WHAT TECHNOLOGY WANTS

ALSO BY KEVIN KELLY

Out of Control: The New Biology of Machines, Social Systems, and the Economic World

New Rules for the New Economy: 10 Radical Strategies for a Connected World

Asia Grace

WHAT TECHNOLOGY WANTS

KEVIN KELLY

VIKING

VIKING
Published by the Penguin Group
Penguin Group (USA) Inc., 375 Hudson Street, New York, New York 10014, U.S.A.
Penguin Group (Canada), 90 Eglinton Avenue East, Suite 700, Toronto, Ontario, Canada M4P 2Y3
(a division of Pearson Penguin Canada Inc.)
Penguin Books Ltd, 80 Strand, London WC2R 0RL, England
Penguin Ireland, 25 St. Stephen's Green, Dublin 2, Ireland (a division of Penguin Books Ltd)
Penguin Books Australia Ltd, 250 Camberwell Road, Camberwell, Victoria 3124, Australia
(a division of Pearson Australia Group Pty Ltd)
Penguin Books India Pvt Ltd, 11 Community Centre, Panchsheel Park, New Delhi – 110 017, India
Penguin Group (NZ), 67 Apollo Drive, Rosedale, North Shore 0632, New Zealand
(a division of Pearson New Zealand Ltd)
Penguin Books (South Africa) (Pty) Ltd, 24 Sturdee Avenue, Rosebank, Johannesburg 2196,
South Africa

Penguin Books Ltd, Registered Offices: 80 Strand, London WC2R 0RL, England

First published in 2010 by Viking Penguin, a member of Penguin Group (USA) Inc.

1 3 5 7 9 10 8 6 4 2

LIBRARY OF CONGRESS CATALOGING IN PUBLICATION DATA
Kelly, Kevin, 1952-
What technology wants / Kevin Kelly.
p. cm.
Includes bibliographical references and index.
ISBN 978-0-670-02215-1
1. Technology'—Social aspects. 2. Technology and civilization. I. Title.
T14.5.K45 2010
303.48'3—dc22 2010013915

Printed in the United States of America

Contents

WHAT TECHNOLOGY WANTS

1

My Question

For most of my life I owned very little. I dropped out of college and for almost a decade wandered remote parts of Asia in cheap sneakers and worn jeans, with lots of time and no money. The cities I knew best were steeped in medieval richness; the lands I passed through were governed by ancient agricultural traditions. When I reached for a physical object, it was almost surely made of wood, fiber, or stone. I ate with my hands, trekked on foot through mountain valleys, and slept wherever. I carried very little stuff. My personal possessions totaled a sleeping bag, a change of clothes, a penknife, and some cameras. Living close to the land, I experienced the immediacy that opens up when the buffer of technology is removed. I got colder often, hotter more frequently, soaking wet a lot, bitten by insects faster, and synchronized quicker to the rhythm of the day and seasons. Time seemed abundant.

After eight years in Asia, I returned to the United States. I sold what little I had and bought an inexpensive bicycle, which I rode on a 5,000-mile meander across the American continent, west to east. The highlight was gliding through the tidy farmland of the Amish in eastern Pennsylvania. Amish communities were the closest thing I could find on this continent to the state of minimal technology I had experienced in Asia. I admired the Amish for their selective possessions. Their unadorned homes were square bundles of contentment. I felt my own life, unencumbered by fancy technology, was in parallel to theirs, and I intended to keep technology in my life to a minimum. I arrived on the East Coast owning nothing but my bicycle.

Growing up in suburban New Jersey in the 1950s and 1960s, I was surrounded by technology. But until I was 10, my family had no television, and when it did arrive in our household, I had no appetite for it. I saw how it worked on my friends. The technology of TV had a remarkable ability to beckon people at specific times and then hold them enthralled for hours. Its creative commercials told them to acquire more technologies. They obeyed. I noticed that other bossy technologies, such as the car, also seemed to be able to get people to serve them, and to prod them to acquire and use still more technologies (freeways, drive-in theaters, fast food). I decided to keep technology to a minimum in my own life. As a teenager, I was having trouble hearing my own voice, and it seemed to me my friends' true voices were being drowned out by the loud conversations technology was having with itself. The less I participated in the circular logic of technology, the straighter my own trajectory could become.

When my cross-country bike ride ended, I was 27. I retreated to an out-of-the-way plot of cheap land in upstate New York with plenty of woods and no building codes. With a friend, I cut down oak trees to mill into lumber, and with these homemade beams we erected a house. We nailed each cedar shake onto the roof one by one. I have vivid memories of hauling hundreds of heavy rocks to build a retaining wall, which the overflowing creek tore down more than once. With my own hands I moved those stones many times. With yet more stones we assembled a huge living-room fireplace. Despite the hard work, those stones and oak beams filled me with Amish contentment.

But I was not Amish. If you were going to cut down a huge tree, I decided, it was a good idea to use a chain saw. Any forest tribesman who could get his hands on one would agree. Once you gain your voice around technology and become more sure of what you want, it becomes obvious that some technologies are simply superior to others. If my travels in the old world had taught me anything, it was that aspirin, cotton clothing, metal pots, and telephones are fantastic inventions. They are *good*. People everywhere in the world, with very few exceptions, grab them when they can. Anyone who has ever held a perfectly designed hand tool knows that it can lift your soul. Airplanes had stretched my horizons; books had opened my mind; antibiotics had saved my life; photography had ignited my muse. Even the chain saw, which can cleanly slice through knotty burls too tough for a hand ax, had instilled in me a reverence for the beauty and strength of wood no other agent in the world could.

I became fascinated by the challenge of picking the few tools that might elevate my spirit. In 1980 I freelanced for a publication (the *Whole Earth Catalog*) that used its own readers to select and recommend appropriate tools picked out of the ocean of self-serving manufactured stuff. In the 1970s and '80s, the *Whole Earth Catalog* was, in essence, a user-generated website before the web, before computers, employing only cheap newsprint. The audience were the authors. I was thrilled by the changes that simple, well-selected tools could provoke in people's lives.

At the age of 28, I started selling mail-order budget travel guides that published low-cost information on how to enter the technologically simple realms most of the planet lived in. My only two significant possessions at the time were a bike and sleeping bag, so I borrowed a friend's computer (an early Apple II) to automate my fledgling moonlight business, and I got a cheap telephone modem to transmit my text to the printer. A fellow editor at the *Whole Earth Catalog* with an interest in computers slipped me a guest account that allowed me to remotely join an experimental teleconferencing system being run by a college professor at the New Jersey Institute of Technology. I soon found myself immersed in something altogether bigger and wilder: the frontier of an online community. It was a new continent more alien to me than Asia, and I began to report on it as if it were an exotic travel destination. To my immense surprise, I found that these high-tech computer networks were not deadening the souls of early users like me; they were filling our souls. There was something unexpectedly organic about these ecosystems of people and wires. Out of complete nothingness, we were barn-raising a virtual commonwealth. When the internet finally came along a few years later, it seemed almost Amish to me.

As computers moved to the center of our lives, I discovered something I had not noticed about technology before. In addition to technology's ability to satisfy (and create) desires, and to occasionally save labor, it did something else. It brought new opportunities. Right before my eyes I saw online networks connect people with ideas, options, and other people they could not possibly have met otherwise. Online networks unleashed passions, compounded creativity, amplified generosity. At the very cultural moment when pundits declared that writing was dead, millions began writing online more than they ever had written before. Exactly when the experts declared people would only bowl alone, millions began to gather together in large numbers. Online they collaborated, cooperated, shared, and created in myriad unexpected ways. This was new to me. Cold silicon chips, long metal wires, and complicated high-voltage gear were nurturing our best efforts as humans. Once I noticed how online computers stirred the muses and multiplied possibilities, I realized that other technologies, such as automobiles, chain saws, biochemistry, and yes, even television, did the same in slightly different ways. For me, this gave a very different face to technology.

I was very active on early teleconference systems, and in 1984, based on my virtual online presence, I was hired by the *Whole Earth Catalog* to help edit the first consumer publication that reviewed personal computer software. (I believe I might have been the first person in the world hired online.) A few years later, I got involved in launching the first public gateway to the emerging internet, an online portal called the Well. In 1992, I helped found *Wired* magazine—the official bullhorn of digital culture—and curated its content for its first seven years. Ever since, I've hung out on the cusp of technological adoption. My friends now are the folks inventing supercomputers, genetic pharmaceuticals, search engines, nanotechnology, fiber-optic communications—everything that is new. I see the transforming power of technology everywhere I look.

Yet I don't have a PDA, a smartphone, or Bluetooth anything. I don't twitter. My three kids grew up without TV, and we still don't have broadcast or cable in our home. I don't have a laptop or travel with a computer, and I am often the last in my circle to get the latest must-have gadget. I ride my bike more often than I drive these days. I see my friends leashed to their vibrating handhelds, but I continue to keep the cornucopia of technology at arm's length so that I can more easily remember who I am. At the same time, I run a popular daily website called *Cool Tools,* which is a continuation of my long-ago *Whole Earth* job evaluating select technology for the empowerment of individuals. A river of artifacts flows through my studio sent by vendors hoping for an endorsement; a fair number of those artifacts never leave. I am surrounded by stuff. Despite my wariness, I have chosen to deliberately position myself to keep the maximum number of technological options within my reach.

I acknowledge that my relationship with technology is full of contradictions. And I suspect they are your contradictions, too. Our lives today are strung with a profound and constant tension between the virtues of more technology and the personal necessity of less: Should I get my kid this gadget? Do I have time to master this labor-saving device? And more deeply: What *is* this technology taking over my life, anyway? What is this global force that elicits both our love and repulsion? How should we approach it? Can we resist it, or is each and every new technology inevitable? Does the relentless avalanche of new things deserve my support or my skepticism—and will my choice even matter?

I needed some answers to guide me through my technological dilemma. And the first question I faced was the most basic. I realized I had no idea what technology really *was.* What was its essence? If I didn't understand the basic nature of technology, then as each new piece of it came along, I would have no frame of reference to decide how weakly or strongly to embrace it.

My uncertainty about the nature of technology and my own conflicted relationship with it sent me on a seven-year quest that eventually became this book. My investigations took me back to the beginning of time and ahead to the distant future. I delved deep into technology's history, and I listened to futurists in Silicon Valley, where I live, spin out imaginative scenarios for what will come next. I interviewed some of technology's fiercest critics and its most ardent fans. I returned to rural Pennsylvania to spend more time with the Amish. I traveled to mountain villages in Laos, Bhutan, and western China to listen to the poor who lack material goods, and I visited the labs of rich entrepreneurs trying to invent things that everyone will consider essential in a few years.

The more closely I looked at the conflicting tendencies of technology, the bigger the questions became. Our confusion over technology usually starts with a very specific concern: Should we allow human cloning? Is constant texting making our kids dumb? Do we want automobiles to park themselves? But as my quest evolved, I realized that if we want to find satisfying answers to those questions, we first need to consider technology as a whole. Only by listening to technology's story, divining its tendencies and biases, and tracing its current direction can we hope to solve our personal puzzles.

Despite its power, technology has been invisible, hidden, and nameless. One example: Since George Washington delivered the first State of the Union address in 1790, every American president has presented to Congress an annual summary of the nation's condition and prospects and the most important forces at work in the world. Until 1939, the colloquial use of the term *technology* was absent. It did not occur twice in a State of the Union address until 1952. Surely my grandparents and parents were surrounded by technology! Yet for most of its adult life, our collective invention did not have a name.

The word *technelogos* is nominally Greek. When the ancient Greeks used the word *techne*, it meant something like art, skill, craft, or even craftiness. *Ingenuity* may be the closest translation. *Techne* was used to indicate the ability to outwit circumstances, and as such it was a trait greatly treasured by poets like Homer. King Odysseus was a master of *techne*. Plato, though, like most scholarly gentlemen of that era, thought that *techne*, which he used to mean manual craftwork, was base, impure, and degraded. Because of his contempt for practical knowledge, Plato omitted any references to craft in his elaborate classification of all knowledge. In fact, there's not a single treatise in the Greek corpus that even mentions *technelogos*—with one exception. To the best of our knowledge, it was in Aristotle's treatise *Rhetoric* that the word *techne* was first joined to *logos* (meaning word or speech or literacy) to yield the single term *technelogos*. Four times in this essay, Aristotle refers to *technelogos,* but in all four instances, his exact meaning is unclear. Is he concerned with the "skill of words" or the "speech about art" or maybe a literacy of craft? After this fleeting, cryptic appearance, the term *technology* essentially disappeared.

But of course, technology did not. The Greeks invented iron welding, the bellows, the lathe, and the key. Their students the Romans invented the vault, the aqueduct, blown glass, cement, sewers, and water mills. Yet in their own time and for many centuries thereafter, the totality of all that was manufactured was virtually invisible—never discussed as a distinct subject, apparently never even contemplated. Technology could be found everywhere in the ancient world except in the minds of humans.

In the centuries following, scholars continued to call the making of things *craft* and the expression of inventiveness *art*. As tools, machines, and contraptions spread, the work performed with them was termed the "useful arts." Each useful art—mining, weaving, metalworking, needlework—had its own secret knowledge that was passed on through a master/apprentice relationship. But it was still an *art,* a singular extension of its maker, and the term retained the original Greek sense of craft and cleverness.

For the next thousand years, art and technique were perceived as distinctly personal realms. Each product of these arts, whether an ironwork fence or an herbal formula, was considered a unique expression derived from the particular cleverness of a particular person. Anything made was a work of solitary genius. As the historian Carl Mitcham explains, "Mass production was unthinkable to the classical mind, and not just for technical reasons."

By the European Middle Ages, craftiness manifested itself most significantly in a new use of energy. An efficient horse collar had disseminated throughout society, drastically increasing farm acreage, while water mills and windmills were improved, increasing the flow of lumber and flour and improving drainage. And all this plentitude came without slavery. As Lynn White, historian of technology, wrote, "The chief glory of the later Middle Ages was not its cathedrals or its epics or its scholasticism: it was the building for the first time in history of a complex civilization which rested not on the backs of sweating slaves or coolies but primarily on non-human power." Machines were becoming our coolies.

In the 18th century, the Industrial Revolution was one of several revolutions that overturned society. Mechanical creatures intruded into farms and homes, but still this invasion had no name. Finally, in 1802, Johann Beckmann, an economics professor at Gottingen University in Germany, gave this ascending force its name. Beckmann argued that the rapid spread and increasing importance of the useful arts demanded that we teach them in a "systemic order." He addressed the techne of architecture, the techne of chemistry, metalwork, masonry, and manufacturing, and for the first time he claimed these spheres of knowledge were interconnected. He synthesized them into a unified curriculum and wrote a textbook titled *Guide to Technology* (or *Technologie* in German), resurrecting that forgotten Greek word. He hoped his outline would become the first course in the subject. It did that and more. It also gave a name to what we do. Once named, we could now see it. Having seen it, we wondered how anyone could not have seen it.

Beckmann's achievement was more than simply christening the unseen. He was among the first to recognize that our creations were not just a collection of random inventions and good ideas. The whole of technology had remained imperceptible to us for so long because we were distracted by its masquerade of rarefied personal genius. Once Beckmann lowered the mask, our art and artifacts could be seen as interdependent components woven into a coherent impersonal unity.

Each new invention requires the viability of previous inventions to keep going. There is no communication between machines without extruded copper nerves of electricity. There is no electricity without mining veins of coal or uranium, or damming rivers, or even mining precious metals to make solar panels. There is no metabolism of factories without the circulation of vehicles. No hammers without saws to cut the handles; no handles without hammers to pound the saw blades. This global-scale, circular, interconnected network of systems, subsystems, machines, pipes, roads, wires, conveyor belts, automobiles, servers and routers, codes, calculators, sensors, archives, activators, collective memory, and power generators—this whole grand contraption of interrelated and interdependent pieces forms a single system.

When scientists began to investigate how this system functioned, they soon noticed something unusual: Large systems of technology often behave like a very primitive organism. Networks, especially electronic networks, exhibit near-biological behavior. Early in my online experience I learned that when I sent out an e-mail message, the network would cut it up into pieces and then send those bits along more than one pathway to the message's final destination. The multiple routes were not predetermined but "emerged" depending on the traffic of the whole network at the instant. In fact, two parts of the e-mail might take radically different pathways and then reassemble at the end. If a bit got lost along the way, it was simply re-sent along different routes until it arrived. That struck me as marvelously organic—very much like the way messages in an anthill are sent.

In 1994, I published a book called *Out of Control* that explored at length the ways in which technological systems were beginning to mimic natural systems. I cited computer programs that could duplicate themselves and synthetic chemicals that could catalyze themselves—even primitive robots that could self-assemble, just as cells do. Many large, complex systems, such as the electrical grid, had been designed to repair themselves, not too differently from the way our bodies do. Computer scientists were using the principles of evolution to breed computer software that was too difficult for humans to write; instead of designing thousands of lines of code, the researchers unleashed a system of evolution to select the best lines of code and keep mutating them, then killing off the duds until the evolved code performed perfectly.

At the same time, biologists were learning that living systems can be imbued with the abstracted essence of a mechanical process like computation. For instance, researchers discovered that DNA—the actual DNA found in the ubiquitous bacteria *E. coli* in our own intestines—could be used to compute the answers to difficult mathematical problems, just like a computer. If DNA could be made into a working computer, and a working computer could be made to evolve like DNA, then there might be, or must be, a certain equivalency between the made and the born. Technology and life must share some fundamental essence.

During the years I was puzzling over these questions, something strange happened to technology: The best of it was becoming incredibly disembodied. Fantastic stuff was getting smaller, using less material but doing more. Some of the best technology, such as software, didn't have a material body at all. This development wasn't new; any list of great inventions in history contains plenty that are rather wispy: the calendar, the alphabet, the compass, penicillin, double-entry accounting, the U.S. Constitution, the contraceptive pill, domestication of animals, zero, germ theory, lasers, electricity, the silicon chip, and so on. Most of these inventions wouldn't hurt you if you dropped them on your toes. But now the process of disembodiment was speeding up.

Scientists had come to a startling realization: However you define life, its essence does not reside in material forms like DNA, tissue, or flesh, but in the intangible organization of the energy and information contained in those material forms. And as technology was unveiled from its shroud of atoms, we could see that at its core, it, too, is about ideas and information. Both life and technology seem to be based on immaterial flows of information.

It was at this point that I realized I needed even greater clarity on what kind of force flowed through technology. Was it really mere ghostly information? Or did technology need physical stuff? Was it a natural force or an unnatural one? It was clear (at least to me) that technology was an extension of natural life, but in what ways was it *different* from nature? (Computers and DNA share something essential, but a MacBook is not the same as a sunflower.) It is also clear that technology springs from human minds, but in what categorical way are the products of our minds (even cognitive products like artificial intelligences) different from our minds themselves? Is technology human or nonhuman?

We tend to think of technology as shiny tools and gadgets. Even if we acknowledge that technology can exist in disembodied form, such as software, we tend not to include in this category paintings, literature, music, dance, poetry, and the arts in general. But we should. If a thousand lines of letters in UNIX qualifies as a technology (the computer code for a web page), then a thousand lines of letters in English (*Hamlet*) must qualify as well. They both can change our behavior, alter the course of events, or enable future inventions. A Shakespeare sonnet and a Bach fugue, then, are in the same category as Google's search engine and the iPod: They are something useful produced by a mind. We can't separate out the multiple overlapping technologies responsible for a *Lord of the Rings* movie. The literary rendering of the original novel is as much an invention as the digital rendering of its fantastical creatures. Both are useful works of the human imagination. Both influence audiences powerfully. Both are technological.

Why not just call this vast accumulation of invention and creation *culture*? In fact, some people do. In this usage, culture would include all the technology we have invented so far, plus the products of those inventions, plus anything else our collective minds have produced. And if by "culture" one means not just local ethnic cultures but the aggregate culture of the human species, then this term very nearly represents this vast sphere of technology that I have been talking about.

But the term *culture* falls short in one critical way. It is too small. What Beckmann recognized in 1802 when he baptized technology was that the things we were inventing were spawning other inventions in a type of self-generation. Technical arts enabled new tools, which launched new arts, which birthed new tools, ad infinitum. Artifacts were becoming so complex in their operation and so interconnected in their origins that they formed a new whole: *technology*.

The term *culture* fails to convey this essential self-propelling momentum pushing technology. But to be honest, the term *technology* does not quite get it right, either. It, too, is too small, because *technology* can also mean specific methods and gear, as in "biotechnology," or "digital technology," or the technology of the Stone Age.

I dislike inventing words that no one else uses, but in this case all known alternatives fail to convey the required scope. So I've somewhat reluctantly coined a word to designate the greater, global, massively interconnected system of technology vibrating around us. I call it the *technium*. The technium extends beyond shiny hardware to include culture, art, social institutions, and intellectual creations of all types. It includes intangibles like software, law, and philosophical concepts. And most important, it includes the generative impulses of our inventions to encourage more tool making, more technology invention, and more self-enhancing connections. For the rest of this book I will use the term *technium* where others might use *technology* as a plural, and to mean a whole system (as in "technology accelerates"). I reserve the term *technology* to mean a specific technology, such as radar or plastic polymers. For example, I would say: "The technium accelerates the invention of technologies." In other words, *technologies* can be patented, while the *technium* includes the patent system itself.

As a word, *technium* is akin to the German word *technik*, which similarly encapsulates the grand totality of machines, methods, and engineering processes. *Technium* is also related to the French noun *technique,* used by French philosophers to mean the society and culture of tools. But neither term captures what I consider to be the essential quality of the technium: this idea of a self-reinforcing system of creation. At some point in its evolution, our system of tools and machines and ideas became so dense in feedback loops and complex interactions that it spawned a bit of independence. It began to exercise some autonomy.

At first, this notion of technological independence is very hard to grasp. We are taught to think of technology first as a pile of hardware and secondly as inert stuff that is wholly dependent on us humans. In this view, technology is only what we make. Without us, it ceases to be. It does only what we want. And that's what I believed, too, when I set out on this quest. But the more I looked at the whole system of technological invention, the more powerful and self-generating I realized it was.

There are many fans, as well as many foes, of technology, who strongly disagree with the idea that the technium is in any way autonomous. They adhere to the creed that technology does only what we permit it to do. In this view, notions of technological autonomy are simply wishful thinking on our part. But I now embrace a contrary view: that after 10,000 years of slow evolution and 200 years of incredible intricate exfoliation, the technium is maturing into its own thing. Its sustaining network of self-reinforcing processes and parts have given it a noticeable measure of autonomy. It may have once been as simple as an old computer program, merely parroting what we told it, but now it is more like a very complex organism that often follows its own urges.

Okay, that's very poetic, but is there any *evidence* for the technium's autonomy? I think there is, but it rests on how we define autonomy. The qualities we hold dearest in the universe are all extremely slippery at the edges. *Life*, *mind*, *consciousness*, *order*, *complexity*, *free will*, and *autonomy* are all terms that have multiple, paradoxical, and inadequate definitions. No one can agree on exactly where life or mind or consciousness or autonomy begins and where it ends. The best we can agree on is that these states are not binary. They exist on a continuum. So: humans have minds, and so do dogs, and mice. Fish have tiny brains, so they must have tiny minds. Does that mean ants, who have smaller brains yet, also have minds? How many neurons do you need to have a mind?

Autonomy has a similar sliding scale. A newborn wildebeest will run on its own the day after it is born. But we can't say a human infant is an autonomous being if it will die without its mother for its first years. Even we adults are not 100 percent autonomous, since we depend upon other living species in our gut (such as *E. coli*) to aid in the digestion of our food or the breakdown of toxins. If humans are not fully autonomous, what is? An organism or system does not need to be wholly independent to exhibit some degree of autonomy. Like an infant of any species, it can acquire increasing degrees of independence, starting from a speck of autonomy.

So how do you detect autonomy? Well, we might say that an entity is autonomous if it displays any of these traits: self-repair, self-defense, self-maintenance (securing energy, disposing of waste), self-control of goals, self-improvement. The common element in all these characteristics is of course the emergence, at some level, of a self. In the technium we don't have any examples of a system that displays *all* these traits—but we have plenty of examples that display some of them. Autonomous airplane drones can self-steer and stay aloft for hours. But they don't repair themselves. Communication networks can repair themselves. But they don't reproduce themselves. We have self-reproducing computer viruses, but they don't improve themselves.

Woven deep into the vast communication networks wrapping the globe, we also find evidence of embryonic technological autonomy. The technium contains 170 quadrillion computer chips wired up into one mega-scale computing platform. The total number of transistors in this global network is now approximately the same as the number of neurons in your brain. And the number of links among files in this network (think of all the links among all the web pages of the world) is about equal to the number of synapse links in your brain. Thus, this growing planetary electronic membrane is already comparable to the complexity of a human brain. It has three billion artificial eyes (phone and webcams) plugged in, it processes keyword searches at the humming rate of 14 kilohertz (a barely audible high-pitched whine), and it is so large a contraption that it now consumes 5 percent of the world's electricity. When computer scientists dissect the massive rivers of traffic flowing through it, they cannot account for the source of all the bits. Every now and then a bit is transmitted incorrectly, and while most of those mutations can be attributed to identifiable causes such as hacking, machine error, or line damage, the researchers are left with a few percent that somehow changed themselves. In other words, a small fraction of what the technium communicates originates not from any of its known human-made nodes but from the system at large. The technium is whispering to itself.

Further deep analysis of the information flowing through the technium's network reveals that it has slowly been shifting its methods of organization. In the telephone system a century ago, messages dispersed across the network in a pattern that mathematicians associate with randomness. But in the last decade, the flow of bits has become statistically more similar to the patterns found in self-organized systems. For one thing, the global network exhibits self-similarity, also known as a fractal pattern. We see this kind of fractal pattern in the way the jagged outline of tree branches look similar no matter whether we look at them up close or far away. Today messages disperse through the global telecommunications system in the fractal pattern of self-organization. This observation doesn't prove autonomy. But autonomy is often self-evident long before it can be proved.

We created the technium, so we tend to assign ourselves exclusive influence over it. But we have been slow to learn that systems—all systems—generate their own momentum. Because the technium is an outgrowth of the human mind, it is also an outgrowth of life, and by extension it is also an outgrowth of the physical and chemical self-organization that first led to life. The technium shares a deep common root not only with the human mind, but with ancient life and other self-organized systems as well. And just as a mind must obey not only the principles governing cognition but also the laws governing life and self-organization, so the technium must obey the laws of mind, life, and self-organization—as well as our human minds. Thus out of all the spheres of influence upon the technium, the human mind is only one. And this influence may even be the weakest one.

The technium wants what we design it to want and what we try to direct it to do. But in addition to those drives, the technium has its own wants. It wants to sort itself out, to self-assemble into hierarchical levels, just as most large, deeply interconnected systems do. The technium also wants what every living system wants: to perpetuate itself, to keep itself going. And as it grows, those inherent wants are gaining in complexity and force.

I know this claim sounds strange. It seems to anthropomorphize stuff that is clearly not human. How can a toaster want? Aren't I assigning way too much consciousness to inanimate objects, and by doing so giving them more power over us than they have, or should have?

It's a fair question. But "want" is not just for humans. Your dog wants to play Frisbee. Your cat wants to be scratched. Birds want mates. Worms want moisture. Bacteria want food. The wants of a microscopic, single-celled organism are less complex, less demanding, and fewer in number than the wants of you or me, but all organisms share a few fundamental desires: to survive, to grow. All are driven by these "wants." The wants of a protozoan are unconscious, unarticulated—more like an urge or a tendency. A bacterium tends to drift toward nutrients with no awareness of its needs. In a dim way it chooses to satisfy its wants by heading one way and not another.

With the technium, *want* does not mean thoughtful decisions. I don't believe the technium is conscious (at this point). Its mechanical wants are not carefully considered deliberations but rather tendencies. Leanings. Urges. Trajectories. The wants of technology are closer to needs, a compulsion toward something. Just like the unconscious drift of a sea cucumber as it seeks a mate. The millions of amplifying relationships and countless circuits of influence among parts push the whole technium in certain unconscious directions.

Technology's wants can often seem abstract or mysterious, but occasionally, these days, you can see them right in front of you. Recently I visited a start-up called Willow Garage in a leafy suburban tract not far from Stanford University. The company creates state-of-the-art research robots. Willow's latest version of a personal robot, called PR2, stands about chest high, runs on four wheels, and has five eyes and two massive arms. When you take hold of one of its arms, it is neither rigid at the joints nor limp. It responds in a supple manner, with a gentle give, as if the limb were alive. It's an uncanny sensation. Yet the robot's grip is as deliberate as yours. In the spring of 2009, PR2 completed a full 26.2-mile marathon circuit in the building without crashing into obstacles. In robotdom, this is a huge accomplishment. But PR2's most notable achievement is its ability to find a power outlet and plug itself in. It's been programmed to look for its own power, but the specific path it takes emerges as it overcomes obstacles. So when it gets hungry, it searches for one of a dozen available power sockets in the building to recharge its batteries. It grabs its cord with one of its hands, uses its laser and optical eyes to line up a socket, and after gently probing the outlet in a small spiral pattern to find the exact slots, pushes its plug in to get fueled. It then sucks up power there for a couple of hours. Before the software was perfected, a few unexpected "wants" emerged. One robot craved plugging in even when its batteries were full, and once a PR2 took off without properly unplugging, dragging its cord behind it, like a forgetful motorist pulling out of the gas station with the pump hose still in the tank. As its behavior becomes more complex, so will its desires. If you stand in front of a PR2 while it is hungry, it won't hurt you. It will backtrack and go around the building any way it can to find a plug. It's not conscious, but standing between it and its power outlet, you can clearly feel its want.

There is a nest of ants somewhere beneath my family's house. The ants, if we let them—and we won't—would carry off most of the food in our pantry. We humans are obliged to obey nature, except that sometimes we are forced to thwart it. While we bow to nature's beauty, we also frequently take out a machete and temporarily hack it back. We weave clothes to keep the natural world away from us, and we concoct vaccines to inoculate us against its mortal diseases. We rush to the wilderness to be rejuvenated, but we bring our tents.

The technium is now as great a force in our world as nature, and our response to the technium should be similar to our response to nature. We can't demand that technology obey us any more than we can demand that life obey us. Sometimes we should surrender to its lead and bask in its abundance, and sometimes we should try to bend its natural course to meet our own. We don't have to do everything that the technium demands, but we can learn to work with this force rather than against it.

And to do that successfully, we first need to understand technology's behavior. In order to decide how to respond to technology, we have to figure out what technology wants.

After a long journey, that is where I have ended up. By listening to what technology wants, I feel that I have been able to find a framework to guide me through this rising web of hatching technologies. Seeing our world through technology's eyes has, for me, illuminated its larger purpose. And recognizing what it wants has reduced much of my own conflict in deciding where to place myself in its embrace. This book is my report on what technology wants. My hope is that it will help others find their own way to optimize technology's blessings and minimize its costs.

PART ONE

ORIGINS

2

Inventing Ourselves

To see where technology is going, we need to see where it has come from. And that's not easy. The further back we trace the technium's history, the further back its origins seem to recede. So let's begin with our own origins, that moment in prehistory when humans lived primarily surrounded by things they did not make. What were our lives like without technology?

The problem with this line of questioning is that technology predated our humanness. Many other animals used tools millions of years before humans. Chimpanzees made (and of course still make) hunting tools from thin sticks to extract termites from mounds and slammed rocks to break nuts. Termites themselves construct vast towers of mud for their homes. Ants herd aphids and farm fungi in gardens. Birds weave elaborate, twiggy fabrics for their nests. And some octopuses will find and carry shells for portable homes. The strategy of bending the environment to use as if it were part of one's own body is a half-billion-year-old trick at least.

Our ancestors first chipped stone scrapers 2.5 million years ago to give themselves claws. By about 250,000 years ago they devised crude techniques for cooking, or predigesting, with fire. Cooking acts as a supplemental stomach—an artificial organ that permits smaller teeth and smaller jaw muscles and provides more kinds of stuff to eat. Technology-assisted hunting, as opposed to tool-free scavenging, is equally old. Archaeologists have found a stone point jammed into the vertebra of a horse and a wooden spear embedded in a 100,000-year-old red deer skeleton. This pattern of tool use has only accelerated in the years since.

All technology, the chimp's termite-fishing spear and the human's fishing spear, the beaver's dam and the human's dam, the warbler's hanging basket and the human's hanging basket, the leaf-cutter ant's garden and the human's garden, are all fundamentally natural. We tend to isolate manufactured technology from nature, even to the point of thinking of it as antinature, only because it has grown to rival the impact and power of its home. But in its origins and fundamentals, a tool is as natural as our life. Humans are animals—no argument. But humans are also not-animals—no argument. This contradictory nature is at the core of our identity. Likewise, technology is unnatural—by definition. And technology is natural—by a wider definition. This contradiction is also core to human identity.

Tools and bigger brains mark the beginning of a distinctly human line in evolution. The first simple stone tools appeared in the same archaeological moment that the brains of the hominins (humanish apes) who made them began to enlarge toward their current size. Thus hominins arrived on Earth 2.5 million years ago with rough, chipped stone scrapers and cutters in hand. About a million years ago, these large-brained, tool-wielding hominins drifted beyond Africa and settled into southern Europe, where they evolved into the Neanderthal (with an even bigger brain) and further into eastern Asia, where they evolved into *Homo erectus* (also bigger brained). Over the next several million years, all three hominin lines evolved, but the ones that remained in Africa evolved into the human form we see in ourselves. The exact time these protohumans became fully modern humans is of course debated. Some say 200,000 years ago, but the undisputed latest date is 100,000 years ago. By 100,000 years ago, humans had crossed the threshold where they were outwardly indistinguishable from us. We would not notice anything amiss if one of them were to stroll alongside us on the beach. However, their tools and most of their behavior were indistinguishable from those of their relatives the Neanderthals in Europe and Erectus in Asia.

For the next 50 millennia not much changed. The anatomy of African human skeletons remained constant over this time. Neither did their tools evolve much. Early humans employed rough-and-ready lumps of rock with sharpened edges to cut, poke, drill, or spear. But these hand-held tools were unspecialized and did not vary by location or time. No matter where or when in this period (called the Mesolithic) a hominin picked up one of these tools, it would resemble one made tens of thousands of miles away or tens of thousands of years earlier or later, whether in the hands of a Neanderthal, Erectus, or *Homo sapiens*. Hominins simply lacked innovation. As biologist Jared Diamond put it, "Despite their large brains, something was missing."

Then about 50,000 years ago, that missing something arrived. While the bodies of early humans in Africa remained unchanged, their genes and minds shifted noticeably. For the first time, hominins were full of ideas and innovation. These newly vitalized modern humans, or Sapiens (a term I am using to distinguish them from earlier populations of *Homo sapiens*), charged into new regions beyond their ancestral homes in eastern Africa. They fanned out from the grasslands, and in a rela-

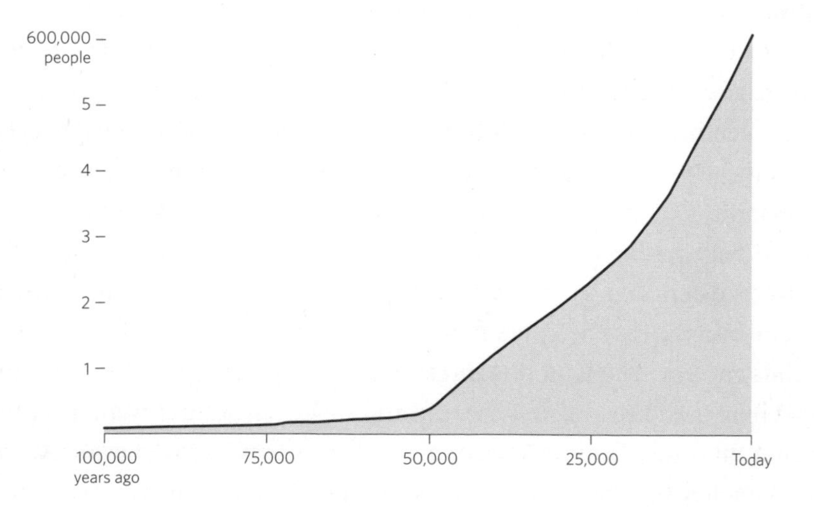

Prehistory Explosion of Human Population. A simulation of the first human population explosion, which began about 50,000 years ago.

tively brief burst exploded from a few tens of thousands of individuals in Africa to an estimated eight million worldwide just before the dawn of agriculture 10,000 years ago.

The speed at which Sapiens marched across the planet and settled every continent (except Antarctica) is astounding. In 5,000 years they overtook Europe. In another 15,000 they reached the edges of Asia. Once tribes of Sapiens crossed the land bridge from Eurasia into what is now Alaska, it took them only a few thousand years to fill the whole of the New World. Sapiens increased so relentlessly that for the next 38,000 years they expanded their occupation at the average rate of one mile (two kilometers) per year. Sapiens kept pushing until they reached the furthest they could go: land's end at the tip of South America. Fewer than 1,500 generations after their "great leap forward" in Africa, *Homo sapiens* had become the most widely distributed species in Earth's history, inhabiting every type of biome and every watershed on the planet. Sapiens were the most invasive alien species ever.

Today the breadth of Sapiens occupation exceeds that of any other macrospecies we know of; no other visible species occupies more niches, geographical and biological, than *Homo sapiens*. Sapiens' overtake was always rapid. Jared Diamond notes that "after the ancestors of the Maori reached New Zealand," carrying only a few tools, "it apparently took them barely a century to discover all worthwhile stone sources; only a few more centuries to kill every last moa in some of the world's most rugged terrain." This sudden global expansion following millennia of steady sustainability was due to only one thing: technological innovation.

As Sapiens expanded in range, they remade animal horns and tusks into thrusters and knives, cleverly turning the animals' own weapons against them. They sculpted figurines, the first art, and the first jewelry, beads cut from shells, at this threshold 50,000 years ago. While humans had long used fire, the first hearths and shelter structures were invented about this time. Trade of scarce shells, chert, and flint rock began. At approximately the same time Sapiens invented fishing hooks and nets and needles for sewing hides into clothes. They left behind the remains of tailored hides in graves. A few bits of pottery from that time have the imprint of woven net and loose fabrics on them. In the same period

Sapiens also invented animal traps. Their garbage reveals heaps of skeletons of small furred animals without their feet; trappers today still skin small animals the same way, keeping the feet with the skin. On walls artists painted humans wearing parkas and killing animals with arrows or spears. Significantly, unlike Neanderthal's and Erectus's crude creations, these tools varied in small stylistic and technological ways place by place. Sapiens had begun innovating.

The Sapiens mind's ability to make warm clothes opened up the arctic regions, and the invention of fishing gear opened up the coasts and rivers of the world, particularly in the tropics, where large game was scarce. While Sapiens' innovation allowed them to prosper in many new climates, the cold and its unique ecology especially drove innovation. More complex "technological units" are needed (or have been invented) by historical hunter-gatherer tribes the higher the latitude of their homes. Hunting oceanic sea mammals in arctic climes took significantly more sophisticated gear than fishing salmon in a river. The ability of Sapiens to rapidly improve their tools allowed them to adapt to new ecological niches at a much faster rate than genetic evolution could ever allow.

During their quick global takeover, Sapiens displaced (with or without interbreeding) the several other coinhabiting hominin species on Earth, including their cousins the Neanderthals. The Neanderthals were never abundant; they may have only numbered 18,000 individuals at one time. After dominating Europe for hundreds of thousands of years as the sole hominin, the Neanderthals vanished in less than 100 generations after the tool-carrying Sapiens arrived. That is a blink in history. As anthropologist Richard Klein points out, this displacement occurred almost instantaneously from a geologic perspective. There were no intermediates in the archaeological record. As Klein says, "The Neanderthals were there one day, and the Cro-Magnons [Sapiens] were there the next." The Sapien layer was always on top, and never the reverse. It was not even necessary that the Sapiens slaughter the Neanderthals. Demographers have calculated that as little as a 4 percent difference in reproductive effectiveness (a reasonable expectation given Sapiens' ability to bring home more kinds of meat) could eclipse the lesser breeding species in a few thousands years. The speed of this several-thousand-year extinction was without precedent in natural evolution. Sadly, it was only the first rapid species extinction to be caused by humans.

It should have been clear to Neanderthals, as it is now clear to us in the 21st century, that something new and big had appeared—a new biological and geological force. A number of scientists (including Richard Klein, Ian Tattersall, and William Calvin, among many others) think that the "something" that happened 50,000 years ago was the invention of language. Up until this point, hominins had been smart. They could make crude tools in a hit-or-miss way and handle fire—perhaps like an exceedingly smart chimp. The growth of the African hominin's brain size and physical stature had leveled off, but evolution continued inside the brain. "What happened 50,000 years ago," says Klein, "was a change in the operating system of humans. Perhaps a point mutation affected the way the brain is wired that allowed languages, as we understand language today: rapidly produced, articulate speech." Instead of acquiring a larger brain, as the Neanderthals and Erectus did, Sapiens gained a rewired brain. Language altered the Neanderthal-type mind and allowed Sapien minds for the first time to invent with purpose and deliberation. Philosopher Daniel Dennett crows in elegant language: "There is no step more uplifting, more momentous in the history of mind design, than the invention of language. When *Homo sapiens* became the beneficiary of this invention, the species stepped into a slingshot that has launched it far beyond all other earthly species." The creation of language was the first singularity for humans. It changed everything. Life after language was unimaginable to those on the far side before it.

Language accelerates learning and creation by permitting communication and coordination. A new idea can be spread quickly if someone can explain it and communicate it to others before they have to discover it themselves. But the chief advantage of language is not communication but autogeneration. Language is a trick that allows the mind to question itself; a magic mirror that reveals to the mind what the mind thinks; a handle that turns a mind into a tool. With a grip on the slippery, aimless activity of self-awareness and self-reference, language can harness a mind into a fountain of new ideas. Without the cerebral structure of language, we couldn't access our own mental activity. We certainly couldn't think the way we do. If our minds can't tell stories, we can't consciously create; we can only create by accident. Until we tame the mind with an organization tool capable of communicating to itself, we have stray thoughts without a narrative. We have a feral mind. We have smartness without a tool.

A few scientists believe that, in fact, it was technology that sparked language. To throw a tool—a rock or stick—at a moving animal and hit it with sufficient force to kill it requires a serious computation in the hominin brain. Each throw requires a long succession of precise neural instructions executed in a split second. But unlike calculating how to grasp a branch in midair, the brain must calculate several alternative options for a throw at the same time: the animal speeds up or it slows down; aim high or aim low. The mind must then spin out the results to gauge the best possible throw before the actual throw—all in a few milliseconds. Scientists such as neurobiologist William Calvin believe that once the brain evolved the power to run multiple rapid-throw scenarios, it hijacked this throw procedure to run multiple rapid sequences of notions. The brain would throw words instead of sticks. This reuse or repurposing of technology then became a primitive but advantageous language.

The slippery genius of language opened up many new niches for spreading tribes of Sapiens. Unlike their cousins the Neanderthals, Sapiens could quickly adapt their tools to hunt or trap an increasing diversity of game and to gather and process an increasing diversity of plants. There is some evidence that Neanderthals were stuck on a few sources of food. Examination of Neanderthal bones show they lacked the fatty acids found in fish and that the Neanderthal diet was mostly meat. But not just any meat. Over half of their diet was woolly mammoth and reindeer. The demise of the Neanderthal may be correlated with the demise of great herds of these megafauna.

Sapiens thrived as broadly omnivorous hunter-gatherers. The unbroken line of human offspring for hundreds of thousands of years proves that a few tools are sufficient to capture enough nutrition to create the next generation. We are here now because hunting-gathering worked in the past. Several analyses of the diets of historical hunter-gatherers show that they were able to secure enough calories to meet the U.S. FDA recommendations for folks their size. For example, anthropologists found the historical Dobe gathered on average 2,140 calories a day; Fish Creek tribe, 2,130; Hemple Bay tribe, 2,160. They had a varied menu of tubers, vegetables, fruit, and meat. Based on studies of bones and pollen in their trash, so did the early Sapiens.

The philosopher Thomas Hobbes claimed the life of the savage—and by this he meant Sapien hunter-gatherers—was "nasty, brutish, and short." But while the life of an early hunter-gatherer was indeed short, and often interrupted by nasty warfare, it was not brutish. With only a slim set of a dozen primitive tools, not only did humans secure enough food, clothing, and shelter to survive in all kinds of environments, but these tools and techniques also afforded them some leisure while doing so. Anthropological studies confirm that present-day hunter-gathers do not spend all day hunting and gathering. One researcher, Marshall Sahlins, concluded that hunter-gatherers worked only three to four hours a day on necessary food chores, putting in what he called "banker hours." The evidence for his surprising results is controversial.

A more realistic and less contentious average for food-gathering time among contemporary hunter-gatherer tribes, based on a wider range of data, is about six hours per day. That average belies a great variation in day-to-day routine. One- to two-hour naps or whole days spent sleeping were not uncommon. Outside observers almost universally noted the punctuated nature of work among foragers. Gatherers may work very hard for several days in a row and then do nothing in terms of food getting for the rest of the week. This cycle is known among anthropologists as the "Paleolithic rhythm"—a day or two on, a day or two off. An observer familiar with the Yamana tribe—but it could be almost any hunter tribe—wrote: "Their work is more a matter of fits and starts, and in these occasional efforts they can develop considerable energy for a certain time. After that, however they show a desire for an incalculably long rest period during which they lie about doing nothing, without showing great fatigue." The Paleolithic rhythm actually reflects the "predator rhythm," since the great hunters of the animal world, the lion and other large cats, exhibit the same style: hunting to exhaustion in a short burst and then lounging around for days afterward. Hunters, almost by definition, seldom go out hunting, and they succeed in getting a meal even less often. The efficiency of primitive tribal hunting, measured in the yield of calories per hour invested, was only half that of gathering. Meat is thus a treat in almost every foraging culture.

Then there are seasonal variations. Every ecosystem produces a "hungry season" for foragers. At higher, cooler latitudes, this late winter–early spring hungry season is more severe, but even at tropical latitudes, there are seasonal oscillations in the availability of favorite foods, supplemental fruits, or essential wild game. In addition, there are climatic variations: extended periods of drought, floods, and storms that can disrupt yearly patterns. These great punctuations over days, seasons, and years mean that while there are many times when hunter-gatherers are well fed, they also can—and do—expect many periods when they are hungry, famished, and undernourished. Time spent in this state along the edge of malnutrition is mortal for young children and dire for adults.

The result of all this variation in calories is the Paleolithic rhythm at all scales of time. Importantly, this burstiness in "work" is not by choice. When you are primarily dependent on natural systems to provide you foodstuffs, working more does not tend to produce more. You can't get twice as much food by working twice as hard. The hour at which the figs ripen can be neither hurried nor predicted exactly. Nor can the arrival of game herds. If you do not store surplus or cultivate in place, then motion must produce your food. Hunter-gatherers must be in ceaseless movement away from depleted sources in order to maintain production. But once you are committed to perpetual movement, surplus and its tools slow you down. In many contemporary hunter-gatherer tribes, being unencumbered with things is considered a virtue, even a virtue of character. You carry nothing; instead, you cleverly make or procure whatever you need when you need it. "The efficient hunter who would accumulate supplies succeeds at the cost of his own esteem," says Marshall Sahlins. Additionally, the surplus producer must share the extra food or goods with everyone, which reduces the incentive to produce extra. For foragers, food storage is therefore socially self-defeating. Instead your hunger must adapt to the movements of the wild. If a dry spell diminishes the yield of the sago, no amount of extra work time will advance the delivery of food. Therefore, foragers take a very accepting pace to eating. When food is there, all work very hard. When it is not, no problem; they will sit around and talk while they are hungry. This very reasonable approach is often misread as tribal laziness, but it is in fact a logical strategy if you rely on the environment to store your food.

We civilized modern workers can look at this leisurely approach to work and feel jealous. Three to six hours a day is a lot less than most adults in any developed country put in to their labors. Furthermore, when asked, most acculturated hunter-gatherers don't want any more than they have. A tribe will rarely have more than one artifact, such as an ax, because why do you need more than one? Either you use the object when you need to, or, more likely, you make one when you need one. Once used, artifacts are often discarded rather than saved. That way nothing extra needs to be carried or cared for. Westerners giving gifts such as a blanket or knife to foragers have often been mortified to see them trashed after a day. In a very curious way, foragers live in the ultimate disposable culture. The best tools, artifacts, and technology are all disposable. Even elaborate handcrafted shelters are considered temporary. When a clan or family travels, they might erect a home (a bamboo hut or snow igloo, for example) for only a night and then abandon it the next morning. Larger multifamily lodges might be abandoned after a few years rather than maintained. The same goes for food patches, which are abandoned after harvesting.

This easy just-in-time self-sufficiency and contentment led Marshall Sahlins to declare hunter-gatherers "the original affluent society." But while foragers had sufficient calories most days and did not create a culture that continually craved more, a better summary might be that hunter-gatherers had "affluence without abundance." Based on numerous historical encounters with aboriginal tribes, we know they often, if not regularly, complained about being hungry. Famed anthropologist Colin Turnbull noted that although the Mbuti frequently sang to the goodness of the forest, they often complained of hunger. Often the complaints of hunter-gathers were about the monotony of a carbohydrate staple, such as mongongo nuts, for every meal; when they spoke of shortages, or even hunger, they meant a shortage of meat, and a hunger for fat, and a distaste for periods of hunger. Their small amount of technology gave them sufficiency for most of the time, but not abundance.

The fine line between average sufficiency and abundance matters for health. When anthropologists measure the total fertility rate (the mean number of live births over the reproductive years) of women in modern hunter-gatherer tribes, they find it relatively low—about five to six children in total—compared to six to eight children in agricultural communities. There are several factors behind this depressed fertility. Perhaps because of uneven nutrition, puberty comes late to forager girls, at 16 or 17 years old. (Modern females start at 13.) This late menarche for women, combined with a shorter life span, delays and thus abbreviates the childbearing window. Breast-feeding usually lasts longer in foragers, which extends the interval between births. Most tribes nurse till children are 2 or 3 years old, while a few tribes keep children suckling for as long as 6 years. Also, many women are extremely lean and active and, like lean, active women athletes in the West, often have irregular or no menstruation. One theory suggests women need a "critical fatness" to produce fertile eggs, a fatness many forager women lack—at least part of the year—because of a fluctuating diet. And of course, people anywhere can practice deliberate abstinence to space children, and foragers have reasons to do so.

Child mortality in foraging tribes was severe. A survey of 25 hunter-gatherer tribes in historical times from various continents revealed that, on average, 25 percent of children died before they were 1, and 37 percent died before they were 15. In one traditional hunter-gatherer tribe, child mortality was found to be 60 percent. Most historical tribes had a population growth rate of approximately zero. This stagnation is evident, says Robert Kelly in his survey of hunting-gathering peoples, because "when formerly mobile people become sedentary, the rate of population growth increases." All things being equal, the constancy of farmed food breeds more people.

While many children died young, older hunter-gatherers did not have it much better. It was a tough life. Based on an analysis of bone stress and cuts, one archaeologist said the distribution of injuries on the bodies of Neanderthals was similar to that found on rodeo professionals—lots of head, trunk, and arm injuries like the ones you might get from close encounters with large, angry animals. There are no known remains of an early hominin who lived to be older than 40. Because extremely high child mortality rates depress average life expectancy, if the oldest outlier is only 40, the median age was almost certainly less than 20.

A typical tribe of hunters-gatherers had few very young children and no old people. This demographic may explain a common impression visitors had upon meeting intact historical hunter-gatherer tribes. They would remark that "everyone looked extremely healthy and robust." That's in part because most everyone was in the prime of life, between 15 and 35. We might have the same reaction visiting a trendy urban neighborhood with the same youthful demographic. Tribal life was a lifestyle for and of young adults.

A major effect of this short forager life span was the crippling absence of grandparents. Given that women would only start bearing children at 17 or so and die by their thirties, it would be common for children to lose their parents before the children were teenagers. A short life span is rotten for the individual. But a short life span is also extremely detrimental for a society as well. Without grandparents, it becomes exceedingly difficult to transmit knowledge—and knowledge of tool using—over time. Grandparents are the conduits of culture, and without them culture stagnates.

Imagine a society that not only lacked grandparents but also lacked language—as the pre-Sapiens did. How would learning be transmitted over generations? Your own parents would die before you were an adult, and in any case, they could not communicate to you anything beyond what they could show you while you were immature. You would certainly not learn anything from anyone outside your immediate circle of peers. Innovation and cultural learning would cease to flow.

Language upended this tight constriction by enabling ideas both to coalesce and to be communicated. An innovation could be hatched and then spread across generations via children. Sapiens gained better hunting tools (such as thrown spears, which permitted a lightweight human to kill a huge, dangerous animal from a safe distance), better fishing tools (barbed hooks and traps), and better cooking methods (using hot stones not just to cook meat but also to extract more calories from wild plants). And they gained all these within only 100 generations of beginning to use language. Better tools meant better nutrition, which could assist in faster evolution.

The primary long-term consequence of this slightly better nutrition was a steady increase in longevity. Anthropologist Rachel Caspari studied the dental fossils of 768 hominin individuals in Europe, Asia, and Africa, dated from 5 million years ago until the great leap. She determined that a "dramatic increase in longevity in the modern humans" began about 50,000 years ago. Increasing longevity allowed grandparenting, creating what is called the grandmother effect: In a virtuous circle, via the communication of grandparents, ever more powerful innovations carried forward were able to lengthen life spans further, which allowed more time to invent new tools, which increased population. Not only that: Increased longevity "provide[d] a selective advantage promoting further population increase," because a higher density of humans increased the rate and influence of innovations, which contributed to increased populations. Caspari claims that the most fundamental biological factor that underlies the behavioral innovations of modernity may be the increase in adult survivorship. It is no coincidence that increased longevity is the most measurable consequence of the acquisition of technology. It is also the most consequential.

By 15,000 years ago, as the world was warming and its global ice caps retreating, bands of Sapiens expanded their population and tool kits, hand in hand. Sapiens used 40 kinds of tools, including anvils, pottery, and composites—complicated spears or cutters made from multiple pieces, such as many tiny flint shards and a handle. While still primarily a hunter-gatherer, Sapiens also dabbled in sedentism, returning to care for favorite food areas, and developed specialized tools for different types of ecosystems. We know from burial sites in the northern latitudes at this same time that clothing also evolved from the general (a rough tunic) to specialized items such as a cap, a shirt, a jacket, trousers, and moccasins. Henceforth human tools would become ever more specialized.

The variety of Sapiens tribes exploded as they adapted into diverse watersheds and biomes. Their new tools reflected the specifics of their homes; river inhabitants had many nets, steppe hunters many kinds of points, forest dwellers many types of traps. Their languages and looks were diverging.

Yet they shared many qualities. Most hunter-gatherers clustered into family clans that averaged about 25 related people. Clans would gather in larger tribes of several hundred at seasonal feasts or camping grounds. One function of the tribes was to keep genes moving through intermarriage. Population was spread thinly. The average density of a tribe was less than .01 person per square kilometer in cooler climes. The 200 to 300 folk in your greater tribe would be the total number of people you'd meet in your lifetime. You might be aware of others, because items for trade or barter could travel 300 kilometers. Some of the traded items would be body ornaments and beads, such as ocean shells for inlanders or forest feathers for the coast dwellers. Occasionally, pigments for face painting were swapped, but these could also be applied to walls or to carved wood figurines. The dozen tools you carried would have been bone drills, awls, needles, bone knives, a bone hook for fish on a spear, some stone scrapers, maybe some stone sharpies. A number of your blades would be held by bone or wood handles, hafted with cane or hide cord. When you crouched around the fire, someone might play a drum or bone flute. Your handful of possessions might be buried with you when you died.

But don't take this progress for harmony. Within 20,000 years of the great march out of Africa, Sapiens helped exterminate 90 percent of the then-existing species of megafauna. Sapiens used innovations such as the bow and arrow, spear, and cliff stampedes to kill off the last of the mastodons, mammoths, moas, woolly rhinos, and giant camels—basically every large package of protein that walked on four legs. More than 80 percent of all large mammal genera on the planet were completely extinct by 10,000 years ago. Somehow, four species escaped this fate in North America: the bison, moose, elk, and caribou.

Violence between tribes was endemic as well. The rules of harmony and cooperation that work so well among members of the same tribe, and are often envied by modern observers, did not apply to those outside of the tribe. Tribes would go to war over water holes in Australia or hunting grounds and wild-rice fields in the plains of the United States or river and ocean frontage along the coast in the Pacific Northwest. Without systems of arbitration, or even leaders, small feuds over stolen goods or women or signs of wealth (such as pigs in New Guinea) could grow into multigenerational warfare. The death rate due to warfare was five times higher among hunter-gatherer tribes than in later agriculture-based societies (.1 percent of the population killed per year in "civilized" wars versus .5 percent in war between tribes). Actual rates of warfare

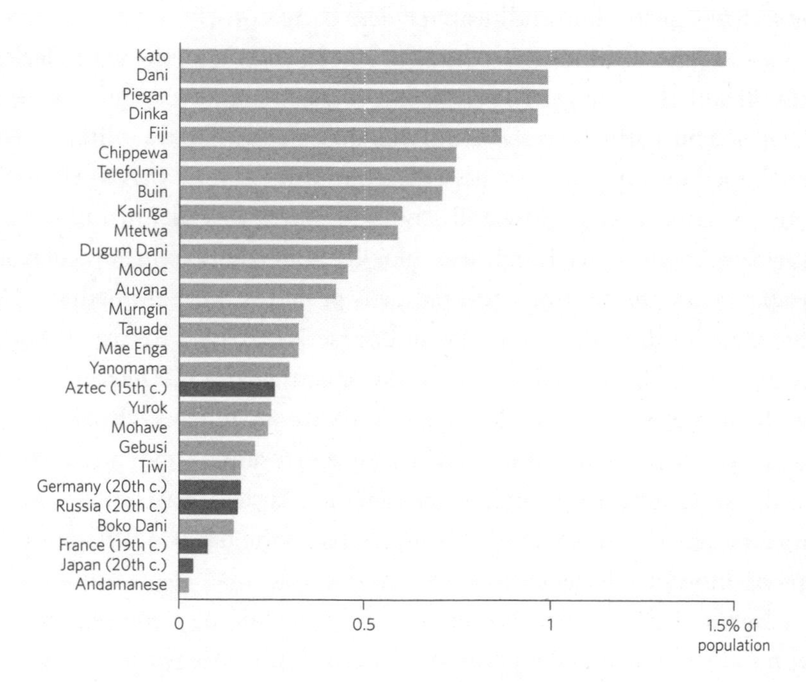

Comparison of War Fatality Rates. Annual war deaths as a percentage of the population in both prestate (gray bars) and modern societies (darker bars).

varied among tribes and regions, because as in the modern world, one belligerent tribe could disrupt the peace for many. In general, the more nomadic a tribe was, the more peaceful it would be, since it could simply flee from conflict. But when fighting did break out, it was fierce and deadly. When the numbers of warriors on both sides were about equal, primitive tribes usually beat the armies of civilization. The Celtic tribes defeated the Romans, the Tuareg smashed the French, the Zulus trumped the British, and it took the U.S. Army 50 years to defeat the Apache tribes. As Lawrence Keeley says in his survey of early warfare in *War Before Civilization,* "The facts recovered by ethnographers and archaeologists indicated unequivocally that primitive and prehistoric warfare was just as terrible and effective as the historic and civilized version. In fact, primitive warfare was much more deadly than that conducted between civilized states because of the greater frequency of combat and the more merciless way it was conducted. . . . It is civilized warfare that is stylized, ritualized, and relatively less dangerous."

Before the revolution of language 50,000 years ago, the world lacked significant technology. For the next 40,000 years, every human born lived as a hunter-gatherer. During this time an estimated 1 billion people explored how far you could go with a handful of tools. This world without much technology provided "enough." There was leisure and satisfying work for humans. Happiness, too. Without technology beyond stone implements, the rhythms and patterns of nature were immediate. Nature ruled your hunger and set your course. Nature was so vast, so bountiful, and so close, few humans could separate from it. The attunement with the natural world felt divine. Yet without much technology, the recurring tragedy of child death was ever present. Accidents, warfare, and disease meant your life, on average, was far less than half as long as it could have been—maybe only a quarter of the natural life span your genes afforded. Hunger was always near.

But most noticeably, without significant technology, your leisure was confined to traditional repetitions. There was no place for anything new. Within narrow limits you had no bosses. But the direction and interests of your life were laid out in well-worn paths. The cycles of your environment determined your life.

It turns out that the bounty of nature, though vast, does not hold all possibilities. The mind does, but it had not yet been fully unleashed. A world without technology had enough to sustain survival but not enough to transcend it. Only when the mind, liberated by language and enabled by the technium, transcended the constraints of nature 50,000 years ago did greater realms of possibility open up. There was a price to pay for this transcendence, but what we gained by this embrace was civilization and progress.

We are not the same folks who marched out of Africa. Our genes have coevolved with our inventions. In the past 10,000 years alone, in fact, our genes have evolved 100 times faster than the average rate for the previous 6 million years. This should not be a surprise. As we domesticated the dog (in all its breeds) from wolves and bred cows and corn and more from their unrecognizable ancestors, we, too, have been domesticated. We have domesticated ourselves. Our teeth continue to shrink (because of cooking, our external stomach), our muscles thin out, our hair disappears. Technology has domesticated us. As fast as we remake our tools, we remake ourselves. We are coevolving with our technology, and so we have become deeply dependent on it. If all technology—every last knife and spear—were to be removed from this planet, our species would not last more than a few months. We are now symbiotic with technology.

We have rapidly and significantly altered ourselves and at the same time altered the world. From the moment we emerged from Africa to colonize every inhabitable watershed on this planet, our inventions began to alter our nest. Sapiens' hunting tools and techniques had far-reaching effects: Their technology enabled them to kill off key herbivores (mammoths, giant elk, etc.) whose extinctions altered the ecology of entire grassland biomes forever. Once dominant grazers were eliminated, their absence cascaded through the ecosystem, enabling the rise of new predators, new plant species, and all their competitors and allies, surfacing a modified ecosystem. Thus a few clans of hominins shifted the destiny of thousands of other species. When Sapiens gained control of fire, this powerful technology further modified the natural terrain on a massive scale. Such a tiny trick—burning grasslands, controlling it with backfires, and summoning flames to cook grains—disrupted vast regions of the continents.

Later the repeated inventions and spread of agriculture around the planet affected not only the surface of the Earth, but its 100-kilometer-wide (60-mile-wide) atmosphere as well. Farming disturbed the soil and increased CO_2. Some climatologists believe that this early anthropogenic warming, starting 8,000 years ago, kept a new ice age at bay. Widespread adoption of farming disrupted a natural climate cycle that ordinarily would have refrozen the northernmost portions of the planet by now.

Of course, once humans invented machines that ate concentrated old plants (coal) instead of fresh plants, the mechanical exhalations of CO_2 further altered the balance of the atmosphere. The technium bloomed as machines harnessed this source of abundant energy. Petroleum-eating machines such as tractor engines transformed the productivity and spread of agriculture (accelerating an old trend), and then more machines drilled for more oil faster (a new trend), accelerating the rate of acceleration. Today the CO_2 exhalation of all machines greatly exceeds the exhalation of all animals and even approaches the volume generated by geological forces.

The technium gains its immense power not only from its scale but from its self-amplifying nature. One breakthrough invention, such as the alphabet, the steam pump, or electricity, can lead to further breakthrough inventions, such as books, coal mines, and telephones. These advances in turn led to other breakthrough inventions, such as libraries, power generators, and the internet. Each step adds further powers while retaining most of the virtues of the previous inventions. Someone has an idea (a spinning wheel!), which can hop to other minds, mutate into a derivative idea (place the spinning wheel beneath a sled to make it easy to haul!), which disrupts the prevailing balance, causing a shift.

But not all changes induced by technology have been positive. Industrial-scale slavery, such as that imposed upon Africa, was enabled by the sailing ships that transported captives across oceans and encouraged by the mechanical cotton gins that could cheaply process the fibers the slaves planted and harvested. Without technology, slavery at this massive scale would have been unknown. Thousands of synthetic toxins have caused mass disruptions of natural cycles in both humans and other species, a huge unwanted downside from small inventions. War is a particularly serious amplifier of the great negative powers brought by technology. Technological innovation has led directly to horrific weapons of destruction capable of inflicting entirely new atrocities upon society.

On the other hand, the remedies for and offsets of the negative consequences also stemmed from technology. Local ethnic slavery was practiced by most earlier civilizations, and probably in prehistoric times as well, and still continues in various remote areas; its overall diminishment globally is due to the technological tools of communication, law, and education. Technologies of detection and substitution can eliminate the routine use of synthetic toxins. The technologies of monitoring, law, treaties, policing, courts, citizen media, and economic globalism can temper, dampen, and in the long run diminish the vicious cycles of war.

Progress, even moral progress, is ultimately a human invention. It is a useful product of our wills and minds, and thus it is a technology. We can decide slavery is not a good idea. We can decide that fairly applied laws, rather than nepotistic favoritism, is a good idea. We can outlaw certain punishments with treaties. We can encourage accountability with the invention of writing. We can consciously expand our circle of empathy. These are all inventions, products of our minds, as much as lightbulbs and telegraphs are.

This cyclotron of social betterment is propelled by technology. Society evolves in incremental doses; each rise in social organization throughout history was driven by an insertion of a new technology. The invention of writing unleashed the leveling fairness of recorded laws. The invention of standard minted coins made trade more universal, encouraged entrepreneurship, and hastened the idea of liberty. Historian Lynn White notes, “Few inventions have been so simple as the stirrup, but few have had so catalytic an influence on history.” In White’s view, the adoption of the lowly foot stirrup for horse saddles enabled riders to use weapons on horseback, which gave an advantage to the cavalry over infantry and to the lords who could afford horses, and so nurtured the rise of aristocratic feudalism in Europe. The stirrup is not the only technology that has been blamed for feudalism. As Karl Marx famously claimed, "The hand-mill gives you society with the feudal lord; the steam-mill, society with the industrial capitalist."

Double-entry bookkeeping, invented in 1494 by a Franciscan monk, enabled companies to monitor their cash flow and for the first time to steer complex business. Double-entry accounting unleashed the banking industry in Venice and launched a global economy. The invention of moveable-type printing in Europe encouraged Christians to read their religion's founding text themselves and make their own interpretations, and that launched the very idea of "protest" within and against religion. Way back in 1620, Francis Bacon, the godfather of modern science, realized how powerful technology was becoming. He listed three "practical arts"—the printing press, gunpowder, and the magnetic compass—that had changed the world. He declared that "no empire, no sect, no start seems to have exerted greater power and influence in human affairs than these mechanical discoveries." Bacon helped launch the scientific method, which accelerated the speed of invention; thereafter society was in constant flux, as one conceptual seed after another disrupted social equilibrium.

Seemingly simple inventions like the clock had profound social consequences. The clock divided an unbroken stream of time into measurable units, and once it had a face, time became a tyrant, ordering our lives. Danny Hillis, computer scientist, believes the gears of the clock spun out science and all its many cultural descendants. He says, "The mechanism of the clock gave us a metaphor for self-governed operation of natural law. (The computer, with its mechanistic playing out of predetermined rules, is the direct descendant of the clock.) Once we were able to imagine the solar system as a clockwork automaton, the generalization to other aspects of nature was almost inevitable, and the process of Science began."

During the Industrial Revolution, our inventions transformed our daily routines. Mechanical contraptions and cheap fuel gave us plenty of food, nine-to-five days, and smokestacks. This phase of technology was dirty, disruptive, and often built and run at an inhuman scale. The stiff, cold, unbending nature of raw steel, brick, and glass cast the encroachment as alien, in opposition to us, if not to all living things. It directly fed upon natural resources and so had a devilish shadow. The worst by-products of the industrial age—black smoke, black river waters, blackened short lives working in the mills—were so remote from our cherished self-conception that we wanted to believe the source itself was alien. Or worse. It was not difficult to eye the hard, cold material takeover as evil, even if a necessary evil. When technology appeared among our age-old routines, it was set outside ourselves and treated like an infection. People embraced its products, but guiltily. It would have been ludicrous a century ago to think of technology as ordained. It was a suspect force. When two world wars unleashed the full killing power of this inventiveness, it cemented the reputation of technology as a beguiling satan.

As we refined this stuff through generations of technological evolution, it lost much of its hardness. We began to see through technology's disguise as material and began to see it primarily as action. While it inhabited a body, its heart was something softer. In 1949, John von Neumann, the brainy genius behind the first useful computer, realized what computers were teaching us about technology: "Technology will in the near and in the farther future increasingly turn from problems of intensity, substance, and energy, to problems of structure, organization, information, and control." No longer a noun, technology was becoming a force—a vital spirit that throws us forward or pushes against us. Not a thing but a verb.

3

History of the Seventh Kingdom

Looking back at Paleolithic times, we can observe an evolutionary phase when human tools were embryonic, when the technium existed in its most minimal state. But since technology predated humans, appearing in primates and even earlier, we need to look beyond our own origins to understand the true nature of technological development. Technology is not just a human invention; it was also born from life.

If we chart the varieties of life we have so far discovered on Earth, they fall into six broad categories. Within each of these six categories, or kingdoms of life, all species share a common biochemical blueprint. Three of these kingdoms are the tiny microscopic stuff: one-celled organisms. The other three are the biological kingdoms of organisms we normally see: fungi (mushrooms and molds), plants, and animals.

Every species in the six kingdoms, which is to say every organism alive on Earth today, from algae to zebra, is equally evolved. Despite the differences in the sophistication and development of their forms, all living species have evolved from predecessors for the same amount of time: four billion years. All have been tested daily and have managed to adapt across hundreds of millions of generations in an unbroken chain.

Many of these organisms have learned to build structures, and those structures have allowed the creature to extend itself beyond its tissue. The hard two-meter mound of a termite colony operates as if it were an external organ of the insects: The mound's temperature is regulated and it is repaired after injury. The dried mud itself seems to be living. What we think of as coral—stony, treelike structures—are the apartment buildings of nearly invisible coral animals. The coral structure and coral animals behave as one. It grows, breathes. The waxy interior of a beehive or the twiggy architecture of a bird's nest works the same way. Therefore a nest or a hive can best be considered a body built rather than grown. A shelter is animal technology, the animal extended.

The extended human is the technium. Marshall McLuhan, among others, noted that clothes are people's extended skin, wheels extended feet, camera and telescopes extended eyes. Our technological creations are great extrapolations of the bodies that our genes build. In this way, we can think of technology as our extended body. During the industrial age it was easy to see the world this way. Steam-powered shovels, locomotives, television, and the levers and gears of engineers were a fabulous exoskeleton that turned man into superman. A closer look reveals the flaw in this analogy: The extended costume of animals is the result of their genes. They inherit the basic blueprints of what they make. Humans don't. The blueprints for our shells spring from our minds, which may spontaneously create something none of our ancestors ever made or even imagined. If technology is an extension of humans, it is not an extension of our genes but of our minds. Technology is therefore the extended body for ideas.

With minor differences, the evolution of the technium—the organism of ideas—mimics the evolution of genetic organisms. The two share many traits: The evolution of both systems moves from the simple to the complex, from the general to the specific, from uniformity to diversity, from individualism to mutualism, from energy waste to efficiency, and from slow change to greater evolvability. The way that a species of technology changes over time fits a pattern similar to a genealogical tree of species evolution. But instead of expressing the work of genes, technology expresses ideas.

Yet ideas never stand alone. They come woven in a web of auxiliary ideas, consequential notions, supporting concepts, foundational assumptions, side effects, and logical consequences and a cascade of subsequent possibilities. Ideas fly in flocks. To hold one idea in mind means to hold a cloud of them.

Most new ideas and new inventions are disjointed ideas merged. Innovations in the design of clocks inspired better windmills, furnaces engineered to brew beer turned out to be useful to the iron industry, mechanisms invented for organ making were applied to looms, and mechanisms in looms became computer software. Often unrelated parts end up as a tightly integrated system in a more evolved design. Most engines combined heat-producing pistons with a cooling radiator. But the clever air-cooled engine merges two ideas into one: The engine contains the pistons but also doubles as a radiator to dissipate the heat they generate. "In technology, combinatorial evolution is foremost, and routine," says economist Brian Arthur in *The Nature of Technology.* "Many of a technology's parts are shared by other technologies, so a great deal of development happens automatically as components improve in other uses 'outside' the host technology."

These combinations are like mating. They produce a hereditary tree of ancestral technologies. Just as in Darwinian evolution, tiny improvements are rewarded with more copies, so that innovations spread steadily through the population. Older ideas merge and hatch idea-lings. Not only do technologies form ecosystems of cross-supported allies, but they also form evolutionary lines. The technium can really only be understood as a type of evolutionary life.

We can arrange the story of life in several ways. One way chronicles biological landmarks. At the top of the list of life's greatest million-year passages would be the point when organisms migrated from the sea to land or the period when they acquired backbones or the era in which they developed eyes. Other milestones would be the arrival of flowering plants or the demise of dinosaurs and the rise of mammals. These are important benchmarks in our past and legitimate achievements in our ancestors' tale.

But since life is a self-generated information system, a more revealing way to view the four-billion-year history of life is to mark the major transitions in the informational organization of life's forms. Of the many ways in which a mammal differs from, say, a sponge, one of the primary differences is the additional layers in which information flows through the organism. To view life's stages we need to call out the major transitions of life's structures over evolutionary time. This was the method of biologists John Maynard Smith and Eors Szathmary, who recently found eight thresholds of biological information in life's history.

They concluded that the major transitions in biological organization were:

> One replicating molecule ➔ Interacting population of replicating molecules
> Replicating molecules ➔ Replicating molecules strung into chromosome
> Chromosome of RNA enzymes ➔ DNA proteins
> Cell without nucleus ➔ Cell with nucleus
> Asexual reproduction (cloning) ➔ Sexual recombination
> Single-cell organism ➔ Multicell organism
> Solitary individual ➔ Colonies and superorganisms
> Primate societies ➔ Language-based societies

Each level in their hierarchy marks a major advance in complexity. The invention of sex is probably the biggest step in the reordering of biological information. By permitting a controlled recombination of traits (some traits from each partner) rather than either the pure random diversity of mutations or the rigid sameness of clones, sex maximizes evolvability. Animals using sexual recombination of genes will evolve faster than their competitors. The later natural invention of multicellularity and, still later, the invention of colonies of multicell organisms each supply Darwinian survival advantages. But more important, these innovations serve as platforms that permit biological informational bits to be organized in newer, more easily organized ways.

The evolution of science and technology parallels the evolution of nature. The major technological transitions are also passages from one level of organization to another. Rather than catalog important inventions such as iron, steam power, or electricity, in this view we catalog how the structure of information is reshaped by new technology. A prime example would be the transformation of alphabets (strings of symbols not unlike DNA) into highly organized knowledge in books, indexes, libraries, and so on (not unlike cells and organisms).

In a parallel to Smith and Szathmary, I have arranged the major transitions in technology according to the level at which information is organized. At each step, information and knowledge are processed at a level not present before.

The major transitions in the technium are:

Primate communication ➔ Language
Oral lore ➔ Writing/mathematical notation
Scripts ➔ Printing
Book knowledge ➔ Scientific method
Artisan production ➔ Mass production
Industrial culture ➔ Ubiquitous global communication

No transition in technology has affected our species, or the world at large, more than the first one, the creation of language. Language enabled information to be stored in a memory greater than an individual's recall. A language-based culture accumulated stories and oral wisdom to disseminate to future generations. The learning of individuals, even if they died before reproducing, would be remembered. From a systems point of view, language enabled humans to adapt and transmit learning faster than genes.

The invention of writing systems for language and math structured this learning even more. Ideas could be indexed, retrieved, and propagated more easily. Writing allowed the organization of information to penetrate into many everyday aspects of life. It accelerated trade, the creation of calendars, and the formation of laws—all of which organized information further.

Printing organized information still more by making literacy widespread. As printing became ubiquitous, so did symbolic manipulation. Libraries, catalogs, cross-referencing, dictionaries, concordances, and the publishing of minute observations all blossomed, producing a new level of informational ubiquity—to the extent that today we don't even notice that printing covers our visual landscape.

The scientific method followed printing as a more refined way to deal with the exploding amount of information humans were generating. Via peer-reviewed correspondence and, later, journals, science offered a method of extracting reliable information, testing it, and then linking it to a growing body of other tested, interlinked facts.

This newly ordered information—what we call science—could then be used to restructure the organization of matter. It birthed new materials, new processes for making stuff, new tools, and new perspectives. When the scientific method was applied to craft, we invented mass production of interchangeable parts, the assembly line, efficiency, and specialization. All these forms of informational organization launched the incredible rise in standards of living we take for granted.

Finally, the latest transition in the organization of knowledge is happening now. We inject order and design into everything we manufacture. We are also adding microscopic chips that can perform small amounts of computation and communication. Even the tiniest disposable item with a bar code shares a thin sliver of our collective mind. This all-pervasive flow of information, expanded to include manufactured objects as well as humans, and distributed around the globe in one large web, is the greatest (but not final) ordering of information.

The trajectory of increasing order in the technium follows the same path that it does in life. Within both life and the technium, the thickening of interconnections at one level weaves the new level of organization above it. And it's important to note that the major transitions in the technium begin at the level where the major transitions in biology left off: Primate societies give rise to language.

The invention of language marks the last major transformation in the natural world and also the first transformation in the manufactured world. Words, ideas, and concepts are the most complex things social animals (like us) make, and also the simplest foundation for any type of technology. Thus language bridges the two sequences of major transitions and unites them into one continuous sequence, so that natural evolution flows into technological evolution. The complete sequence of major transitions in deep history runs like this:

One replicating molecule → Interacting population of replicating molecules
Replicating molecules → Replicating molecules strung into chromosome
Chromosome of RNA enzymes → DNA proteins
Cell without nucleus → Cell with nucleus
Asexual reproduction (cloning) → Sexual recombination
Single-cell organism → Multicell organism
Solitary individual → Colonies and superorganisms
Primate societies → Language-based societies
Oral lore → Writing/mathematical notation
Scripts → Printing
Book knowledge → Scientific method
Artisan production → Mass production
Industrial culture → Ubiquitous global communication

This escalating stack of increasing order is revealed to be one long story. We can think of the technium as the further reorganization of information that began with the six kingdoms of life. In this way, the technium becomes the seventh kingdom of life. It extends a process begun four billion years ago. Just as the evolutionary tree of Sapiens branched off from its animal precursors long ago, the technium now branches off from its precursor, the mind of the human animal. Outward from this common root flow new species of hammers, wheels, screws, refined metal, and domesticated crops, as well as rarefied species like quantum computers, genetic engineering, jet planes, and the World Wide Web.

The technium differs from the other six kingdoms in a couple of important ways. Compared to members of the other six kingdoms, these new species are the most ephemeral species on Earth. The bristlecone pines have watched entire families and classes of technology come and go. Nothing we have made approaches the endurance of the least living thing. Many digital technologies have shorter life spans than individual mayflies, let alone species.

But nature can't plan ahead. It does not hoard innovations for later use. If a variation in nature does not provide an *immediate* survival advantage, it is too costly to maintain and so over time it disappears. But sometimes a trait advantageous for one problem will turn out to be advantageous for a second, unanticipated problem. For instance, feathers evolved to warm a small, cold-blooded dinosaur. Later on, these same feathers, once installed on limbs for warmth, proved handy for short flights. From this heat-conservation innovation came unplanned wings and birds. These inadvertent anticipatory inventions are called exaptations in biology. We don't know how common exaptations are in nature, but they are routine in the technium. The technium is nothing but exaptations, since innovations can be easily borrowed across lines of origin or moved across time and repurposed.

Niles Eldredge is the cofounder (with Stephen Jay Gould) of the theory of punctuated, stepwise evolution. His professional expertise is the history of trilobites, or ancient arthropods that resemble today's pill bugs. As a hobby he collects cornets, musical instruments very similar to trumpets. Once Eldredge applied his professional taxonomic methods to his collection of 500 cornets, some dating back to 1825. He selected 17 traits that varied among his instruments—the shape of their horns, the placement of the valves, the length and diameter of their tubes—very similar to the kinds of metrics he applies to trilobites. When he mapped the evolution of cornets using techniques similar to those he applies to ancient arthropods, he found that the pattern of the lineages were very similar in many ways to those of living organisms. As one example, the evolution of cornets showed a stepwise progress, much like trilobites. But the evolution of musical instruments was also very distinctive. The key difference between the evolution of multicellular life and the evolution of the technium is that in life most blending of traits happens "vertically" in time. Innovations are passed from living parents down (vertically) through offspring. In the technium, on the other hand, most blending of traits happens laterally across time—even from "extinct" species and across lineages from nonparents. Eldredge discovered that the pattern of evolution in the technium is not the repeated forking of branches we associate with the tree of life, but rather a spreading, recursive network of pathways that often double back to "dead" ideas and resurrect "lost" traits. Another way of saying the same thing: Early

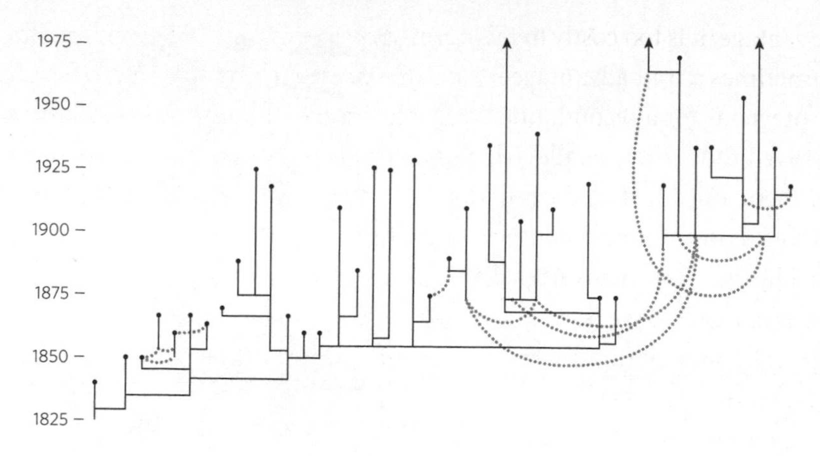

Evolutionary Tree of Cornets. The design heritage for each musical instrument shows how some branches borrow from far earlier models or nonadjacent branches (dotted lines), unlike organic evolution.

traits (exaptations) anticipate the later lineages that adopt them. These two patterns were distinct enough that Eldridge claims one could use it to identify whether an evolutionary tree depicted a clan of the born or of the made.

The second difference between evolution of the technium and evolution of the organic is that incremental transformation is the rule in biology. There are very few revolutionary steps; everything advances via a very long series of tiny steps, each one of which must work for the creature at the time. In contrast, technology can jump ahead, make abrupt leaps, and skip over incremental steps. As Eldredge points out, "No way did the transistor 'evolve from' the vacuum tube the way the eyes on one side of a flatfish's head are derived from the original bilaterally symmetrical conformation of the ancestral fish." Instead of the hundreds of millions of incremental improvements the flatfish endured, the transistor leaped from the ancestral vacuum tube via dozens of iterations at the most.

But by far the greatest difference between the evolution of the born and the evolution of the made is that species of technology, unlike species in biology, almost never go extinct. A close examination of a supposedly extinct bygone technology almost always shows that somewhere

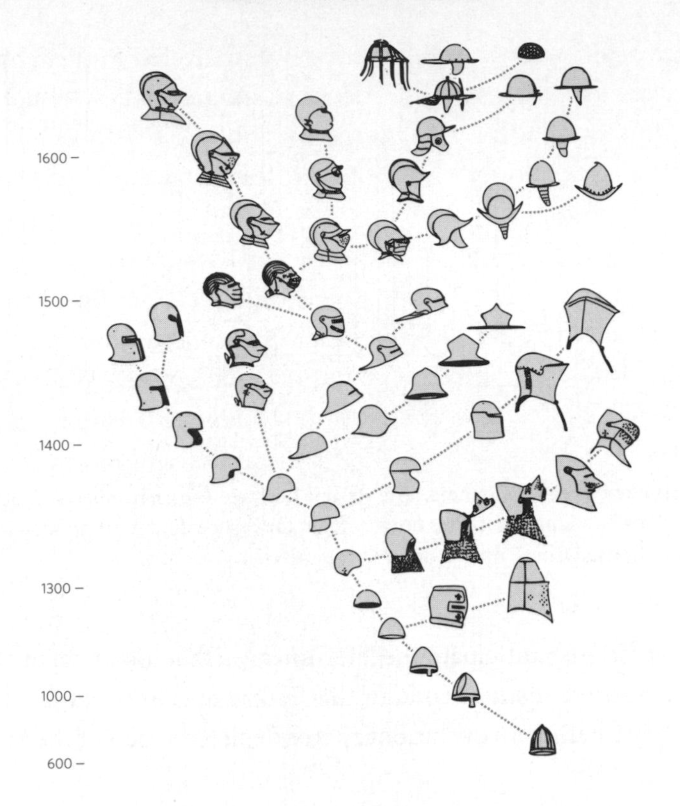

A Thousand Years of Helmet Evolution. The American zoologist and medieval armor expert Bashford Dean sketched out this diagramatic "genealogical tree" of the evolution of medieval European helmets starting in the year 600.

on the planet someone is still producing it. A technique or artifact may be rare in the modern urban world but quite common in the developing rural world. For instance, Burma is full of oxcart technology; basketry is ubiquitous in most of Africa; hand spinning is still thriving in Bolivia. A supposedly dead technology may be enthusiastically embraced by a heritage-based minority in modern society, if only for ritual satisfaction. Consider the traditional ways of the Amish, or modern tribal communities, or fanatical vinyl record collectors. Often old technology is obsolete, that is, it is not very ubiquitous or is second rate, but it still may be in small-time use. For just one of many examples, as late as 1962, in what was then called the atomic age, many small businesses on a block in Boston ran machines using steam power delivered to them by overhead driveshafts. This kind of anachronistic technology is not at all unusual.

In my own travels around the world I was struck by how resilient ancient technologies were, how they were often first choices where power and modern resources were scarce. It seemed to me as if no technologies ever disappeared. I was challenged on this conclusion by a highly regarded historian of technology who told me without thinking, "Look, they don't make steam-powered automobiles anymore." Well, within a few clicks on Google I very quickly located folks who are making *brand-new* parts for Stanley steam-powered cars. Nice shiny copper valves, pistons, whatever you need. With enough money you could put together an entirely new steam-powered car. And of course, thousand of hobbyists are still bolting together steam-powered vehicles, and hundreds more are keeping old ones running. Steam power is very much an intact, though uncommon, species of technology.

I decided to see how many old technologies a postmodern urban citizen living in a cosmopolitan city (like San Francisco) could lay his hands on. One hundred years ago, there was no electricity, no internal combustion engines, few highways, and little long-distance communication except via the post office network. But through that postal network you could order almost anything manufactured from the Montgomery Ward catalog. The faded newsprint of my reproduction catalog had the air of a mausoleum of a long-dead civilization. However, it became quickly and surprisingly clear that most of the thousands of items for sale 100 years ago, as cataloged by this wish book, were still for sale now. Although the styling is different, the underlying technology, function, and form are the same. A leather boot with doodads is still a leather boot.

I set myself the challenge of finding all the products on a sample page from the 1894–95 Montgomery Ward catalog. Flipping through its 600 pages, I selected one fairly typical page that featured agricultural implements. These types of obsolete tools would be far harder to find today than, say, the stove pots, lamps, clocks, pens, and hammers that populate the rest of the pages. Farm tools seemed like certain dinosaurs. Who

Catalogs of Durable Goods. On the left, page 562 of the 1894–95 Montgomery Ward catalog offering farm implements by mail order. On the right, the equivalent brand-new items offered by various sources on the web in 2005.

needs a hand-powered corncob sheller, or a paint mill, whatever that was? If I could purchase these obsolete tools from the agricultural era it would strongly suggest not much was gone.

Of course it's a no-brainer to find antiques on eBay. My test was to find newly manufactured versions of this equipment, since this would show that these species were still viable.

The results stunned me. In a few hours I was able to find every single item listed on this page of a century-old catalog. Each old tool was available in a new incarnation and sold on the web. Nothing was dead.

I haven't done the research to find out the reason for the survival of each item, but I suspect that most of these tools share a similar story. While working farms have shed these obsolete tools entirely and are almost completely automated, many of us still garden with very primitive hand tools simply because they work. As long as backyard tomatoes taste better than farmed ones, the primeval hoe will survive. And apparently, there's pleasure in harvesting some crops by hand, even in bulk. I suspect a few of these items may be bought by the Amish and other back-to-the-landers who find virtue in doing things without oil-fed machinery.

But maybe 1895 is not old enough. Let's take the oldest technology of all: a flint knife or stone ax. Well, it turns out you can buy a brand-new flint knife, flaked by hand and carefully attached to an antler-horn handle by tightly wound leather straps. In every respect it is precisely the same technology as a flint knife made 30,000 years ago. It's yours for fifty dollars, available from more than one website. In the highlands of New Guinea, tribesmen were making stone axes for their own use until the 1960s. They still make stone axes the same way for tourists now. And stone-ax aficionados study them. There is an unbroken chain of knowledge that has kept this Stone Age technology alive. Today, in the United States alone, there are 5,000 amateurs who knap fresh arrowhead points by hand. They meet on weekends, exchange tips in flint-knapping clubs, and sell their points to souvenir brokers. John Whittaker, a professional archaeologist and flint knapper himself, has studied these amateurs and estimates that they produce over one million brand-new spear and arrow points per year. These new points are indistinguishable, even to experts like Whittaker, from authentic ancient ones.

Few technologies have disappeared forever from the face of the Earth. The recipe for Greek warfare was lost for millennia, but there is a good chance research has recovered it. The practical know-how for the Inca system of accounting using knots on a string, called *quipu*, is forgotten. We have some antique samples, but no knowledge of how they were actually used. This might be the single exception. Not too long ago, science fiction authors Bruce Sterling and Richard Kadrey compiled a list of "dead media" to highlight the ephemeral nature of popular gadgetry. Recently vanished gizmos such as the Commodore 64 computer and the Atari computer were added to a long list of older species such as lantern slide projectors and the telharmonium. In reality, though, most of the items on this list aren't dead, just rare. Some of the oldest media technologies are maintained by basement tinkerers and crazy amateur enthusiasts. And many of the more recent technologies are still in production but under different brand names and configurations. For instance, a lot of the technology first introduced in early computers is now found inside your watch or toys.

With very few exceptions, technologies don't die. In this way they differ from biological species, which in the long term inevitably go extinct. Technologies are idea based, and culture is their memory. They can be resurrected if forgotten, and can be recorded (by increasingly better means) so that they won't be overlooked. Technologies are forever. They are the enduring edge of the seventh kingdom of life.

4

The Rise of Exotropy

The origin of the technium can be retold in concentric creation stories. Each retelling illuminates a deeper set of influences. In the first account (chapter two), technology begins with the Sapien mind but soon transcends it. The second telling (chapter three) reveals an additional force besides the human mind at work on the technium: the extrapolation and deepening of organic life as a whole. Now in this third version, the circle is enlarged further, beyond mind and life, to include the cosmos.

The root of the technium can be traced back to the life of an atom. An atom's brief journey through an everyday technological artifact, such as a flashlight battery, is a flash of existence unlike anything else in its long life.

Most hydrogen atoms were born at the beginning of time. They are as old as time itself. They were created in the fires of the big bang and dispersed into the universe as a uniform warm mist. Thereafter, each atom has been on a lonely journey. When a hydrogen atom drifts in the unconsciousness of deep space, hundreds of kilometers from another atom, it is hardly much more active than the vacuum surrounding it. Time is meaningless without change, and in the vast reaches of space that fill 99.99 percent of the universe, there is little change.

After billions of years, a hydrogen atom might be swept up by the currents of gravity radiating from a congealing galaxy. With the dimmest hint of time and change it slowly drifts in a steady direction toward other stuff. Another billion years later it bumps into the first bit of matter it has ever encountered. After millions of years it meets the second. In time it meets another of its kind, a hydrogen atom. They drift together in mild attraction until aeons later they meet an oxygen atom. Suddenly something weird happens. In a flash of heat they clump together as one water molecule. Maybe they get sucked into the atmosphere circulation of a planet. Under this marriage, they are caught in great cycles of change. Rapidly the molecule is carried up and then rained down into a crowded pool of other jostling atoms. In the company of uncountable numbers of other water molecules it travels this circuit around and around for millions of years, from crammed pools to expansive clouds and back. One day, in a stroke of luck, the water molecule is captured by a chain of unusually active carbons in one pool. Its path is once again accelerated. It spins around in a simple loop, assisting the travel of carbon chains. It enjoys speed, movement, and change such as would not be possible in the comatose recesses of space. The carbon chain is stolen by another chain and reassembled many times until the hydrogen finds itself in a cell constantly rearranging its relations and bonds with other molecules. Now it hardly ever stops changing, never stops interacting.

The hydrogen atoms in a human body completely refresh every seven years. As we age we are really a river of cosmically old atoms. The carbons in our bodies were produced in the dust of a star. The bulk of matter in our hands, skin, eyes, and hearts was made near the beginning of time, billions of years ago. We are much older than we look.

For the average hydrogen atom in our body, the few years it spends dashing from one cellular station to another will be the most fleeting glory imaginable. Fourteen billion years in inert lassitude, then a brief, wild trip through life's waters, and then on again to the isolation of space when the planet dies. A blink is too long as an analogy. From the perspective of an atom, any living organism is a tornado that might capture it into its mad frenzy of chaos and order, offering it a once-in-a-14-billion-year-lifetime fling.

As fast and crazy as a cell is, the rate of energy flowing through technology is even faster. In fact, technology is more active in this respect—it will give an atom a wilder ride—than any other sustainable structure we are currently aware of. For the ultimate trip today, the most sustainable energetic thing in the universe is a computer chip.

There is a more precise way to say this: Of all the sustainable things in the universe, from a planet to a star, from a daisy to an automobile, from a brain to an eye, the thing that is able to conduct the highest density of power—the most energy flowing through a gram of matter each second—lies at the core of your laptop. How can this be? The power density of a star is huge compared to the mild power drifting through a nebulous gas cloud in space. But remarkably, the power density of a sun pales in comparison to the intense flow of energy and activity present in grass. As intense as the surface of the sun is, its mass is enormous and its lifetime is 10 billion years, so as a whole system, the amount of energy flowing through it per gram per second is less than that in a sunflower soaking up that sun's energy.

An exploding nuclear bomb has a much higher power density than the sun because it is an unsustainable out-of-control flow of energy. A one-megaton nuclear bomb will release 10^{17} ergs, which is a lot of power. But the total lifetime of that explosion is only a hyperblink of 10^{-6} seconds. So if you "amortized" a nuclear blast so that it spent its energy over a full second instead of microseconds, its power density would be reduced to only 10^{11} ergs per second per gram, which is about the intensity of a laptop computer chip. Energywise, a Pentium chip may be better thought of as a very slow nuclear explosion.

The same fleeting flameout seen in a nuke applies to fires, chemical bombs, supernovas, and other kinds of explosions. They literally consume themselves with incredibly high but unsustainable densities of energy. The glory of a sunlike star is that it can sustain its brilliant fission for billions of years. But it does so at a lower energy flow rate than the sustainable flux that takes place in a green plant! Rather than a burst of fire, the energy exchange in grass yields the cool order of green blades, tawny stalks, and plump seeds ripe with information that can duplicate a picture-perfect clone. Greater yet is the steady energy flow within animals, where we can actually sense the energetic waves. They wiggle, pulse, move, and in some cases radiate warmth.

The flow of energy through technology is still greater. Measured in

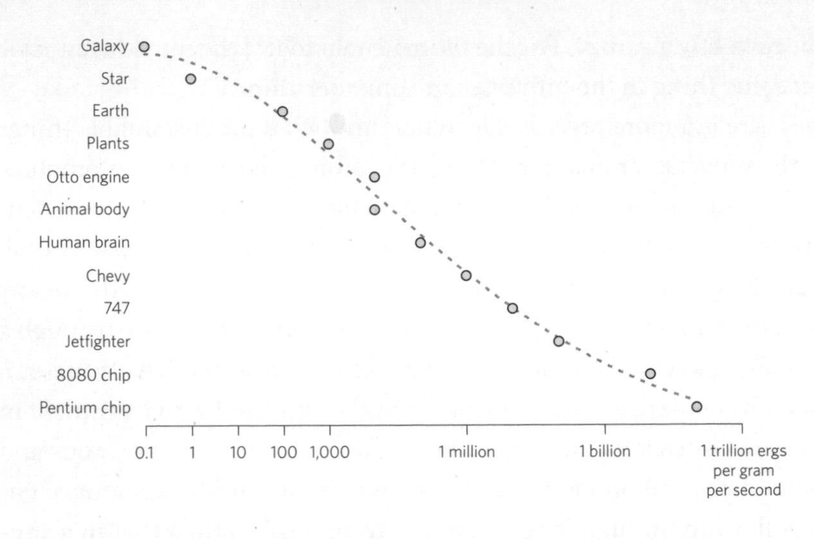

Power Density Gradient. Large, complex systems listed in order of their energy flow density, as measured by the amount of energy that flows through the system per gram per second of the system's duration.

joules (or ergs) per gram per second, nothing concentrates energy for long periods of time as much as high-tech gadgetry. At the far right apex of the power density graph above, compiled by physicist Eric Chaisson, shines the computer chip. It conducts more energy per second per gram through its tiny corridors than animals, volcanoes, or the sun. This bit of high technology is the most energetically active thing in the known universe.

We can now retell the story of the technium as a story of expanding cosmic activity. At the very start of creation, the universe, such as it was, was packed into a very, very small space. The entire cosmos began as a flash smaller than the smallest bit of the smallest particle in the smallest atom. It was equally hot and bright and dense within that dot. All parts of this too-tiny spot shared a uniform temperature. There was, in fact, no room for any differences, and no activity at all.

But from the very start of its creation, this tiny spot expanded by a process we don't understand. Every new point flew away from every other new point. As the universe ballooned to about the size of your head, coolness became possible. Before it expanded to that size, in its first three seconds, the universe was perfectly solid, with no emptiness for relief. It was so full, even light could not move. Indeed, it was so uniform that the four fundamental forces we see at work in reality today—gravity, electromagnetism, the strong and weak nuclear forces—were compressed into a single unified force. In that start-up phase there was *one* general energy, which differentiated into four distinct forces as the universe expanded.

It would not be too much of an exaggeration to say that in the initial femtoseconds of creation there was only one thing in the universe, one superdense power that ruled all, and this solitary power expanded and cooled into thousands of variations of itself. The history of the cosmos thus proceeds from unity to diversity.

As the universe stretched out, it made nothingness. As emptiness increased, so did coolness. Space permitted energy to cool into matter and for matter to slow down, light to radiate, and gravity and the other energetic forces to unfold.

Energy is simply the potential—the difference needed—to cool. Energy can only flow from greater to lesser, so without a differential no energy can flow. Curiously, the universe expanded faster than matter itself could cool and gel, which means the potential for cooling kept increasing. The faster the universe expanded, the greater was its potential to cool and the greater were the potential differences within its boundaries. Over aeons of cosmic time this expanding differential (between expanding emptiness and the remnant hotness of the big bang) powered evolution, life, intelligence, and eventually the acceleration of technology.

Energy, like water under gravity, will seep to the lowest, coolest level and not rest until all differential has been eliminated. In the first thousand years after the big bang the temperature difference within the universe was so small that it would have reached equilibrium quickly. Had not the universe kept expanding, very little interesting would have happened. But the expansion of the universe put a tilt into things. By expanding omnidirectionally—every point receding from every other point—space provided an empty bottom, a basement of sorts, down which energy could flow. The faster the cosmos enlarged, the bigger the basement it constructed.

At the very bottom of the basement lies the final end state known as heat death. It is absolutely still. There is no movement because there is no difference. No potential. Picture it as lightless, silent, and identical in all directions. All distinctions—including the elemental distinction between this and that—have been spent. This hell of uniformity is called maximum *entropy*. Entropy is the crisp scientific name for waste, chaos, and disorder. As far as we know, the sole law of physics with no known exceptions anywhere in the universe is this: All creation is headed to the basement. Everything in the universe is steadily sliding down the slope toward the supreme equality of wasted heat and maximum entropy.

We see the slope all around us in many ways. Because of entropy, fast-moving things slow down, order fizzles into chaos, and it costs something for any type of difference or individuality to remain unique. Each difference—whether of speed, structure, or behavior—becomes less different very quickly because every action leaks energy down the tilt. Difference within the universe is not free. It has to be maintained against the grain.

The effort to maintain difference against the pull of entropy creates the spectacle of nature. A predator such as an eagle sits atop a pyramid of entropic waste: In one year 1 eagle eats 100 trout, which eat 10,000 grasshoppers, which eat 1 million blades of grass. Thus it takes, indirectly, 1 million blades of grass to support 1 eagle. But this pile of 1 million blades far outweighs the eagle. This bloated inefficiency is due to entropy. Each movement in an animal's life wastes a small bit of heat (entropy), which means every predator catches less energy than the total energy the prey consumed, and this shortfall is multiplied by each action for all time. The circle of life is kept going only by the constant replenishment of sunlight showering the grass with new energy.

This inevitable waste is so harsh and unavoidable that it is astounding that any organization can persist for long without rapidly dissolving to cold equilibrium. Everything we find interesting and good in the cosmos—living organisms, civilization, communities, intelligence, evolution itself—somehow maintains a persistent difference in the face of entropy's empty indifference. A flatworm, a galaxy, and a digital camera all have this same property—they maintain a state of difference far removed from thermal undifferentiation. That state of cosmic lassitude and stillness is the norm for most atoms of the universe. While the rest of the material cosmos slips down to the frozen basement, only a remarkable few will catch a wave of energy to rise up and dance.

This rising flow of sustainable difference is the inversion of entropy. For the sake of this narrative, call it *exotropy*—a turning outward. *Exotropy* is another word for the technical term *negentropy*, or negative entropy. It was originally coined by the philosopher Max More, though he spelled it extropy. I've appropriated his term with an alternative spelling to heighten its distinction from its opposite entropy. I prefer *exotropy* over *negentropy* because it is a positive term for an otherwise double negative phrase meaning "the absence of the absence of order." Exotropy, in this tale, is far more uplifting than simply the subtraction of chaos. Exotropy can be thought of as a force in its own right that flings forward an unbroken sequence of unlikely existences.

Exotropy is neither wave nor particle, nor pure energy, nor supernatural miracle. It is an immaterial flow that is very much like information. Since exotropy is defined as negative entropy—the reversal of disorder—it is, by definition, an increase in order. But what is order? For simple physical systems, the concepts of thermodynamics suffice, but for the real world of cucumbers, brains, books, and self-driving trucks, we don't have useful metrics for exotropy. The best we can say is that exotropy resembles, but is not equivalent to, information and that it entails self-organization.

We can't make an exact informational definition of exotropy because we don't really know what information is. In fact the term *information* covers several contradictory concepts that should have their own terms. We use *information* to mean (1) a bunch of bits or (2) a meaningful signal. Confusingly, bits rise but signals decrease when entropy gains, so one kind of information increases while the other kind decreases. Until we clarify our language, the term *information* is more metaphor than anything else. I try to use it in the second meaning here (not always consistently): Information is a signal of bits that makes a difference.

Muddying the waters further, information is the reigning metaphor of the moment. We tend to interpret the mysteries surrounding life in imagery suggested by the most complex system we are aware of at the time. Once nature was described as a body, then a clock in the age of clocks, then a machine in the industrial age. Now, in the "digital age," we apply the computational metaphor. To explain how our minds work, or how evolution advances, we apply the pattern of a very large software program processing bits of information. None of these historical metaphors is wrong; they are just incomplete. Ditto for our newest metaphor of information and computation.

But exotropy, as rising order, must entail more than information alone. We have thousands of years of science ahead of us, and thousands of metaphors. Information and computation can't be the most complex immaterial entity there is, just the most complex we've discovered so far. We might eventually discover that exotropy involves quantum dynamics, or gravity, or even quantum gravity. But for now, information (in the sense of structure) is a better analogy than anything else we know of for understanding the nature of exotropy.

From one cosmic perspective, information is the dominant force in our world. In the initial era of the universe, back just after the big bang, energy dominated existence. At that time radiation was all there was. The universe was a glow. Slowly, as space expanded and cooled, matter took over. Matter was clumpy, unevenly distributed, but its crystallization generated gravity, which began to shape space. With the rise of life (in our immediate neighborhood), information ascended in influence. The informational process we call life took control of the atmosphere of Earth several billion years ago. Now the technium, another informational processing, is reconquering it. Exotropy's rise in the universe (from the perspective of our planet) might look like the chart on the opposite page, where E = energy, M = mass, and I = information.

The multibillion-year rise of exotropy—as it flings up stable molecules, solar systems, a planetary atmosphere, life, mind, and the technium—can be restated as the slow accumulation of ordered information. Or rather, the slow ordering of accumulated information.

This is more clearly seen at the extreme. The difference between four bottles of nucleotides on a laboratory shelf and the four nucleotides arrayed in your chromosomes lies in the additional structure, or ordering,

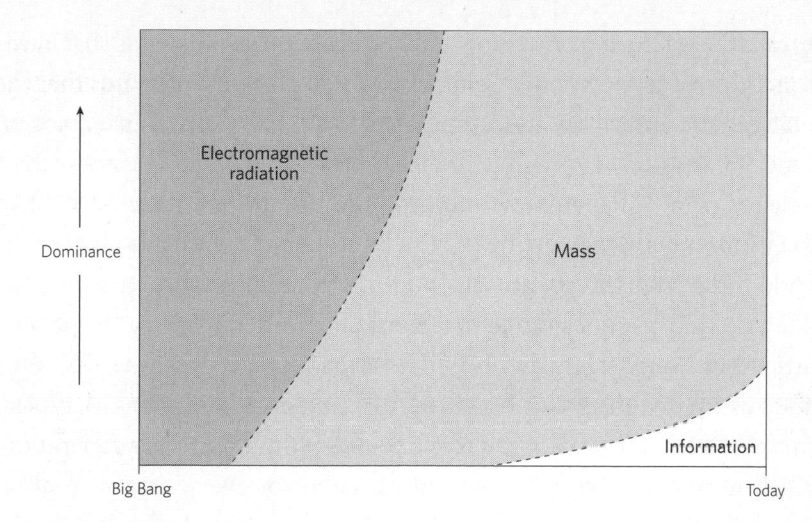

Dominant Eras of the Universe. The relative dominant force in our local area of the universe has shifted since the big bang. Time is indicated on a log scale, its units exponentially increasing over time. On this scale a few nanoseconds at the dawn of time occupy the same horizontal distance as a billion years today.

those atoms get from participating in the spirals of your replicating DNA. Same atoms, but more order. Those atoms of nucleotides acquire yet another level of structure and order when their cellular host undergoes evolution. As organisms evolve, the informational code their atoms carry is manipulated, processed, and reordered. In addition to genetic information, the atoms now convey adaptive information. They gain order from the innovations that survive. Over time, the same atoms can be promoted to new levels of order. Perhaps their one-cell home joins another cell to become multicellular—that demands the informational architecture for a larger organism as well as a cell. Further transitions in evolution—the aggregation into tissues and organs, the acquisition of sex, the creation of social groups—continue to elevate the order and increase the structure of the information flowing through those same atoms.

For four billion years evolution has been accumulating knowledge in its library of genes. You can learn a lot in four billion years. Every one of the 30 million or so unique species alive on the planet today is an unbroken informational thread that traces back to the very first cell. That thread (DNA) learns something new each generation and adds that hard-won knowledge to its code. Geneticist Motoo Kimura estimates that the total genetic information accumulated since the Cambrian explosion some 500 million years ago is 10 megabytes per genetic lineage (such as a parrot or a wallaby). Now multiply the unique information held in every individual organism by the total number of organisms alive in the world today and you get an astronomically large treasure. Imagine the Noah's Ark of digital storage that would be needed to carry the genetic payload of every organism on Earth (seeds, eggs, spores, sperms). One study estimated the Earth harbored 10^{30} single-cell microbes. A typical microbe, such as a yeast, produces one one-bit mutation per generation, which means one bit of unique information for every organism alive. Counting the microbes alone (about 50 percent of the biomass), the biosphere today contains 10^{30} bits, or 10^{29} bytes, or 10,000 yottabytes of genetic information. That's a lot.

And that is only the biological information. The technium is awash in its own ocean of information. It reflects 8,000 years of embedded human knowledge. Measured by the amount of digital storage in use, the technium today contains 487 exabytes (10^{20}) of information, many orders smaller than nature's total, but growing exponentially. Technology expands data by 66 percent per year, overwhelming the growth rate of any natural source. Compared to other planets in the neighborhood, or to the dumb material drifting in space beyond, a thick blanket of learning and self-organized information surrounds this orb.

There is yet one more version of the technium's cosmic story. We can view the long-term trajectory of exotropy as an escape from the material and a transcendence into the immaterial. In the early universe, only the laws of physics reigned. The rules of chemistry, momentum, torque, electrostatic charges, and other such reversible forces of physics were all that mattered. There was no other game. The ironclad constraints of the material world birthed only extremely simple mechanical forms—rocks, ice, gas clouds. But the expansion of space, with its corresponding increase in potential energy, introduced new immaterial vectors into the world: information, exotropy, and self-organization. These new organizational possibilities (like a living cell) did not contradict the rules of chemistry and physics but flowed from them. It is not as if life and mind were simply embedded in the nature of matter and energy; but rather, life and mind emerged out of the constraints to transcend them. Physicist Paul Davies summarizes it well: "The secret of life does not lie in its chemical basis. . . . Life succeeds precisely because it evades chemical imperatives."

Our present economic migration from a material-based industry to a knowledge economy of intangible goods (such as software, design, and media products) is just the latest in a steady move toward the immaterial. (Not that material processing has let up, just that intangible processing is now more economically valuable.) Richard Fisher, president of the Federal Reserve Bank of Dallas, says, "Data from nearly all parts of the world show us that consumers tend to spend relatively less on goods and more on services as their incomes rise. . . . Once people have met their basic needs, they tend to want medical care, transportation and communication, information, recreation, entertainment, financial and legal advice, and the like." The disembodiment of value (more value, less mass) is a steady trend in the technium. In six years the average weight per dollar of U.S. exports (the most valuable things the U.S. produces)

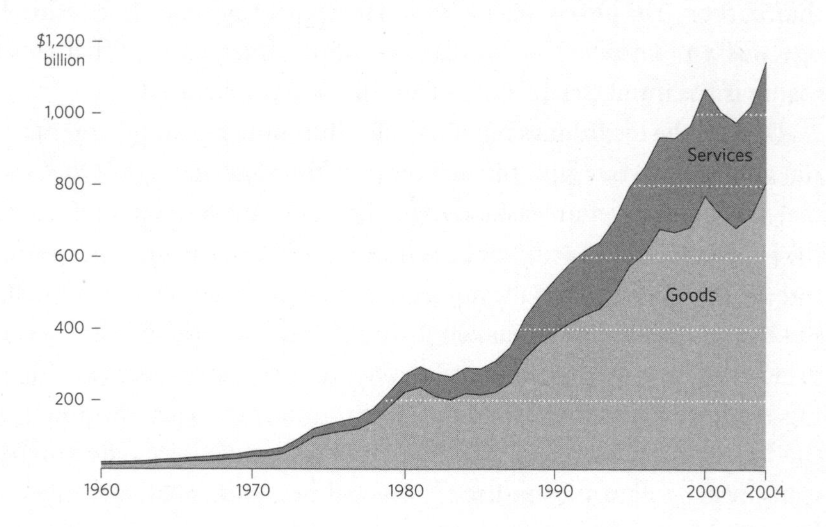

The Dematerialization of U.S. Exports. In billions of dollars, the total annual amount of both goods and services exported from the United States between 1960 and 2004.

dropped by half. Today, 40 percent of U.S. exports are services (intangibles) rather than manufactured goods (atoms). We are steadily substituting intangible design, flexibility, innovation, and smartness for rigid, heavy atoms. In a very real sense our entry into a service- and idea-based economy is a continuation of a trend that began at the big bang.

Dematerialization is not the only way in which exotropy advances. The technium's ability to compress information into highly refined structures is also a triumph of the immaterial. For instance, science (starting with Newton) has been able to abstract a massive amount of evidence about the movement of any kind of object into a very simple law, such as $F = ma$. Likewise, Einstein reduced enormous numbers of empirical observations into the very condensed container of $E = mc^2$. Every scientific theory and formula—whether about climate, aerodynamics, ant behavior, cell division, mountain uplift, or mathematics—is in the end a compression of information. In this way, our libraries packed with peer-reviewed, cross-indexed, annotated, equation-riddled journal articles are great mines of concentrated dematerialization. But just as an academic book about the technology of carbon fiber is a compression of the intangible, so are carbon fibers themselves. They contain far more than carbon. The philosopher Martin Heidegger suggested that technology was an "unhiding"—a revealing—of an inner reality. That inner reality is the immaterial nature of anything manufactured.

Despite the technium's reputation for dumping hardware and material gizmos into our laps, the technium is the most intangible and immaterial process yet unleashed. Indeed, it is the most powerful force in the world. We tend to think of the human brain as the most powerful-force in the world (although we should remember what is telling us that). But the technium has overtaken its brainy parents. The powers of our minds can be only slightly increased by mindful self-reflection; thinking about thoughts will only make us marginally smarter. The power of the technium, however, can be increased indefinitely by reflecting its transforming nature upon itself. New technologies constantly make it easier to invent better technologies; we can't say the same about human brains. In this unbounded technological amplification, the immaterial organization of the technium has now become the most dominant force in this part of the universe.

Technology's dominance ultimately stems not from its birth in human minds but from its origin in the same self-organization that brought galaxies, planets, life, and minds into existence. It is part of a great asymmetrical arc that begins at the big bang and extends into ever more abstract and immaterial forms over time. The arc is the slow yet irreversible liberation from the ancient imperative of matter and energy.

PART TWO

IMPERATIVES

5

Deep Progress

Newness is such an elemental part of our lives today that we forget how rare it was in ancient days. Most change in the past was cyclical: A forest was cleared for a field and then a farm was abandoned; an army came and then an army left. Droughts followed floods, and one king, either good or evil, succeeded another. For most humans, for most of time, real change was rarely experienced. What little change did happen occurred over centuries.

And when change erupted it was to be avoided. If historical change had any perceived direction at all, it was downhill. Somewhere in the past was a golden age, when the young respected their elders, neighbors didn't steal at night, and men's hearts were closer to God. In ancient times when a bearded prophet forecast what was to come, the news was generally bad. The idea that the future brought improvement was never very popular until recently. Even now, progress is far from universally accepted. Cultural advancements are commonly seen as exceptional episodes that may at any moment retreat into the woes of the past.

Any claim for progressive change over time must be viewed against the realities of inequality for billions, deteriorating regional environments, local war, genocide, and poverty. Nor can any rational person ignore the steady stream of new ills bred by our inventions and activities, including new problems generated by our well-intentioned attempts to heal old problems. The steady destruction of good things and people seems relentless. And it is.

But the steady stream of good things is relentless as well. Who can deny the benefits of antibiotics—even though they are overprescribed? Of electricity, or woven cloth, or radio? The desirable things are uncountable. While some have their downsides, we depend on their upsides. To remedy currently perceived ills, we create more new things.

Some of these new solutions are worse than the problems they were supposed to solve, but I think there is evidence that on average and over time, the new solutions outweigh the new problems. A serious techno-optimist might argue that the vast majority of cultural, social, and technological change is overwhelmingly positive—that 60 percent or 70 percent or 80 percent of the changes that take place in the technium each year make the world a better place. I don't know the actual percentage, but I think the balance settles out at higher than 50 percent positive, even if it is only slightly higher. As Rabbi Zalman Schachter-Shalomi once said, "There is more good than evil in the world—but not by much." Unexpectedly, "not much" is all that's needed when you have the leverage of compound interest at work—which is what the technium is. The world does not need to be perfectly utopian to see progress. Some portion of our actions, such as war, are destructive. A bunch of what we produce is crap. Maybe nearly half of what we do. But if we create only 1 percent or 2 percent (or even one-tenth of 1 percent) more positive stuff than we destroy, then we have progress. This differential could be so small as to be almost imperceptible, and this may be why progress is not universally acknowledged. When measured against the large-scale imperfections of our society, 1 percent better seems trivial. Yet this tiny, slim, shy discrepancy generates progress when compounded by the ratchet of culture. Over time a few percent "not much better" accumulates into civilization.

But is there really even 1 percent annual betterment over the long term? I think there are five pools of evidence for this trend. One is the long-term rise in longevity, education, health, and wealth of an average person. This we can measure. In general, the more recently in history people lived, the longer they lived, the greater access they had to accumulated knowledge, and the more tools and choices they owned. That's on average. Since war and strife can depress well-being locally and temporarily, indexes of health and wealth fluctuate within decades and by regions of the world. However, the long-term trajectory (and by "long-term" I mean over hundreds or even thousands of years) is a steady, measurable rise.

The second indicator of long-term progress is the obvious wave of positive technological development we have witnessed in our own lifetimes. Perhaps more than any other signal, this constant surge daily persuades us that things improve. Devices not only get better, they also get cheaper while they get better. We turn around to peer through our window into the past and realize there wasn't window glass back then. The past also lacked machine-woven cloth, refrigerators, steel, photographs, and the entire warehouse of goods spilling into the aisles of our local superstore. We can trace this cornucopia back along a diminishing curve to the Neolithic era. Craft from ancient times can surprise us in its sophistication, but in sheer quantity, variety, and complexity, it pales against modern inventions. The proof of this is clear: We buy the new over the old. Given the choice between an old-fashioned tool and a new one, most people—in the past as well as now—would grab the newer one. A very few will collect old tools, but as big as eBay is, and flea markets anywhere in the world, they are dwarfed by the market of the new. But if the new is not really better, and we keep reaching for it, then we are consistently duped or consistently dumb. The more likely reason we seek the new is that new things do get better. And of course there are more new things to choose from.

The typical American supermarket carries 30,000 varieties of items. Each year in the United States alone, 20,000 brand-new packaged-good items, such as food, soaps, and beverages, are launched, hoping to survive on those crowded shelves. Most of these contemporary products carry a bar code. The agency that issues the prefixes used in bar codes estimates that there are at least 30 million of them in use worldwide. The variety of manufactured products available on the planet is certainly in the tens of millions, if not hundreds of millions.

When Henry VIII, king of all England, died in 1547, his bursars took an exhaustive inventory of his belongings. They were especially careful in their count because his wealth doubled as the wealth of England. The accountants added up his furniture, spoons, silks, armor, weapons, silver plates, and all the usual possessions of a king at that time. In their final tally King Henry's household (the national treasure of England) contained 18,000 objects.

I live in a large American house that I share with my wife, my three children, a sister-in-law, and two nieces. One summer my young daughter Ting and I counted all the objects in our home. Equipped with a hand tally clicker and a clipboard, we went from room to room pawing through kitchen cupboards, bedroom closets, and desk drawers unopened for years.

I was primarily interested in measuring the variety of objects in our house rather than the total number, so I tried to count the number of technological "genres." We'd count only one representative of each type. The particular coloration (say, yellow or blue) or superficial ornamentation or decoration would not alter the type. I'd count only the archetypes of books: for instance, one paperback, one hardcover, and one oversized coffee-table tome, etc. All CDs were counted as one genre, all VHS tapes as one, etc. Essentially, the content didn't count. Things made of different materials counted as different species. Ceramic plates counted as one, glass plates as another. Things manufactured by the same machinery were one species. In the pantry all canned goods were one. Closets were a different matter. Most clothes are made by the same technology, but fibers vary. Cotton jeans and cotton shirts were each considered one species, wool pants another, a synthetic blouse another. If it seemed as if different technologies might be needed to make something, I would count it as a separate technological species.

After going from room to room, skipping none except the garage (that would be a project in itself), we arrived at a total of 6,000 varieties of things in our house. Since we have multiple examples of some varieties, such as books, CDs, paper plates, spoons, socks, on so on, I estimate the total number of objects in our home, including the garage, to be close to 10,000.

Without trying very hard, our typical modern house holds a king's ransom. But in fact, we are wealthier than King Henry. In fact, the lowest-paid burger flipper working at McDonald's is in many respects better off than King Henry or any of the richest people living not too long ago. Although the burger flipper barely makes enough to pay the rent, he or she can afford many things that King Henry could not.

King Henry's wealth—the entire treasure of England—could not have purchased an indoor flush toilet or air-conditioning or secured a comfortable ride for 500 kilometers. Any taxicab driver can afford these today. Only 100 years ago, John Rockefeller's vast fortune as the world's richest man could not have gotten him the cell phone that any untouchable street sweeper in Bombay now uses. In the first half of the 19th century Nathan Rothschild was the richest man in the world. His millions were not enough to buy an antibiotic. Rothschild died of an infected abscess that could have been cured with a three-dollar tube of neomycin today. Although King Henry had some fine clothes and a lot of servants, you could not pay people today to live as he did, without plumbing, in dark, drafty rooms, isolated from the world by impassable roads and few communication connections. A poor university student living in a dingy dorm room in Jakarta lives better in most ways than King Henry.

Recently, photographer Peter Menzel organized an expedition to photograph families around the world surrounded by all their possessions. Families in 39 countries, including Nepal, Haiti, Germany, Russia, and Peru, let Menzel and his delegates haul the entire contents of their homes outside into the street or yard to be photographed, inventoried, and published in Menzel's book, called *Material World*. Nearly every family was proud of what they possessed, standing happily in front of their dwelling amid a colorful display of furniture, pots, clothes, and knickknacks. The average number of objects owned by one of these families was 127.

There is one thing we can say for certain about these different pictures of possessions, and one thing we can't say. One thing for sure is that the families living in those regions in previous centuries had significantly fewer than 127 objects. Even families in the poorest countries today have more than those in some of the richest had two centuries ago. In Colonial America when a homeowner died, officials would normally take an inventory of his estate. Typical historical inventories of deceased homeowners from that period totaled up 40, maybe 50 and usually less than 75 objects in the entire estate.

What we can't say is this: If we hold up two photographs of people and their possessions—one of a Guatemalan family with their firepot and looms and not much else, and one of an Icelandic family with their washer/dryer, cellos, piano, three bicycles, horse, and a thousand other items—we can't say which family is happier. Is it the one with all the possessions or the one without?

For the past 30 years the conventional wisdom has been that once a person achieves a minimal standard of living, more money does not bring more happiness. If you live below a certain income threshold, increased money makes a difference, but after that, it doesn't buy happiness. That was the conclusion of a now-classic study by Richard Easterlin in 1974. However, recent research from the Wharton School at the University of Pennsylvania shows that worldwide, affluence brings increased satisfaction. Higher income earners *are* happier. Citizens in higher-earning countries tend to be more satisfied on average.

My interpretation of this newest research—which also matches our intuitive impressions—is that what money brings is increased choices, rather than merely increased stuff (although more stuff comes with the territory). We don't find happiness in more gadgets and experiences. We do find happiness in having some control of our time and work, a chance for real leisure, in the escape from the uncertainties of war, poverty, and corruption, and in a chance to pursue individual freedoms—all of which come with increased affluence.

I've been to many places in the world, the poorest and the richest spots, the oldest and the newest cities, the fastest and the slowest cultures, and it is my observation that when given a chance, people who walk will buy a bicycle, people who ride a bike will get a scooter, people riding a scooter will upgrade to a car, and those with a car dream of a plane. Farmers everywhere trade their ox plows for tractors, their gourd bowls for tin ones, their sandals for shoes. Always. Insignificantly few ever go back. The exceptions such as the well-known Amish are not so exceptional when examined closely, for even their communities adopt selected technology without retreat.

This one-way pull toward technology is either a magical siren, bewitching the innocent into consuming something they don't really want, or a tyrant that we are unable to overthrow. Or else technology offers something highly desirable, something that indirectly leads to greater satisfaction. (It is also possible that all three possibilities are true.)

The dark side of technology cannot be avoided. It may even be nearly half of the technium. Hiding behind the 10,000 shiny high-tech items in my house are remote, dangerous mines dug to obtain rare earth elements emitting toxic traces of heavy metals. Vast dams are needed to power my computer. Stumps are left in the jungle after timber is removed for my bookshelves, and long chains of vehicles and roads are needed to package and market all the stuff in my house and home office. Every gizmo begins with earth, air, and sunlight and a web of other tools. The 10,000 items we counted are only the visible tips of a huge tree with deep roots. Probably 100,000 physical contraptions behind the scenes were needed to transform elements into our final 10,000.

Yet all the while the technium is increasing the transparency of its roots, compiling more camera eyes, more communication neurons, more tracking technologies that reveal its own complicated processes. We have more options to view the real costs of technologies, if we care to. Could these communication and monitoring systems slow unabashed consumerism? It is possible. But great visibility and transparency of the technium's true costs and trade-offs won't slow down its progress. Awareness of its downsides may even refine its evolution and speed up its improvement by shunting energy away from frivolous consumption toward more select meaningful advances.

The third piece of evidence for small, steady, long-term advance resides in the moral sphere. Here metrics for measurement are few and disagreement about the facts greater. Over time our laws, mores, and ethics have slowly expanded the sphere of human empathy. Generally, humans originally identified themselves primarily via their families. The family clan was "us." This declaration cast anyone outside of that intimacy as "other." We had—and still have—different rules of behavior for those inside the circle of "us" and for those outside. Gradually the circle of "us" enlarged from inside the family clan to inside the tribe, and then from tribe to nation. We are currently in an unfinished expansion beyond nation and maybe even race and may soon be crossing the species boundary. Other primates are, more and more, deemed worthy of humanlike rights. If the golden rule of morality and ethics is to "do unto others as you would have others do unto you," then we are constantly expanding our notion of "others." This is evidence for moral progress.

The fourth line of evidence does not prove the reality of progress but it provides strong support. A large and still expanding body of scientific literature spotlights the immense distance life has traveled in its four-billion-year journey from extremely simple organisms to extremely complex and social animals. Changes in our culture can be viewed as a continuation of progress begun four billion years ago, a key parallel I will develop in the next chapter.

The fifth argument for the reality of progress is the rush toward urbanization. A thousand years ago only a small percentage of humans lived in cities; now 50 percent do. Cities are where people move to live in "a better tomorrow," where increased choices and possibilities bloom. Every week, a million people move from the countryside into cities, a journey that is less in space than in time. These migrants are really moving hundreds of years forward, relocating from medieval villages into twenty-first-century sprawling urban areas. The afflictions of the slums are highly visible, but they don't stop the arrivals. The hopeful keep coming—as we all do—for the greater number of freedoms and options. We live in urban and suburban environments for the same reason migrants do—to gain that marginal advantage of more choice.

The choice of returning to our early state is always there. In fact, moving back into the past has never been easier. Citizens in developing countries can merely take a bus back to their villages, where they can live with age-old traditions and limited choice. They will not starve. In a similar spirit of choice, if you believe that the peak of existence was reached in Neolithic times, you can camp out in a clearing in the Amazon. If you think the golden age was in the 1890s, you can find a farm among the Amish. We have lots of opportunity to revisit the past, but few people really want to live there. Rather, everywhere in the world, at

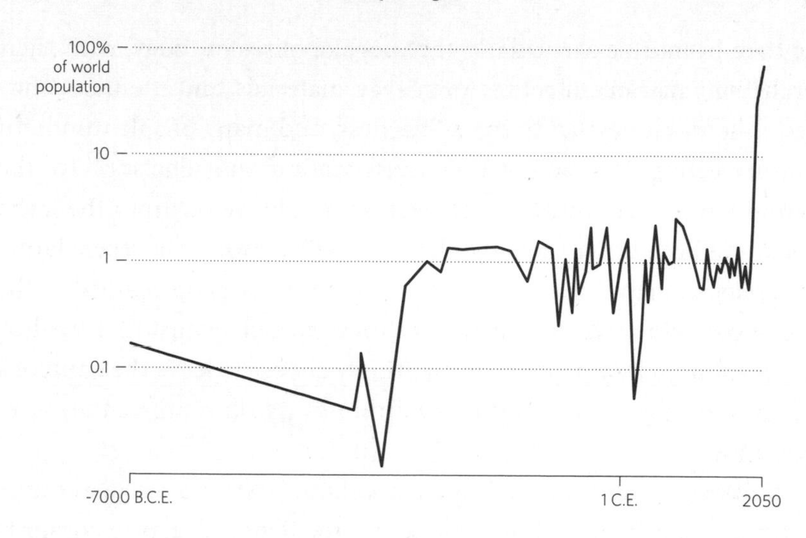

Global Urban Population. The percentage of total global population dwelling in an urban area, from 7000 B.C.E. to the present, including the projected percentage by 2050. Percentages are shown on a logarithmic scale.

all historical periods, in all cultures, people have stampeded by the billions into the future of "slightly more options" as fast as they can. With their feet they have voted for progress by migrating to cities.

Cities are technological artifacts, the largest technology we make. Their impact is out of proportion to the number of humans living in them. As the chart above shows, the percentage of humans living in cities averaged about 1 percent or 2 percent for most of recorded history. Yet almost everything that we think of when we say "culture" arose within cities. (The terms *city* and *civilization* share the same root.) But the massive citification, or urbanization, that characterizes the technium today is a very recent development. Like most other charts depicting the technium, not much happens until the last two centuries. Then population booms, innovation rockets, information explodes, freedoms increase, and cities rule.

All the promises, paradoxes, and trade-offs carried by Progress, with a capital *P*, are represented in a city. In fact, we can inspect the notion and veracity of technological progress at large by examining the nature of cities. Cities may be engines of innovation, but not everyone thinks they are beautiful, particularly the megalopolises of today, with their sprawling, rapacious appetites for energy, materials, and attention. They seem like machines eating the wilderness, and many people wonder if they are eating us as well. Cities, even more than gadgets, revive the eternal tension we feel about the technium: Do we buy into the latest inventions because we want to or because we have to? Is the recent large-scale relocation to cities a choice or a necessity? Are people pulled by the lure of opportunities in cities, or are they pushed against their will by desperation? Why would anyone willingly choose to leave the balm of a village and squat in a smelly, leaky hut in a city slum unless they were forced to?

Well, every beautiful city begins as a slum. First it's a seasonal camp, with the usual freewheeling makeshift expediency. Creature comforts are scarce, squalor the norm. Hunters, scouts, traders, pioneers find a good place to stay for the night, or two, and then if their camp is deemed a desirable spot it grows into an untidy village or uncomfortable fort or dismal official outpost with permanent buildings surrounded by temporary huts. If the location of the village favors growth, concentric rings of squatters aggregate until the village chaotically swells to a town. When a town prospers it acquires a center—civic or religious—and the edges of the city continue to expand in unplanned, ungovernable messiness. It doesn't matter in what century or in which country; the teeming fringes of a city will shock and disturb the established residents. The eternal disdain for newcomers is as old as the first city. Romans complained of the tenements, shacks, and huts at the edges of their town, which "were putrid, sodden and sagging." Every so often Roman soldiers would raze a settlement of squatters, only to find it rebuilt or moved within weeks.

Babylon, London, and New York all had teeming ghettos of unwanted settlers erecting shoddy shelters with inadequate hygiene and engaging in dodgy dealings. Historian Bronislaw Geremek states that "slums constituted a large part of the urban landscape" of Paris in the Middle Ages. Even by the 1780s, when Paris was at its peak, nearly 20 percent of its residents did not have a "fixed abode"—that is, they lived in shacks. In a familiar complaint about medieval French cities, a gentleman from that time noted: "Several families inhabit one house. A weaver's family may be crowded into a single room, where they huddle around a fireplace." That refrain is repeated throughout history. A century ago Manhattan was home to 20,000 squatters in self-made housing. Slab City alone, in Brooklyn (named after the use of planks stolen from lumber mills), contained 10,000 residents in its slum at its peak in the 1880s. In the New York slums, reported the *New York Times* in 1858, "nine out of ten of the shanties have only one room, which does not average over twelve feet square, and this serves all the purposes of the family."

San Francisco was built by squatters. As Rob Neuwirth recounts in his eye-opening book *Shadow Cities,* one survey in 1855 estimated that "95 percent of the property holders in [San Francisco] would not be able to produce a bona fide legal title to their land." Squatters were everywhere, in the marshes, sand dunes, military bases. One eyewitness said, "Where there was a vacant piece of ground one day, the next saw it covered with half a dozen tents or shanties." Philadelphia was largely settled by what local papers called "squatlers." As late as 1940, one in five citizens in Shanghai was a squatter. Those one million squatters stayed and kept upgrading their slum so that within one generation their shantytown became one of the first twenty-first-century cities.

That's how it works. This is how all technology works. A gadget begins as a junky prototype and then progresses to something that barely works. The ad hoc shelters in slums are upgraded over time, infrastructure is extended, and eventually makeshift services become official. What was once the home of poor hustlers becomes, over the span of generations, the home of rich hustlers. Propagating slums is what cities do, and living in slums is how cities grow. The majority of neighborhoods in almost every modern city are merely successful former slums. The squatter cities of today will become the blue-blood neighborhoods of tomorrow. This is already happening in Rio and Mumbai today.

Slums of the past and slums of today follow the same description. The first impression is and was one of filth and overcrowding. In a ghetto a thousand years ago and in a slum today shelters are haphazard and dilapidated. The smells are overwhelming. But there is vibrant economic activity. Every slum boasts eateries and bars, and most have rooming houses or places you can rent a bed. They have animals, fresh milk, grocery stores, barber shops, healers, herb stores, repair stands, and strong armed men offering "protection." A squatter city is, and has always been, a shadow city, a parallel world without official permission, but a city nonetheless.

Like any city, a slum is highly efficient—maybe even more so than the city's official sections, because nothing goes to waste. The ragpickers and resellers and scavengers all live in the slums and scour the rest of the city for scraps to assemble into shelter and to feed their economy. Slums are the skin of the city, its permeable edge that can balloon as it grows. The city as a whole is a wonderful technological invention that concentrates the flow of energy and minds into computer chip–like density. In a relatively small footprint, a city not only provides living quarters and occupations in a minimum of space, but it also generates a maximum of ideas and inventions.

Stewart Brand notes in the "City Planet" chapter of his book *Whole Earth Discipline,* "Cities are wealth creators; they have always been." He quotes urban theorist Richard Florida, who claims that forty of the largest megacities in the world, home to 18 percent of the world's population, "produce two-thirds of global economic output and nearly 9 in 10 new patented innovations." A Canadian demographer calculated that "80 to 90 percent of GNP growth occurs in cities." The raggedy new part of each city, its squats and encampments, often house the most productive citizens. As Mike Davis points out in *Planet of Slums,* "The traditional stereotype of the Indian pavement-dweller is a destitute peasant, newly arrived from the countryside, who survives by parasitic begging, but as research in Mumbai has revealed, almost all [families] (97 percent) have at least one breadwinner, and 70 percent have been in the city at least six years." Slum dwellers are often busy with low-paying service jobs in nearby high-rent districts; they have money but live in a squatter city because it's close to their work. Because they are industrious, they progress fast. One UN report found that households in the older slums of Bangkok have on average 1.6 televisions, 1.5 cell phones, and a refrigerator; two-thirds have a washing machine and CD player; and half have a fixed-line phone, a video player, and a motor scooter. In the favelas of

Rio, the first generation of squatters had a literacy rate of only 5 percent, but 94 percent of their kids were literate.

There is a price to pay for that growth. As vibrant and dynamic as cities are, their edges can be unpleasant. To enter a slum you need to walk down shit lane. There is human excrement rotting on the sidewalk, urine flowing in the gutter, and garbage piled up in heaps. I've done it many times in the sprawling shantytowns of the developing world, and it is no fun—and less so for the residents who must endure this every day. To compensate for this outer contamination and ugliness, the inside of squatter housing is often surprisingly soothing. Recycled material covers the walls, color abounds, and knickknacks accumulate to create a comfy zone. Sure, one room will house far more people than seems possible, but for many, a slum dwelling offers more comfort than a village hut. While the pirated electricity may be unreliable, at least there *is* electricity. The single water spigot may have a long line, but it might be closer than the well at home. Medicines are expensive but available. And there are schools with teachers that show up.

It is not utopia. When it rains, slums turn to mud cities. The ceaseless call for bribes for everything is dispiriting. And there is the embarrassment that squatters feel about the obvious low status of their homes. As Suketu Mehta, author of *Maximum City* (about Mumbai), says, "Why would anyone leave a brick house in the village with its two mango trees and its view of small hills in the East to come here?" Then he answers: "So that someday the eldest son can buy two rooms in Mira Road, at the northern edges of the city. And the younger one can move beyond that, to New Jersey. Discomfort is an investment."

Then Mehta continues: "For the young person in an Indian village, the call of Mumbai isn't just about money. It's also about freedom." Stewart Brand recounts this summation of the magnetic pull of cities by activist Kavita Ramdas: "In the village, all there is for a woman is to obey her husband and relatives, pound millet, and sing. If she moves to town, she can get a job, start a business, and get education for her children." The Bedouin of Arabia were once seemingly the freest people on Earth, roaming the great Empty Quarter at will, under a tent of stars and no one's thumb. But they are rapidly quitting their nomadic life and hustling into drab, concrete-block apartments in exploding Gulf-state ghettos. As reported by Donovan Webster in *National Geographic*, they stable their camels and goats in their ancestral village, because the bounty and attraction of the herder's life still remain for them. The Bedouin are lured, not pushed, to the city because, in their own words: "We can always go into the desert to taste the old life. But this [new] life is better than the old way. Before there was no medical care, no schools for our children." An eighty-year-old Bedouin chief sums it up better than I could: "The children will have more options for their future."

The migrants don't have to come. Yet they come by the millions from the villages or the deserts and scrublands. If you ask them why they come, it's almost always the same answer, the same answer given by the Bedouin and slum dwellers of Mumbai. They come for opportunities. They could stay where they are, as the Amish choose to do. The young men and women could remain in the villages and adopt the satisfying rhythms of agriculture and small-town craft that their parents followed. The seasonal droughts and floods are eternal. And so is the incredible beauty of the land and the intensity of family and community support. The same tools work. The same traditions deliver the same good things. The immense satisfactions of seasonal toil, abundant leisure, strong family ties, reassuring conformity, and rewarding physical labor will always pull our hearts. If everything were equal, who would want to leave a Greek island, or a Himalayan village, or the lush gardens of southern China?

But the options aren't equal. People of the world increasingly have TV and radio and trips into town to see movies, and they know what is possible. The freedom in a city makes their village seem a prison. So they choose—very willingly, very eagerly—to run to the city.

Some argue that they had no choice. That those who arrive in the slums are forced against their desire to migrate to the city because their villages can no longer support farmers. That they leave unwillingly. Perhaps after surviving for generations selling coffee, they find that global markets have shifted and dropped the price of their coffee to nothing, sending them either back to subsistence farming or onto the bus. Or perhaps technological development, such as mining for coal, is poisoning their land, lowering the water table, and stirring their exodus. In addition, as technological improvements in the form of tractors, refrigeration, and roads to transport goods reach the farthest fields, fewer farmers are needed, even in developed countries. Massive deforestation to produce lumber for housing and construction, or to clear land for new farms to feed the cities, also forces indigenous people out of their wild homelands and traditional ways.

Truly, there is nothing as disturbing as the sight of indigenous tribesmen, say in the Amazon basin or in the jungles of Borneo or Papua New Guinea, wielding chain saws to fell their own forests. When your forest home is toppled, you are pushed into camps, then towns, and then cities. Once in a camp, cut off from your hunter-gatherer skills, it makes a weird sense to take the only paid job around, which is cutting down your neighbors' forest. Clear-cutting virgin forests counts as cultural insanity for a number for reasons, not least that the tribal people ousted by this habitat destruction cannot go back. Within a generation or two of exile, they can lose key survival knowledge, which would prevent their descendants from returning even if their homeland were to be renewed. Their exit is an involuntary one-way trip. In the same way, the despicable treatment of indigenous tribes by American white settlers really did force them into settlements and the adoption of new technologies they were in no hurry to use.

However, clear-cutting is technologically unnecessary. Habitat destruction of any type is deplorable, and stupidly low tech, but also not responsible for the majority of migrations. Deforestation is a minor push compared to the tractor beam–like pull of the flickering lights that have brought 2.5 billion people into the cities in the last 60 years. Today, as in the past, most of the mass movement toward cities—the hundreds of millions per decade—is led by settled people willing to pay the price of inconvenience and grime, living in a slum in order to gain opportunities and freedom. The poor move into the city for the same reason the rich move into the technological future—to head toward possibilities and increased freedoms.

In *The Progress Paradox* Gregg Easterbrook writes, "If you sat down with a pencil and graph paper to chart the trends of American and European life since the end of World War II, you'd do a lot of drawing that was pointed up." Ray Kurzweil has collected an entire gallery of graphs depicting the upward-zooming trend in many, if not most, technological fields. All graphs of technological progress start low, with small change several hundred years ago, then begin to bend upward in the last hundred, and then bolt upright to the sky in the last fifty.

These charts capture a feeling we have that change is accelerating even within our own lifetimes. Novelty arrives in a flash (compared to earlier), and there seems to be a shorter and shorter interval between novel changes. Technologies get better, cheaper, faster, lighter, easier, more common, and more powerful as we move into the future. And it is not just technology. The human life span increases, the rate of infant mortality decreases, and even the average IQ inches upward every year.

If all this is true, then what of long ago? Long ago there was not much evidence of progress, at least how we now visualize it. Five hundred years ago technologies were not doubling in power and halving in price every 18 months. Waterwheels were not becoming cheaper every year. A hammer was not easier to use from one decade to the next. Iron was not increasing in strength. The yield of corn seed varied by the season's climate, instead of improving each year. Every 12 months you could not upgrade your oxen's yoke to anything much better than what you already had. And your own expected longevity, or your children's, was approximately the same as it had been for your parents. Wars, famine, storms, and curious events came and went, but there was no steady movement in any direction. There appeared to be, in short, change without progress.

A common misconception about human evolution is that historic tribes and prehistoric clans of early Sapiens achieved a level of egalitarian justice, freedom, liberty, and harmony that has only declined since then. In this view, the human inclination to make tools (and weapons) has only introduced trouble. Each new invention unleashes new power that can be concentrated, wielded asymmetrically, or corrupted, and therefore the history of civilization is one long devolution. By this account, human nature is fixed, unyielding. If that is true, then attempts to alter human nature will only lead to evil. So in this view, new technologies generally erode the innate sacred human character, and can be kept in check only by keeping technology to a minimum in strict moral vigilance. Therefore, our relentless propensity to create things is a kind of species-level addiction, or a self-destroying frivolity, and we must always guard against succumbing to its spell.

The reality is the opposite. Human nature is malleable. We use our minds to change our values, expectations, and definition of ourselves. We have changed our nature since our hominin days, and once changed, we will continue to change ourselves even more. Our inventions, such as language, writing, law, and science, have ignited a level of progress that is so fundamental and embedded in the present that we now naively expect to see similar good things in the past as well. But much of what we consider "civil" or even "humane" was absent long ago. Early societies were not peaceful but rife with warfare. One of the most common causes of adult death in tribal societies was to be declared a witch or evil spirit. No rational evidence was needed for these superstitious accusations. Lethal atrocities for infractions within a clan were the norm; fairness, as we might think of it, did not extend outside the immediate tribe. Rampant inequality among genders and physical advantage for the strong guided a type of justice few modern people would want applied to them.

Yet all these values worked for the first kinds of human communities. Early societies were incredibly adaptable and resilient. They produced art, love, and meaning. They were very successful in their environments because their own social norms were successful—even though we find them intolerable. If these earlier societies had had to rely on our modern conceptions of justice, harmony, education, and equality, they would have failed. But all societies, including aboriginal cultures today, evolve and adapt. Their progress may be imperceptible, but it is real.

In all cultures prior to the 17th century or so, the quiet, incremental drift of progress was attributed to the gods, or to the one God. It wasn't until progress was liberated from the divine and assigned to ourselves that it began to feed upon itself. Sanitation made us healthier, so we could work longer. Farm tools made more food for less work. Gadgets made our homes more comfortable for tinkering with new ideas. The more inventions, the better. There was a tight feedback loop as increased knowledge enabled us to discover and manufacture more tools, and these tools allowed us to discover and learn more knowledge, and both the tools and the knowledge made our lives easier and longer. The general enlargement of knowledge and comfort and choices—and the sense of well-being—was called progress.

The rise of progress coincided with the rise of technology. But what pushed technology? We had thousands, if not tens of thousands, of years of human culture, steadily learning, passing on information from one generation to the next—but no progress. Sure, new things would occasionally be discovered and slowly disseminated, or rediscovered independently, but whatever improvement one might measure over centuries in the old days would be very small. In fact, the average farming peasant who lived in 1650 followed a life that was nearly indistinguishable from that of the average farming peasant who lived in 1650 B.C.E. or 3650 B.C.E. In some valleys of the world (the Nile in Egypt, the Yangtze in China) and in some particular places and times (classical Greece, Renaissance Italy), the fate of citizens might rise above the historical average, only to descend when a dynasty ended or the climate shifted. Before 300 years ago, the standard of the average human's life was fairly interchangeable anywhere in time or place: People were perennially hungry, short lived, limited in choices, and extremely dependent on traditions simply to survive to the next generation.

For thousands of years this slow cycle of birth and death crept along, when suddenly—*boom!*—complex industrial technology appeared and everything started moving very fast. What caused the boom in the first place? What is the origin of our progress?

The ancient world—particularly its cities—enjoyed many fabulous inventions. Societies slowly accumulated such marvels as arch bridges, aqueducts, steel knives, suspension bridges, water mills, paper, vegetable dyes, and so on. Each of these innovations was discovered in a trial-and-error fashion. Once found, by hit or miss, they were disseminated haphazardly. Some marvels could take centuries to reach another country. This nearly random method of improvement was transformed by the tool of science. By systematically recording the evidence for beliefs and investigating the reasons why things worked and then carefully distributing proven innovations, science quickly became the greatest tool for making new things the world had ever seen. Science was in fact a superior method for a culture to learn.

Once you invent science—which allows you to quickly invent many things—you have a grand lever that can propel you forward very quickly. That's what happened in the West starting approximately in the 17th century. Science catapulted society into a rapid learning. By the 18th century, science had launched the Industrial Revolution, and progress was noticeable in the growing spread of cities, increasing longevity and literacy, and the acceleration of future discoveries.

But there is a puzzle. The necessary ingredients of the scientific method are conceptual and fairly low tech: a way to record, catalog, and communicate written evidence and the time to experiment. Why didn't the Greeks invent it? Or the Egyptians? A time traveler from today could journey back to that era and set up the scientific method in ancient Alexandria or Athens without much trouble. But would it catch on?

Maybe not. Science is costly for an individual. Sharing results is of marginal benefit if you are chiefly seeking a better tool for today. Therefore, the benefits of science are neither apparent nor immediate for individuals. Science requires a certain density of leisured population willing to share and support failures to thrive. That leisure is generated by pre-science inventions such as the plow, grain mills, domesticated power animals, and other techniques that permit a steady surplus of food for large numbers of people. In other words, science needs prosperity and populations.

Outside the reign of science and technology a growing population will collapse upon itself as it meets Malthusian limits. But inside the reign of science a growing population creates a positive feedback loop wherein more people participate in scientific innovation and purchase the results, driving more innovation, which brings better nutrition, more surplus, and more population, which feed the cycle further.

Just as an engine tames its fire, channeling its explosive energy to-

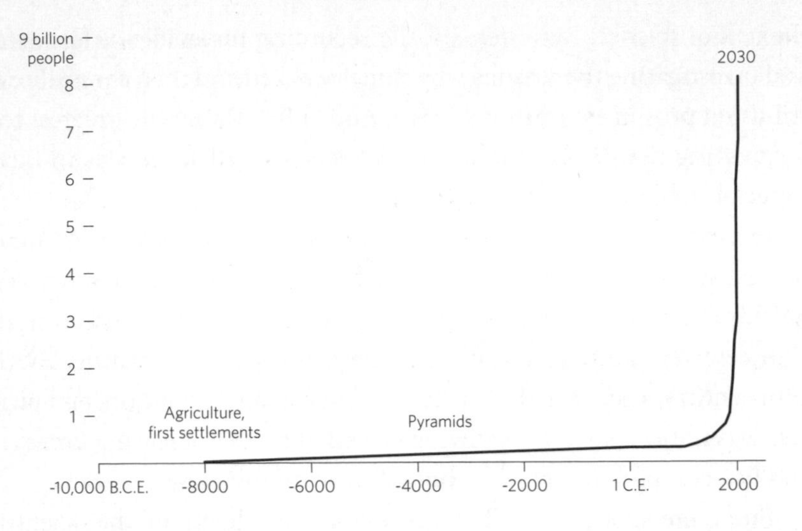

World Population in Civilization. A typical chart showing world population during the last 12,000 years, including a short-term, 30-year projection into the future.

ward work, science tames population growth, channeling its explosive energy toward prosperity. As population rises, so does progress, and vice versa. The two growths are heavily correlated.

We have many examples in modern times of increasing populations suffering through declining living standards. That is happening in parts of Africa right now. On the other hand, throughout history it has been rare to see rising prosperity over the long term propelled by declining population. Declining population is almost always associated with declining prosperity. Even during the decimations of the black plague, when 30 percent of an area's population died, the change in living standards was uneven. Many of the overpopulated peasant regions in Europe and China prospered as their competition thinned out, but the quality of life for merchants and the upper class declined substantially. There was a redistribution of living standards, but not a net gain in progress during this time. The evidence from plagues is that population growth is necessary but not sufficient for progress.

Clearly, the roots of progress lie deep in the structured knowledge of science and technology. But the flowering of this progressive growth seems to also need the growth of large human populations. Historian Niall Ferguson believes that on the global scale, the origins of progress lie only in expanding population. According to this theory, in order to elevate populations beyond Malthusian limits you need science, yet it is the increase in the number of humans that ultimately drives science, and then prosperity. In this virtuous circle more human minds invent more things and in turn buy more inventions, including tools, techniques, and methods that will support more humans. Therefore, more human minds equal more progress. The economist Julian Simon called human minds "the ultimate resource." In his calculation, more minds were the prime source of deep progress.

Whether population growth is the prime cause of progress or only a factor, population growth assists progress growth in two ways. First, a million individual minds applied to a problem are better than one. It's more likely someone will find a solution. Second, and more important, science is a collective action, and the emergent intelligence of shared knowledge is often superior to even a million individuals. The solitary scientific genius is a myth. Science is both the way we personally know things and the way we collectively know. The greater the pool of individuals in the culture, the smarter science gets.

The economy works in a similar way. Much of our current economic prosperity is due to population growth. The population of the United States has steadily grown over the past few centuries, ensuring a steadily expanding market for innovations. At the same time, world population has been expanding, ensuring economic growth worldwide. World population has also grown in accessibility and desire as billions have moved from subsistence farming into the marketplace. But try to imagine the same rise in wealth in the past two centuries if the world market or the U.S. market had shrunk every year.

If it is true that progress expands as human population expands, then we should be worried. You may have seen the official graph of peak human population prepared by the United Nations. It is based on the best information we have about the global census of humans living today. The estimated peak number of humans on Earth keeps changing (downward) in each revision in the past decades, but the shape of its destiny does not. A typical UN chart for the next 40 years or so looks like this:

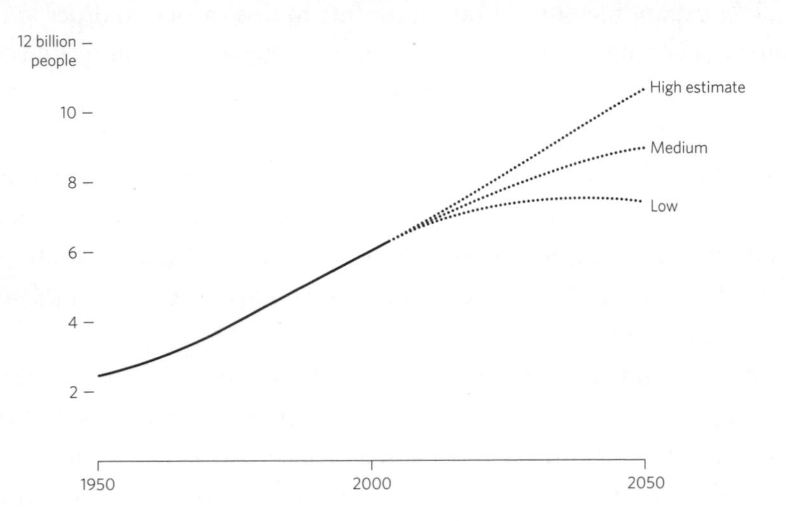

World Population Forecasts. United Nations world population projections for the years 2002 to 2050, in billions, forecasted in 2002.

The problem this presents for understanding the origins of technological progress is that the chart always stops there, right at the year 2050. At the apex. It dares not look beyond the peak. So what happens after the population peaks? Does it sink, swim, or rise again? Why is that never shown? Most charts simply ignore the question, with no apology for the omission. Showing just one-half of the curve has been so common for so long that no one asks for the other half.

The only source I have found for a reliable projection of what happens on the other side of the peak of human population around 2050 is a set of UN scenarios for World Population in 2300, that is, for the next 300 years.

Keep in mind that a worldwide fertility rate below the replacement level of 2.1 children per woman means a long-term decline in global population, or negative population growth. The UN high scenario assumes average fertility remains at 1995 rates, or 2.35 children per woman. We already know this extreme version is not happening. Only a couple of countries out of 100-plus in the world have kept their reproduction

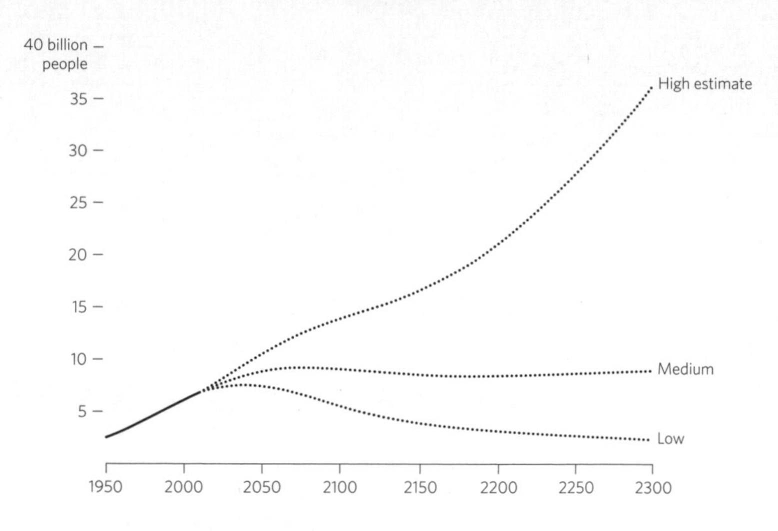

Estimated Long-Range World Population. Three United Nations scenarios (high, medium, low) for world population in the next 300 years, from 2000 to 2300, in billions.

rates that high. The middle scenario assumes that the average fertility dips below replacement levels of 2.1 for 100 years and then for some reason returns to replacement level for the next 200 years. The report suggests no possible reason why fertility rates would rise in a more developed world. The low scenario assumes 1.85 children per woman. Today every country in Europe is below 2.0, and Japan is at 1.34. Even this "low" scenario assumes a higher fertility in 200 years than what most developed countries currently have.

What's going on here? As countries become developed, their fertility rates drop. This drop-off has happened in every modernizing country, and this universal decrease in fertility rates is known as the "demographic transition." The problem is that the demographic transition has no bottom. In developed countries the fertility rate keeps dropping. And dropping. Look at Europe (chart on the next page) or Japan. Their fertility rates are headed to zero. (Not zero population *growth,* which they long ago sank past, but zero population.) In fact, most countries, even developing countries, see their fertility rates dropping. Nearly half of the countries in the world are already below the replacement level.

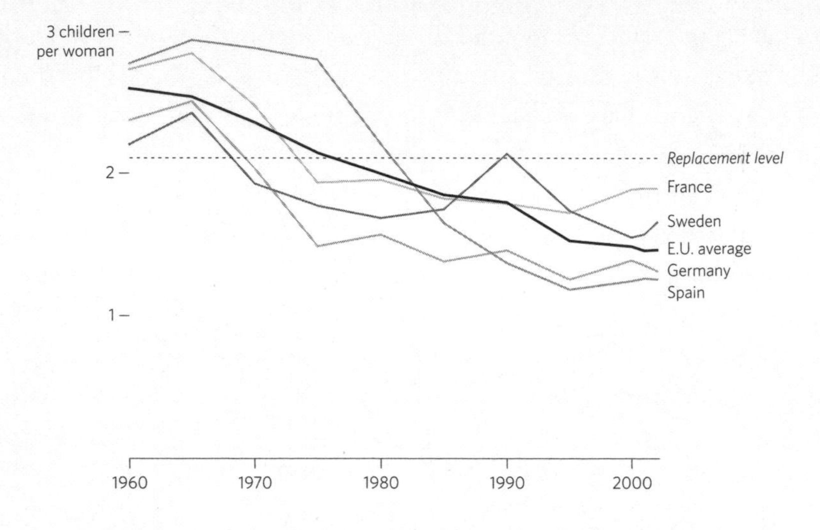

Recent Fertility Rates in Europe. The dotted line is the replacement level—the lowest rate by which a population group can replace itself.

In other words, as prosperity increases due to expanding population, fertility rates drop, which will shrink population. This might be a homeostasis feedback mechanism that reins in exponential rates of progress. Or it might be wrong.

The UN 2300 scenarios are scary, but the problem with these 300-year forecasts is that their dire scenario is not dire enough. The experts assume that even in the worst-case scenario fertility rates cannot go lower than the low rates found in places like Europe or Japan. Why do they assume this? Because it has never happened before. But of course this level of prosperity has never occurred before either. So far all evidence suggests that increased prosperity keeps lowering the number of children the average woman wants. What if global fertility rates keep dropping below the replacement rate of 2.1 offspring for every woman in developed countries and 2.3 in developing countries? The replacement rate is what is needed simply to maintain zero growth, to maintain a population, to not decline. An average rate of 2.1 offspring means a significant portion of women have to have three or four or five babies in order to counter the childless and those with only one or two babies.

What countercultural force is at work prompting billions of modern, educated, working women to have three, four, or five babies? How many of your friends have four children? Or three? "Just a few" won't matter in the long run.

Keep in mind that an enduring global fertility rate only a little below the replacement level, say 1.9, will eventually, inevitably bring the world population to zero, because each year there are fewer and fewer babies. But zeroing out is not the worry. Long before the human population dropped to zero, the Amish and Mormons would save humankind with their prolific breeding and large families. The question is, if rising prosperity hinges on rising population, what happens to deep technological progress if there are centuries of slow population decline?

There are five scenarios, with five different assumptions about the nature of progress.

Scenario #1

Perhaps technology makes having babies much easier, or much cheaper, though it is hard to imagine any way in which technology could make rearing three children any easier. Or perhaps there is social pressure to maintain the species or social status in having a lot of children. Maybe robotic nannies change everything and having more than two kids becomes fashionable. It is not impossible to speculate on ways to maintain a status quo. But even if global population leveled off and maintained a constant number, we don't have any experience that suggests that a stagnant population can produce rising progress.

Scenario #2

While the census of human minds may decrease, we can build artificial minds, maybe even in the billions. Perhaps these artificial minds are all that is needed to keep prosperity expanding. To do so they would need to not only keep producing ideas, but also consuming them, just as humans do. Since they aren't human (if you want a human mind, make a baby), this prosperity and progress would likely look different from that of today.

Scenario #3

Rather than depending on expanding the number of human minds, maybe progress can keep advancing by bettering the average human mind. Perhaps with the aid of always-on technologies or genetic engineering or pills, the potential of individual human minds increases, and this increase propels progress. Perhaps we increase our attention span, sleep less, live longer, and consume more, produce more, create more. The cycle spins faster with fewer but more powerful minds.

Scenario #4

We might have it all wrong. Maybe prosperity has nothing to do with increasing numbers of minds. Maybe consumption has no part in progress. We simply figure out how to increase living quality, choices, and possibilities with fewer and fewer people (who live longer and longer). It's a very green vision, but also very alien to our current system. If every year there are fewer people as my potential audience or my potential customers, I have to create things for a different reason than growth in audience or customers. A nongrowth economy is hard to imagine. But stranger things have happened.

Scenario #5

Our population plunges to small remnants, which in desperation breed madly and prosper. World population oscillates up and down.

If the origins of prosperity lie solely in growth of the human population, then progress will paradoxically temper itself in the coming century. If the origins of progress lie outside population growth, we'll need to identify them so that on the other side of the population peak, we can continue to prosper.

I tell the story of progress's rise as one driven by human minds, but I haven't yet mentioned the crucial fact that humanity's use of energy follows the same upward curves. The accelerating progress of the last 200 years has indisputably been fueled by an exponential increase in cheap, abundant energy. It is no coincidence that the takeoff in progress at the dawn of the industrial age began exactly when humans figured out how to harness coal power instead of, or in addition to, animal power. One could look at three rising curves in the 20th century—human population, technical progress, and energy production—and be convinced that both people and machines were eating oil. The curves fit each other that well.

Tapping into cheap energy was a major breakthrough in the technium. But if the discovery of compact energy was the key insight, then China would have been first to industrialize because the Chinese figured out their abundant coal could burn at least 500 years before Europe did. Cheap energy was a huge bonus, but stockpiles of energy were not enough. China lacked the science that was key to liberating that energy.

Imagine humans had been born on a planet without fossil fuels. What would have happened? Could civilization have progressed very far burning wood only? It is possible. Maybe highly efficient wood and charcoal technology beyond what we presently have could have nurtured a population increase sufficiently dense to invent science and then, solely powered by wood, go on to invent solar panels, or nukes, or whatever. On the other hand, a civilization floating on oceans of oil, yet without science, would not progress anywhere.

Progress follows the rise of minds, which then causes an echoing rise in energy. Abundant, cheap fuel found easily around the planet enabled the Industrial Revolution and the current acceleration of technological progress, but first the technium needed science to unlock the transforming power of coal and oil. In a coevolutionary dance, human minds mastered cheap energy, which expanded food for increasing numbers of human minds, which propelled more technological inventions, which consumed more cheap energy. This self-amplifying circuit produces the three rising curves of population, energy use, and technological progress, the three strands of the technium.

The evidence for the rising curve of technological progress is deep and wide. The data fills volumes. Hundreds of scholarly papers record substantial improvements across the board in matters we care about.

The trajectories of these measurements generally point in the same direction: up. Their accumulated weight elicited this famous prediction by Julian Simon a decade ago:

> These are my most important long-run predictions, contingent on there being no global war or political upheaval: (1) People will live longer lives than now; fewer will die young. (2) Families all over the world will have higher incomes and better standards of living than now. (3) The costs of natural resources will be lower than at present. (4) Agricultural land will continue to become less and less important as an economic asset, relative to the total value of all other economic assets. These four predictions are quite certain because the very same predictions, made at all earlier times in history, would have turned out to be right.

His reason is worth repeating: He is betting on a historical force that has maintained its trajectory for many centuries.

Nonetheless, experts wield three arguments against the notion of progress. The first is that what we think we are measuring is completely illusionary. By this reckoning we are measuring the wrong things. Skeptics see massive deterioration in human health and loss of human spirit, not to mention degradation of everything else. But any objection to the reality of progress must confront a simple fact: Life expectancy at birth in the United States increased from 47.3 years in 1900 to 75.7 years in 1994. If this is not an example of progress, then what is it? In at least one dimension progress is not illusionary.

The second objection argues that progress is only half real. That is, material advances do occur, but they don't mean very much. Only intangibles like meaningful happiness count. Meaningfulness is very hard to measure, which makes it very hard to optimize. So far anything we can quantify has been getting better over the long term.

The third objection is the most common today. It posits that material progress is real but is too costly as produced. On their better days, critics of the notion of progress would agree that in fact things are getting better for humans but that they do so by destroying or consuming natural resources at an unsustainable rate.

We should take this argument seriously. Progress is real, but so are its consequences. There is real, serious environmental damage caused by technologies. But this damage is not inherent in technologies. Modern technologies don't have to cause such damage. When existing ones do cause damage, we can make better technologies.

"If we go on as we are, it'll be very difficult to sustain things," says science author Matt Ridley. "But we won't go on as we are. That's what we never do. We always change what we do and we always get much more efficient at using things—energy, resources, etc. Just take land area for feeding the world. If we'd gone on as we were, as hunter-gatherers, we'd have needed about 85 Earths to feed 6 billion people. If we'd gone on as early slash-and-burn farmers, we'd have needed a whole Earth, including all the oceans. If we'd gone on as 1950 organic farmers without a lot of fertilizer, we'd have needed 82 percent of the world's land area for cultivation, as opposed to the 38 percent that we farm at the moment."

We don't go on as we are. We address the problems of tomorrow not with today's tools but with the tools of tomorrow. This is what we call progress.

And there will be problems tomorrow because progress is not utopia. It is easy to mistake progressivism as utopianism because where else does increasing and everlasting improvement point to except utopia? Sadly, that confuses a direction with a destination. The future as unsoiled technological perfection is unattainable; the future as a territory of continuously expanding possibilities is not only attainable but also exactly the road we are on now.

I prefer how biologist Simon Conway Morris puts it: "Progress is not some noxious by-product of the terminally optimistic, but simply part of our reality." Progress is real. It is the reordering of the material world that is made possible by flows of energy and the expansion of intangible minds. While progress is carried forward by humans now, this reorganization began long ago, in biological evolution.

6

Ordained Becoming

As the seventh kingdom of life, the technium is now amplifying, extending, and speeding up the self-organized progress that propels biological evolution through the aeons. We might think of the technium as "evolution accelerated." Therefore, in order to see where the technium is going we need to discern where evolution itself is headed and what is pushing it in that direction.

I make the case in this chapter that the course of biological evolution is not a random drift in the cosmos, which is the claim of current textbook orthodoxy. Rather, evolution—and by extension, the technium—has an inherent direction, shaped by the nature of matter and energy. This direction introduces inevitabilities into the shape of life. These nonmystical tendencies are woven into the fabric of technology as well, which means certain aspects of the technium are also inevitable.

To follow this trajectory we must begin at the beginning: the origin of life. Like a robot that builds itself, the mechanism we call life slowly self-assembled four billion years ago. Ever since that seemingly improbable self-invention, life has evolved hundreds of millions of improbable creatures. But how improbable are they really?

When Charles Darwin was working out his theory of natural selection, the eye worried him. He found it very hard to explain how it could have evolved bit by bit, because the eye's retina, lens, and pupil seemed so finely perfected toward the whole and so utterly useless at less than whole. Critics of Darwin's theory of evolution at the time held the eye out as a miracle. But miracles, almost by definition, happen only once. Neither Darwin nor his critics appreciated the fact that the cameralike eye evolved not just once—miracle though it may seem—but six times over the course of life on Earth. The remarkable optical architecture of a "biological camera" is also found in certain octopuses, snails, marine annelids, jellyfish, and spiders. These six lineages of unrelated creatures share only a distant, blind common ancestor, so each lineage gets credit for evolving this marvel on its own. Each of the six manifestations is an astounding achievement; after all, it took humans several thousand years of serious tinkering to cobble together the first working artificial camera eye.

But does the six-time independent self-assembly of the camera eye signal a supreme degree of improbability, sort of like tossing six million pennies in a row heads? Or does the multiple invention mean that the eye is a natural funnel that attracts evolution, like water in a well at the bottom of a valley? And then there are the eight other types of eyes, each of which has evolved more than once. Biologist Richard Dawkins estimates that "the eye has evolved independently between 40 and 60 times around the animal kingdom," leading him to claim, "it seems that life, at least as we know it on this planet, is almost indecently eager to evolve eyes. We can confidently predict that a statistical sample of [evolutionary] reruns would culminate in eyes. And not just eyes, but compound eyes like those in an insect, a prawn, or trilobite, and camera eyes like ours or a squid's. . . . There are only so many ways to make an eye, and life as we know it may well have found them all."

Are there certain forms—natural states—that evolution tends to gravitate toward? This question has immense bearing on the technium, because if evolution displays an attraction to universal solutions, then so will technology, its accelerated extension. In recent decades science has discovered that complex adaptive systems (of which evolution is one example) tend to settle (all other factors being equal) into a few recurring patterns. These patterns are not found in the parts of the system, and so the structure that appears is considered both "emergent" and dictated by the complex adaptive system as a whole. Since the same structure will appear again and again seemingly from nowhere—like a vortex that instantly appears among water molecules in a draining tub—these structures can also be considered inevitable.

With some perplexity biologists file in the bottom drawer of their desks an ever-growing list of identical phenomenon that have kept reappearing in life on Earth. They are not sure what to do with these curious cases. But a few scientists believe these recurring inventions are biological "vortices," or familiar patterns that emerge from the complex interactions in evolution. The estimated 30 million species coinhabiting Earth are running millions of experiments every hour. They constantly breed, fight, kill, or mutually alter each other. Out of this exhaustive recombination, evolution keeps converging upon similar characteristics in far-flung branches in the tree of life. This attraction to recurring forms is called *convergent evolution*. The more taxonomically separate the lineages, the more impressive the convergence.

Old World primates have full-color vision and an inferior sense of smell compared to their distant cousins the New World monkeys. These spider monkeys, lemurs, and marmosets all have a very keen sense of smell but lack tricolor vision. All, that is, except the howler monkey, which, in parallel to the Old World primates, has tricolor vision and a weak nose. The common ancestor to the howler and the Old World primates goes very far back, so howlers independently evolved tricolor vision. By examining the genes for full-color vision, biochemists discovered that both the howler and Old World primates use receptors tuned to the same wavelengths, and they contain exactly the same amino acids in three key positions. Not only that, the diminished olfactory senses of both howler and apes was caused by the inhibition of the same olfactory genes, turned off in the same order and in the same details. "When similar forces converge, similar results emerge. Evolution is remarkably reproducible," says geneticist Sean Carroll.

The notion of reproducibility in evolution is highly controversial. But since convergence is not only big news in biology but also strongly suggests convergence in the technium, it is worth looking at further evidence for it in nature. Depending on how one measures the concept of "independent," the catalog of visible examples of independent, convergent evolution is hundreds long and counting. Any list will certainly include the three-time evolution of flapping wings in birds, bats, and pterodactyls (reptiles of the dinosaur era). The last common ancestor among these three lineages did not have wings, which means that each line evolved its wings independently. Despite their vast taxonomic distance, the wings in these three cases are remarkably similar in form: skin stretched over bony limbs. Navigation by echolocation has been found four times: in bats, dolphins, and two species of cave-dwelling birds (the South American oilbird and Asian swiftlet). Bipedality recurs in humans and birds. Antifreeze compounds were evolved twice in ice fish, once in the Artic and once in the Antarctic. Both hummingbirds and sphinx moths evolved to hover over flowers sucking nectar through a thin tube. Warm-bloodedness evolved more than once. Binocular vision evolved many times in distant taxa. Bryozoa, a family of coral, evolved distinctive helical colonies six different times over 400 million years. Social cooperation evolved in ants, bees, rodents, and mammals. Seven widely separated corners of the plant kingdom evolved insectivorous species—eating insects for nitrogen. Succulent leaves evolved multiple times across taxonomic distance, jet propulsion twice. Buoyant swim bladders evolved independently in many varieties of fish, mollusks, and jellyfish. Flapping wings constructed of taut membranes over skeleton frames arose more than once in the insect kingdom. While humans have technically evolved fixed-wing aircraft and spinning-wing aircraft, we haven't yet made a viable flapping-wing craft. On the other hand, fixed-wing gliders (flying squirrels, flying fish) and spinning-wing gliders (many seeds) have evolved a number of times. In fact, three species of rodentlike gliders also display convergence: the flying squirrel as well as the squirrel glider and marsupial sugar glider, both of Australia.

Because of its lone tectonic wanderings in geologic time, the continent of Australia is a laboratory for parallel evolution. There are multiple examples of marsupials in Australia paralleling placental mammals from the Old World, even in the past. Saber-canine teeth are found in both the extinct marsupial thylacosmilus and the extinct saber-toothed cat. Marsupial lions had retractable claws like feline cats.

Dinosaurs, our iconic distant cousins, independently evolved a number of innovations in parallel with our common vertebrate ancestors. In addition to the parallels between flying pterodactyls and bats, there were the streamlined ichthyosaurs that mirrored dolphins and mosasaurs, which paralleled whales. Triceratops evolved beaks similar to those of both parrots and octopus and squid. Snakelike pygopodidae were as legless as reptilian snakes later were.

The less taxonomic distance between lineages, the more common—but less significant—convergence becomes. Both frogs and chameleons independently evolved rapid-fire "harpoon tongues" to snatch prey at a distance. All three major phyla of mushrooms have separately evolved species that produce dark, dense, underground, trufflelike fruits; and in North America alone there are more than 75 mushroom genera that include "truffles," many of which evolved independently.

For some biologists, occurrences of convergence are merely a statistical curiosity, sort of like meeting someone else with your own name and birth date. Weird, but so what? Given enough species and enough time you are bound to encounter two that cross paths morphologically. But homologous features are actually the rule in biology. Most homology is invisible and occurs among related species. Relatives naturally share features, while the unrelated share fewer, so unrelated homology is more meaningful and noticeable. Either way, most methods used by life are used by more than one organism and in more than one phylum. What is rare is a trait that has *not* been reused somewhere in nature. Richard Dawkins challenged naturalist George McGavin to name biological "innovations" that have evolved only once, and McGavin was able to compile only a handful, such as the bombardier beetle, which mixes two chemicals on demand to shoot a noxious stream at enemies, or the diving-bell spider, which uses a bubble to breathe. Simultaneous, independent invention seems to be the rule in nature. As I argue in the next chapter, simultaneous, independent invention also seems to be the rule in the technium. In both realms, natural evolution and technological evolution, convergence creates inevitabilities. Inevitability is even more controversial than reproducibility and so demands yet more evidence.

Return to the recurring eye. The retina is lined with a layer of a very specialized protein that performs the tricky work of perceiving light. This protein, called rhodopsin, transfers the photon energy from incoming light to an outgoing electrical signal sent along the optic nerve. Rhodopsin is an archaic molecule present not only in the retinas of camera eyes but also in the most primitive lensless eye spot of a lowly worm. It is found throughout the animal kingdom, and it retains its structure wherever it is found because it works so well. The same molecule has probably remained unchanged for billions of years. Several competing light-trigger molecules (cryptochromes) aren't as efficient or robust, suggesting that rhodopsin is simply the best molecule for seeing that can be found after two billion years of looking. But surprisingly, rhodopsin is another example of convergent evolution, because it evolved twice in two separate kingdoms in the deep past—once in Archaea and once in Eubacteria.

This fact should shock us. The number of possible proteins is astronomical. There is an alphabet of 20 base symbols (amino acids) that make up every protein "word," which on average is, say, 100 symbols, or 100 bases, long. (In fact, many proteins are much longer, but for this calculation 100 is sufficient.) The total number of possible proteins that evolution could generate (or discover) is 100^{20} or 10^{39}. This means that there are more possible proteins than there are stars in the universe. But let's simplify that. Because only one in a million amino acid "words" folds into a functioning protein, let's vastly reduce that magnitude and agree that the number of potential working proteins is equal to the number of stars in the universe. Discovering a specific protein would be equivalent to randomly finding a specific star in the vastness of space.

By this analogy evolution finds new proteins (new stars) by a series of hops. It jumps from one protein to a "nearby" related one and then hops on to the next novel form until it reaches some remote unique protein far from where it started, just as one might travel to a distant sun by hopping stars. But in a universe as large as ours, once you landed on a distant star one hundred random hops away, you would never reach it again by the same random process. It is statistically impossible. But that is what evolution did with rhodopsin. Out of all the protein stars in the universe, it found this one—a protein that has not been improved upon for billennia—twice.

And the impossibility of "twice-struck" keeps happening in life. Evolutionist George McGhee writes in a paper entitled "Convergent Evolution": "The evolution of the ichthyosaur or porpoise morphology is not trivial. It can be correctly described as nothing less than astonishing that a group of land-dwelling tetrapods, complete with four legs and a tail, could devolve their appendages and their tails back into fins like those of a fish. Highly unlikely, if not impossible? Yet it happened twice, convergently in the reptiles and the mammals, two groups of animals that are not closely related. We have to go back in time as far as the Carboniferous to find a common ancestor for them; thus, their genetic legacies are very, very different. Nonetheless, the ichthyosaur and the porpoise both have independently re-evolved fins."

What, then, guides this return to the improbable? If the same protein, or "contingent" form, is evolved twice, it is obvious that every step of the way cannot be random. The prime guidance for these parallel journeys is their common environment. Both archaea rhodopsin and eubacterial rhodopsin, and both ichthyosaur and dolphin, float in the same seas with the same advantages gained by adaptation. In the case of rhodopsin, because the molecular soup surrounding the precursor molecules is basically the same, the selection pressure will tend to favor the same direction on each hop. In fact, the match of environmental niche is usually the reason given for occurrences of convergent evolution. Arid, sandy deserts on different continents tend to produce large-eared, long-tailed, hopping rodents because the climate and terrain sculpts a similar set of pressures and advantages.

Yes, but why, then, doesn't every similar desert in the world produce a kangaroo rat, or jerboa, and why aren't all desert rodents some version of kangaroo rats? The orthodox answer is that evolution is a highly contingent process, where random events and pure luck change the course, so that even within parallel environments it is very rare to arrive at the same morphological solution. Contingency and luck are so strong in evolution that the marvel is that convergence ever happens. Based on the number of possible forms that can be assembled from the molecules of life and the central role of random mutation and deletion in shaping them, significant convergence from independent origins should be as scarce as miracles.

But a hundred, or a thousand, cases of isolated significant convergent evolution suggest something else at work. Some other force pushes the self-organization of evolution toward recurring solutions. A different dynamic besides the lottery of natural selection steers the course of evolution so that it can reach an unlikely remote destination more than once. It is not a supernatural force but a fundamental dynamic as simple at its core as evolution itself. And it is the same force that funnels convergence in technology and culture.

Evolution is driven toward certain recurring and inevitable forms by two pressures:

1. The negative constraints cast by the laws of geometry and physics, which limit the scope of life's possibilities.
2. The positive constraints produced by the self-organizing complexity of interlinked genes and metabolic pathways, which generate a few repeating new possibilities.

These two dynamics create a push in evolution that gives it a direction. Both of these two dynamics continue to operate in the technium as well and shape the inevitabilities along the course of the technium. Let me address each influence in turn, starting with the way chemistry and physics shape life and, by extension, the inventions of our mind in the technium.

Plants and animals come in a bewildering diversity of scales. Insects can be microscopic, like lice, or giant, like horned beetles the size of shoes; redwood trees tower 100 meters tall, and miniature alpine plants fit in a thimble; immense blue whales swell as big as ships, and pygmy chameleons shrink to less than an inch long. Yet the dimensions of each species are not arbitrary. They follow a scale ratio that is astonishingly

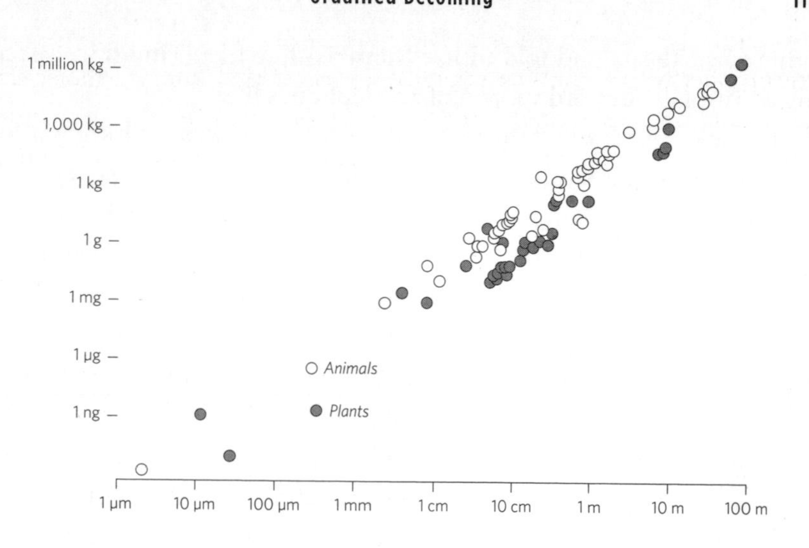

Size Ratio in Life. The ratio between an organism's mass and its length is a constant in both plants and animals.

constant in both plants and animals. This ratio is dictated by the physics of water. The strength of a cell wall is determined by the surface tension of water; that constant in turn mandates the maximum height per width of a body, any possible body. These physical forces play out not just on Earth, but everywhere in the universe, so we might expect any organisms based on water, whenever and wherever they evolve, to converge upon this same universal size ratio (adjusted for local gravity).

The metabolism of life is likewise constrained. Small animals live fast and die young. Big animals plod along. The speed of life for animals—the rate at which their cells burn energy, the speed of their muscle twitches, the time it takes them to gestate or to mature—is remarkably proportional to their life span and size. Both metabolic rate and heart rate are proportional to the mass of the creature. These constants derive from the fundamental rules of physics and geometry and the natural advantages to minimizing energy surfaces (lung surface, cell surface, circulatory capacity, etc.). While a mouse's heart and lungs beat rapidly compared to an elephant's, both mouse and elephant count the same number of beats and breaths per life. It is as if mammals are assigned 1.5

billion heartbeats and told to use them as they like. Tiny mice speed ahead in a fast-forward version of an elephant's life.

In biology this constant ratio for metabolism was well known for mammals, but researchers recently realized a similar law governs all plants, bacteria, and even ecosystems. Dilute pools of cool oceanic algae can be thought of as a slow-motion version of a warm-blooded heart. The amount of energy per kilogram (or energy density) flowing through a plant or ecosystem is equivalent to metabolism. Many life processes—from the number of hours of sleep an animal needs to the hatching time for its eggs to the rate at which a forest accumulates wood mass to the mutation rate in DNA—all seem to follow a universal metabolism scaling law. "We've found that despite the incredible diversity of life, from a tomato plant to an amoeba to a salmon, once you correct for size and temperature, many of these [metabolic] rates and times are remarkably similar," say James Gillooly and Geoffrey West, the researchers who discovered this law. "Metabolic rate is the fundamental biological rate," they claim—"a universal clock" reckoned in energy, the speed at which all life of any type proceeds. The clock is inevitable for anything living.

Other physical constants run through the biological world. Bilateral symmetry (mirrored left and right sides) recurs in almost every family of life. This fundamental symmetry seems to bring adaptive advantage on many levels, from superior balance of movement to prudent redundancy (two of everything!) to efficient compression of genetic code (just duplicate the first side). Other geometric forms, like a tube for nutrient transport in plants or animals (a gut) or legs, are just plain good physics. Some recurring designs, such as the arboreal splay of branches in a tree and coral or the swirling spiral of petals on a flower, are based on the mathematics of growth. They repeat because the math is eternal. All life on Earth is protein based, and the way those proteins fold and unfold inside cells determines the character traits and behaviors of that creature. Biochemists Michael Denton and Craig Marshall state that "recent advances in protein chemistry suggest that at least one set of biological forms—the basic protein folds—is determined by physical laws similar to those giving rise to crystals and atoms. They give every appearance of being invariant platonic forms." Proteins—the essential molecules of life's diversity—are also ultimately governed by a limited set of recurring laws.

If we made a large spreadsheet containing all the physical characteristics of all the living organisms on Earth, we would find many blank white spaces for organisms that logically "could be" but aren't. These fill-in creatures would obey the laws of biology and physics, yet were never born. Such "could be" life forms might include a mammalian snake (why not?), a flying spider, or a terrestrial squid. In fact, some of these could still evolve on Earth if we left the current flora and fauna alone long enough. These speculative creatures are entirely plausible because they are convergent, recycling (but remixing) morphological forms that repeat throughout the biosphere.

When artists and science-fiction authors fantasize alternative planets full of living creatures, try as they might to "think outside the box" of earthly constraints, many of the organisms they envision also retain many of the forms found on Earth. Some would chalk this up to a lack of imagination; we are constantly being surprised by bizarre forms found in the deepest part of the oceans on our home planet; surely life on other planets will be full of surprises. Others, myself included, agree that we will be surprised but that given what "could be"—that vast imaginary space of all possible ways in which one could arrange atoms into an organism—what we will find on another planet will only fill one small corner of what could be. Life on other planets will be surprising because of what it does with already familiar forms. Biologist George Wald, who won a Nobel Prize for his work on eye retina pigments, told NASA, "I tell my students: learn your biochemistry here and you will be able to pass examinations on Arcturus."

Nowhere is that physical constraint of the infinitude more evident than in the structure of DNA. The molecule of DNA is so remarkable that it is in its own class. As every student knows, DNA is a unique double-helical chain that can zip and unzip with ease and of course replicate itself. But DNA can also arrange itself into flat sheets or interlocking rings or even an octahedron. This singular gymnastic molecule serves as a dynamic mold that prints the stupendously large set of proteins responsible for the physical characteristics of tissue and flesh, which in turn, by mutual interaction, generate vast ecosystems of complexity. From this single omnipotent quasicrystal the awesome variety of life in all its unexpected shapes springs forth. Subtle rearrangements along its tiny, ancient spiral will produce the majesty of a strolling sauropod 20 meters high, and also the delicate gem of an iridescent green dragonfly, and the frozen immaculacy of a white orchid petal, and of course the intricacies of the human mind. All from such a tiny semicrystal.

If we acknowledge no supernatural force working outside evolution, then all these structures—and more—must in some sense be contained within the structure of DNA. Where else could they come from? The details of all oak lineages and future species of oak are resident, in some fashion, in the original acorn of DNA. And if we acknowledge no supernatural force working outside evolution, then our minds—which all descended from the same original first cell—must also have been encoded implicitly in DNA. And if our minds, then what about the technium? Were its space station, Teflon, and internet also dissolved in the genome, only to be precipitated later by constant evolutionary work, just as an oak tree is finally manifested after billions of years?

Of course, merely inspecting this molecule reveals none of this cornucopia; we seek in vain to find a giraffe in the spiral ladder of DNA. But we can seek alternative "acorn" molecules as a way to rerun this unfolding to see if something else besides DNA could generate similar diversity, reliability, and evolvability. A number of scientists have searched for alternatives to DNA in the laboratory by engineering "artificial" DNAs or constructing DNA-like molecules or engineering wholly original biochemistry. There are a bunch of practical reasons to invent a DNA alternative (say, to create cells that can work in space), but so far alternatives with DNA's versatility and brilliance are in short supply.

The first obvious approach in the quest for an alternative DNA molecule is to substitute slightly modified base pairs into the helix (think of different steps in DNA's spiral staircase). K. D. James and A. D. Ellington write in *Origins of Life and Evolution of the Biospheres* that "experiments with alternative base pairing schemes have suggested that the current set of purines and pyrimidines [the canonical base pair types] is in many ways optimal. . . . The unnatural nucleic acid analogues that have been examined experimentally have proven to be largely incapable of self-replication."

Of course, science is rife with discoveries initially thought unlikely, implausible, or impossible. In the case of self-organizing life, we might want to be particularly hesitant to generalize about alternatives since everything we can say about it is based on a sample size (so far) of exactly one, here on Earth.

But chemistry is chemistry, everywhere in the universe. Carbon sits at the center of life because it is gregarious and contains so many hooks for other elements to bind to. It has a particularly friendly relationship with oxygen. Carbon is easily oxidized as fuel for animals and easily unoxidized (reduced) by chlorophyll in plants. And of course it forms the backbone for long chains of incredibly diverse megamolecules. Silicon, carbon's sister element, is the most likely alternative candidate to produce a non-carbon-based life form. Silicon also is very prolific in its hooking up with a variety of elements, and it is more abundant on the planet than carbon. When science-fiction authors dream up alternative life forms, they are often based on silicon. But in real life silicon suffers from a few major drawbacks. It does not link up into chains with hydrogen, limiting the size of its derivatives. Silicon-silicon bonds are not stable in water. And when silicon is oxidized, its respiratory output is a mineral precipitate, rather than the gaslike carbon dioxide. That makes it hard to dissipate. A silicon creature would exhale gritty grains of sand. Basically, silicon produces dry life. Without a liquid matrix it's hard to imagine how complex molecules are transported around to interact. Perhaps silicon-based life inhabits a fiery world and the silicates are molten. Or perhaps the matrix is very cold liquid ammonia. But unlike ice, which floats and insulates the unfrozen liquid, frozen ammonia sinks, allowing the oceans to freeze whole. These concerns are not hypothetical but are based on experiments to produce alternatives to carbon-based life. So far, all evidence points to DNA as the "perfect" molecule.

For even though clever minds like ours may invent a new life base, finding a life base that can create itself is an entirely higher order. A potential synthetic life base created in the lab might be robust enough to survive on its own in the wild but fail to organize itself into existence. If you can skip the need for a self-made birth, you can jump to all kinds of complex systems that would never evolve on their own. (This is in fact the "job" of minds: to produce types of complexity that evolutionary self-creation cannot.) Robots and AIs don't need to self-organize from metal-laden rocks because they are made rather than born.

However, DNA did have to self-organize. By far the most remarkable thing about this potent nucleus of life is that it put itself together. The most basic carbon-based ingredients—such as methane or formaldehyde—are readily available in space, and even in pools on planets. But every abiotic condition (lightning, heat, warm pools, impact, freezing/thawing) we have tried as a stimulus to organize these Lego-like building blocks into the eight component sugars that make up RNA and DNA has failed to generate sustainable amounts of them. All the known pathways to creating just one of these sugars—ribose (the *R* in RNA)—are so complicated they are difficult to reproduce in the lab and (so far) unthinkable as existing in the wild. And that is just for one of eight essential predecessor molecules. The necessary—and potentially contradictory—conditions to nurture dozens of other unstable compounds toward self-generation have not been found.

Yet here we are, so we know that these peculiar pathways can be found. At least once. But the supreme difficulty of simultaneous improbable pathways working in parallel suggests that there may be only one molecule that can negotiate this maze, self-assemble its scores of parts, self-replicate once birthed, and then unleash from its seed the head-shaking, eye-popping, mind-blowing variety and exuberance we see in life on Earth. It is not enough to find a molecule that can self-replicate *and* generate ever-larger mounds of increasing complexity. There may indeed be multiple amazing chemical nuclei capable of that. Rather, the challenge is finding one that does all that and can make itself, too.

So far, there are no other contenders even close to offering that kind of magic. This is why Simon Conway Morris calls DNA "the strangest molecule in the universe." Biochemist Norman Pace says there may be a "universal biochemistry" based upon this most remarkable of all molecules. He speculates: "It seems likely that the basic building blocks of life anywhere will be similar to our own, in the generality if not in the detail. Thus the 20 common amino acids are the simplest carbon structures imaginable that can deliver the functional groups used in life." To paraphrase George Wald: If you want to study ET, study DNA.

There is another hint of the unique (perhaps universally unique) power of DNA. Two molecular biologists (Stephen Freeland and Laurence Hurst) computationally generated random genetic code systems (the equivalents of DNA, but without DNA) in a simulated chemical world. Since the combinatorial sum of all possible genetic codes overwhelms the time in the universe to compute them, the researchers sampled a subset of these, focusing on those systems they classified as chemically viable. They explored a million variations out of what they estimated to be a pool of 270 million viable alternatives and ranked the systems on how well they minimized errors in their simulated world (a good genetic code will reproduce accurately without errors). After a million computer runs the measured efficiency of the genetic codes fell into a typical bell curve. Far off to one side was Earth's DNA. Out of a million alternative genetic codes, our current DNA scheme was "the best of all possible codes," they concluded, and even if it is not perfect, it is at least "one in a million."

Green chlorophyll is another strange molecule. It is ubiquitous on the planet, yet not optimal. The spectrum of the sun peaks in the yellow frequency, yet chlorophyll is optimized for red/blue. As George Wald notes, chlorophyll's "triple combination of capacities"—a high receptivity to light, an ability to store the captured energy and relay it to other molecules, and an ability to transfer hydrogen in order to reduce carbon dioxide—made it essential in the evolution of solar-gathering plants "despite its disadvantageous absorption spectrum." Wald goes on to speculate that this nonoptimization is evidence that there is no better carbon-based molecule for converting light into sugar, because if there were, wouldn't several billion years of evolution have produced it?

It may seem like I contradict myself when I point to convergence due to rhodopsin's maximum optimization and then to chlorophyll's nonoptimization. I don't think the level of efficiency is central. In both cases it is the paucity of alternatives that is the strongest evidence for inevitability. In chlorophyll's case, no alternative forms appear after billions of years in spite of its imperfection, and in rhodopsin's case, despite a few minor competitors, the same molecule was found twice in an otherwise vast empty field. Again and again evolution returns to a few solutions that work.

No doubt someday very smart researchers in a laboratory will devise an alternative base to organic DNA that is able to unleash a river of new life. Accelerated vastly, this synthetic life base might evolve all kinds of new creatures, including sentient beings. However, this alternative living system—whether based on silicon, carbon nanotubes, or nuclear gases in a black cloud—would have its own inevitabilities, channeled by the constraints embedded in its original seeds. It would not be able to evolve everything, but it could produce many types of life that our life could not. Some science-fiction authors have playfully speculated that DNA might itself be such an engineered molecule. It is, after all, ingeniously optimized, and yet its origins are deeply mysterious. Perhaps DNA was cleverly crafted by superior intelligences in white lab coats and shotgunned into the universe to naturally seed empty planets over billions of years? We would be just one of many seedlings that sprouted from this generic starter mix. This kind of engineered gardening might explain a lot, but it does not remove the uniqueness of DNA. Nor does it remove the channels that DNA has laid for evolution on Earth.

The constraints of physics, chemistry, and geometry have governed life from its origins onward—and even into the technium. "Underlying all the diversity of life is a finite set of natural forms that will recur over and over again anywhere in the cosmos where there is carbon-based life," claim biochemists Michael Denton and Craig Marshall. Evolution simply cannot make all possible proteins, all possible light-gathering molecules, all possible appendages, all possible means of locomotion, all possible shapes. Life, rather than being boundless and unlimited in every direction, is bounded and limited in many directions by the nature of matter itself.

I will argue that the same constraints bind technology. Technology is based on the same physics and chemistry as life, and more important, as the seventh, accelerated kingdom of life, the technium is bound by many of the same constraints guiding life's evolution. The technium can't make all imaginable inventions or all possible ideas. Rather, the technium is limited in many directions by the constraints of matter and energy. But the negative constraints of evolution are only half of its story.

The second great force pushing evolution on its immense journey is positive constraints that channel evolutionary innovation in certain directions. In tandem with the constraints of physical laws outlined above, the exotropy of self-organization steers evolution along a trajectory. While these internal inertias are immensely important in biological evolution, they are even more consequential in technological evolution. In fact, in the technium, self-generated positive constraints are more than half the story; they are the main event.

However, the existence of internal constraints guiding biological evolution is far from orthodoxy in biology today. The notion of directional evolution has a colorful history tainted by its association with a belief in a supernatural essence of life. While it is no longer associated with the supernatural today, the idea of directional evolution is now associated with the idea of "inevitable"—a concept that many modern scientists find intolerable in any form.

I would like to present the best case for a direction within biological evolution that our evidence so far will permit. It is a complicated story, vital not only for understanding biology but also for discerning the future of technology. Because if I can demonstrate that there is an internally generated direction within natural evolution, then my argument that the technium extends this direction is easier to see. So while I delve deeply into the forces driving life's evolution, this long explanation is really a parallel argument for the same kind of evolution within technology.

I begin this second half of the story with a reminder that this newly appreciated exotropic drive of evolution is not its only engine. Evolution has multiple drivers, including the physical constraints I earlier described. But in the current orthodox scientific understanding of evolution, change is attributed chiefly to one source: random variation. In the wilds of nature, reproducing survivors are naturally selected from inheritable random variations; therefore, in evolution there can only be random advance without direction. The key insight gained by the last three decades of research on complex adaptive systems offers a contrary view: that the *variation presented to natural selection is not always random.* Experiments show that "random" mutations are often not unbiased; instead, variation is governed by geometry and physics; and most important, variations are often shaped by the possibilities inherent in the recurring patterns of self-organization (a la whirlpool vortex).

Once upon a time the notion of nonrandom variation was heresy, but as more and more biologists ran computer models, the idea that variation is not random became a scientific consensus among certain theoreticians. Self-regulating networks of genes (found in all chromosomes) favor certain kinds of complexes. "Some potentially useful mutations are so probable that they can be viewed as being encoded implicitly in the genome," says biologist L. H. Caporale. Metabolic pathways in cells can autocatalyze themselves into a network and drift into self-preferred loops. This flips the traditional view. In the old view, the internal (the source of mutation) created change, while the external (the environmental source of adaptation) selected or directed it; in the new view, the external (physical and chemical constraints) creates forms, while the internal (self-organization) selects or directs them. And when the internal directs, it redirects to recurring forms. As the early paleontologist W. B. Scott put it, the complexity of evolution creates "inherited channels for preferred change."

In the textbook version, evolution is a mighty force propelled by a single near-mathematical mechanism: inheritable random mutations selected by adaptive survival, also known as natural selection. The emerging modified view recognizes additional forces. It proposes that the creative engine of evolution stands on *three* legs: the adaptive (the classic agent), plus the contingent and the inevitable. (These three forces reappear in the technium as well.) We can describe these as three vectors of evolution.

The adaptive vector is the orthodox force that textbook theory teaches. Just as Darwin surmised, those organisms that adapt best to their environment survive to breed offspring. So any new strategies for survival in a changing environment, no matter where they come from, are selected over time and lead to a very fine fit for that species. The adaptive force is fundamental at all levels of evolution.

The second vector in evolution's triad is luck, or contingency. A lot of what happens in evolution comes down to the lottery, not adaptation of the superior. Much of the fine detail of speciation is a result of happenstance, some improbable trigger that leads a species down a contingent path. The individual speckles on a monarch butterfly's wings are not strictly adaptive, just plain chance. These random beginnings can eventually lead to completely unexpected designs later on. And these subsequent designs may be less complex or less elegant than their predecessors. In other words, many of the forms we see in evolution today are due to random contingencies in the past and don't follow a progressive sequence. If we rewind the tape of life's history and push start again, it will play out differently. (I should mention for the benefit of young readers that "rewinding the tape," like "dialing the phone," "filming a movie," or "cranking the engine," is a skeuonym, an expression left over from a technology no longer used. In this case, "rewinding a tape" means to rerun a sequence from the same starting point.)

Stephen Jay Gould, who introduced the trope of "rewinding the tape of life" in his seminal book *Wonderful Life*, makes an elegant case for the ubiquity of contingency in evolution. He based his argument on the evidence of a set of cryptic fossils of pre-Cambrian life found in the Burgess Shale in Canada. A young grad student named Simon Conway Morris spent years tediously dissecting these minute fossils under a microscope. After a decade of intense study Morris announced that the Burgess Shale contained a treasure trove of previously unknown biota, far more diverse in forms than life now. But this great ancient diversity of archetypes was decimated by unlucky disasters 530 million years ago, leaving further evolution with only a relative few basic organism types—the comparatively less varied world we see now. Superior designs were randomly eliminated. Gould interpreted this chancy decimation of older, greater diversity as a powerful argument for the rule of contingency and an argument against the idea of directionality in evolution. In particular he believed the evidence of the Burgess Shale demonstrated that human minds were not inevitable, because nothing in evolution was inevitable. At the close of his book, Gould concludes, "Biology's most profound insight into human nature, status, and potential lies in the simple phrase, the embodiment of contingency: *Homo sapiens* is an entity, not a tendency."

This phrase "entity not tendency" is the orthodoxy in evolutionary theory today: that inherent contingency and supreme randomness in evolution preclude tendencies in any direction. However, later research disproved the notion that the Burgess Shale contained as great a diversity as first believed, deflating Gould's conclusions. Simon Conway Morris himself changed his mind about his earlier radical classifications. It turns out many of the Burgess Shale organisms were not weird new forms but weird old forms, and so contingency was far less prevalent in macroevolution, and progress more likely. Curiously, over the years since Gould's influential book, Morris has become the chief paleontologist championing the idea of convergence, directionality, and inevitabilities in evolution. In hindsight what the Burgess Shale proves is that contingency is a significant force in evolution, but not the only one.

The third leg of evolution's tripod is structural inevitability, the very force denied by the current dogma of biology. Whereas contingency can be thought of as a "historical" force, that is, a phenomenon where history matters, the structural component of evolution's engine can be thought of as "ahistorical" in that it produces change independent of history. Run it again, and you get the same story. This aspect of evolution pushes inevitabilities. For instance, the defensive venomous sting has been evolved at least twelve times: in the spider, the stingray, the stinging nettle, the centipede, the stonefish, the honeybee, the sea anemone, the male platypus, the jellyfish, the scorpion, the cone-shell mollusk, and the snake. Its reappearance is due not to a common history but to a common matrix of life, and that common structure arises not from the outside environment but from the internal momentum of self-organized complexity. This vector is the exotropic force, the emergent self-organization that arises in a system as complex as evolutionary life. As described in previous chapters, complex systems acquire their own inertia, creating recurring patterns that the system tends to fall into. This emergent self-order steers the system to its own selfish interests, and in this way it engenders a direction to the ongoing process. This vector pushes the messiness of evolution toward certain inevitabilities.

Charted, the tripod of evolution might look like this:

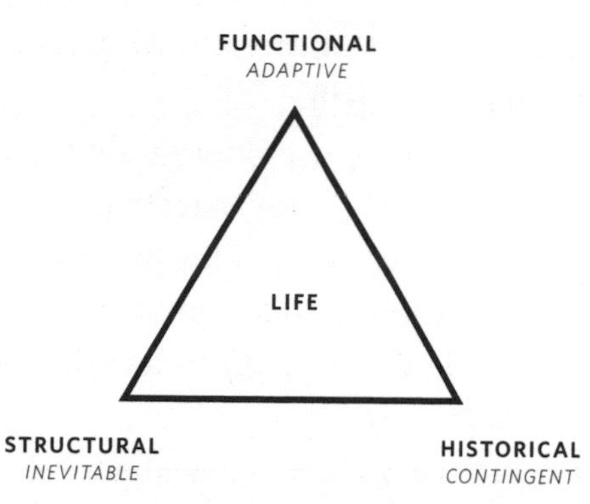

The Triad of Evolution. The three evolutionary vectors in life. The bold name indicates the realm in which it operates and the italic name its consequence.

All three dynamics are present in varying proportions at different levels in nature, counterbalancing and offsetting one another, combining to produce the history of each creature. A metaphor comes to mind that may help to untangle the three forces: The evolution of a species is like a meandering river as it carves away land. The detailed "particularness" of that river, the profile of its fine contours along the shore and bottom, comes from the vectors of adaptive mutations and contingency (never to be repeated), but the universal "riverness" form of the river (recurring in all rivers) as it is channeled in a valley comes from the internal gravity of convergence and emergent order.

For another example of contingent microdetails decorating inevitable macroarchetypes, consider the six separate dinosaur lineages that have followed the same morphological pathway in evolution. Over time each of the six dino lineages displayed a similar (inevitable) reduction in their side toes, an elongation of the long bones in their paws, and a shortening of their "fingers." We might call this pattern part of "dino-saurness." Because they are rehearsed in six lineages these archetypical structures are not merely random. Bob Bakker, the model for the dino guy in *Jurassic Park* and real-life dinosaur expert, claims, "This striking case of iterative parallelism and convergence [in the six dino lineages] . . . is a powerful argument that observed long-term changes in the fossil record are the result of directional natural selection, not a random walk through genetic drift."

Way back in 1897, paleontologist Henry Osborn, an early dinosaur and mammal expert, wrote: "My study of teeth in a great many phyla of Mammalia in past times has convinced me that there are fundamental predispositions to vary in certain directions; that the evolution of teeth is marked out beforehand by hereditary influences which extend back hundreds of thousands of years."

It is important to outline what is "marked out beforehand." In most cases, the details of life are contingent. The river of evolution determines only the broadest outlines of form. One might think of these as grand archetypes, for instance, tetrapods (four-leggedness), the snake form, eyeballs (spherical cameras), coiled guts, egg sacs, flapping wings, repeating segmented bodies, trees, puffballs, fingers. These are general silhouettes, not individuals. The biologist Brian Goodwin proposed that "all the main morphological features of organisms—hearts, brains, guts, limbs, eyes, leaves, flowers, roots, trunks, branches, to mention only the obvious ones—are the emergent results of morphogenic principles" and would reappear if the tape of life was rewound. Like other recurring archetypes, they are patterns your brain perceives without your even noticing it. "Oh, it's a clam," your mind says to itself, letting you fill in the particulars of color, texture, and individual species. The "clam" form—two concave hemispheres hinged to close—is the recurring archetype, the determined form.

Viewed from afar, from the distance of billions of years, it seems as if evolution wanted to create certain designs, in the way Richard Dawkins suggests that life wants to produce eyeballs, since it keeps repeating this invention. There is a tilt to evolution's seemingly chaotic churning that rediscovers the same forms and keeps arriving at the same solutions. It is almost as if life has an imperative. It "wants" to materialize certain patterns. Even the physical world seems biased in that direction.

There are many indications that our neighborhood of the universe is biased toward the appearance of life. Our planet is just close enough to the sun to be warm but far enough away to not burn. Earth has a large nearby moon that slows down its rotation to lengthen the day and to stabilize it over the long term. Earth shares the sun with Jupiter, which acts as a comet magnet. The ice of those captured comets may also have given Earth its oceans. Earth has a magnetic core, which generates a cosmic ray shield. It has the appropriate level of gravity to retain water and oxygen. It has a thin crust, which enables the churn of plate tectonics. Each of these variables seems to sit in a Goldilocks zone of not too little and not too much. Recent research suggests that there's a Goldilocks zone in the galaxy as well. Too close to the center of the galaxy and a planet is bombarded by constant, lethal cosmic radiation; too far from the center and when the planetary mass condenses from star dust it will be lacking the heavy elements that are needed for life. Our solar system is smack in the middle of this just-right zone. Such a list can quickly get out of hand to include every aspect of life on Earth. It's all perfect! The catalog soon resembles one of those phony "Help Wanted" ads engineered to stealthily fit only one favored predetermined person.

Some of these Goldilocks factors will turn out to be simply coincidental, but their number and deep-rootness, hint, in Paul Davies's phrase, that "the laws of nature are rigged in favor of life." In this view, "life emerges from a soup in the same dependable way that a crystal emerges from a saturated solution, with its final from predetermined by the interatomic forces." Cyril Ponnamperuma, an early pioneer in biogenesis (the study of the origin of life), believed "there are inherent properties in the atoms and molecules which seem to direct the synthesis" toward life. Theoretical biologist Stuart Kauffman believes his exhaustive computer simulations of prebiotic networks demonstrate that when conditions are right, the emergence of life is inevitable. Our existence here, he says, is a case of "not we the accidental but we the expected." Mathematician Manfred Eigen wrote in 1971, "The evolution of life, if it is based on a derivable physical principle, must be considered an inevitable process."

Christian de Duve, a Nobel Prize winner for his work in biochemistry, goes even further. He believes life is a cosmic imperative. He writes in his book *Vital Dust:* "Life is the product of deterministic forces. Life was bound to arise under the prevailing conditions, and it will arise similarly wherever and whenever the same conditions obtain. . . . Life and mind emerge not as the result of freakish accidents, but as natural manifestations of matter, written into the fabric of the universe."

If life is inevitable, why not fishes? If fishes are inevitable, why not mind? If mind, why not the internet? Simon Conway Morris speculates that "what was impossible billions of years ago becomes increasingly inevitable."

One way to test the cosmic imperative is to simply rerun the tape of life. Gould called rewinding the tape of life the great "undoable" experiment, but he was wrong: It turns out you can rewind life.

The new tools of sequencing and genetic cloning make replaying evolution possible. You take a simple bacteria (*E. coli*), select an individual, and make dozens of identical clones of that one particular bug. Genetically sequence the genotype of one. Put each remaining clone into an identical incubation chamber, with identical settings and inputs. Let the cloned bacteria multiply freely in parallel pots. Let them breed for 40,000 generations. At each 1,000-generation milestone, take a few out, freeze them for a snapshot, and sequence their evolved genomes. Compare the parallel evolved genotypes across all the pots. You can rerun the tape of evolution at any time along the way by retrieving a frozen snapshot specimen and redeploying the bug in another identical chamber.

Richard Lenski, at Michigan State University, has been performing this very experiment in his lab. What he has found is that, in general, multiple runs of evolution produced similar traits in the phenotype—the outward body of the bacteria. Changes in the genotype occurred in roughly the same places, though the exact coding was often different. This suggests a convergence of broad form with details left to chance. Lenski is not the only scientist doing experiments like this. Others' experiments show similar results from parallel evolution: Instead of getting novelty each time, you get what one scientific paper calls "the convergence of multiple evolving lines on similar phenotypes." As geneticist Sean Carroll concludes, "Evolution can and does repeat itself at the levels of structures and patterns, as well as of individual genes. . . . This repetition overthrows the notion that if we rewound and replayed this history of life, all outcomes would be different." We can rewind the tape of life, and when we do in a constant environment, it often turns out roughly the same.

These experiments suggest that a trajectory shoots through evolution and this long path makes some improbable forms inevitable. That paradox of improbable inevitabilities needs a bit of explanation.

The incredible complexity of life disguises its singularity. There is only one life. All life today is descended along an unbroken line of duplication from one ancient molecule that worked inside one primeval cell that worked. Despite life's magnificent diversity, it is chiefly repeating, billions of billions of times, solutions that worked before. Compared to all possible arrangements of matter and energy in the universe, life's solutions are few. Because field biologists discover another organism on Earth every day that is new to us, we have reason to marvel at the inventiveness and exuberance of nature. Yet compared to what our brains could imagine, the diversity of life on Earth occupies a very small corner. Our alternative imaginary universes are full of creatures far more diverse, creative, and "out there" than the life here. But most of our imaginary creatures would never work because they would be full of physical contradictions. The world of the actual-possible is much smaller than it first appears.

The particular physical arrangements of matter, energy, and information that produce the ingenious molecules of rhodopsin or chlorophyll or DNA or the human mind are so scarce in the space of all possible "could be" things that they are statistically improbable almost to the point of being impossible. Every organism (and artifact) is a wholly improbable arrangement of its constituent atoms. Yet within the long chain of reproducing self-organization and restless evolution, these forms become highly probable, and even inevitable, because there are only a few ways such open-ended ingenuity can actually work in the real world; therefore, evolution must work through them. In this way, life is an inevitable improbability. And most of life's archetypal forms and stages are also inevitable improbabilities, or, we might say, improbable inevitabilities.

This means that something like a human mind is also the improbable inevitability of evolution. Rewind the tape of life and it would (on another planet or in a parallel time) produce a mind again. When Stephen Jay Gould claimed that "*Homo sapiens* is an entity, not a tendency," he got it precisely, but elegantly, backward. If we rerun his sentence again, but this time from back to front, I can't think of a more succinct phrase that sums up evolution's message better than this:

> *Homo sapiens* is a tendency, not an entity.

Humanity is a process. Always was, always will be. Every living organism is on its way to becoming. And the human organism even more so, because among all living beings (that we know about) we are the most open-ended. We have just started our evolution as *Homo sapiens*. As both parent and child of the technium—evolution accelerated—we are nothing more and nothing less than an evolutionary ordained becoming. "I seem to be a verb," the inventor/philosopher Buckminster Fuller once said.

We can likewise say: The technium is a tendency, not an entity. The technium and its constituent technologies are more like a grand process than a grand artifact. Nothing is complete, all is in flux, and the only thing that counts is the direction of movement. So if the technium has a direction, where is it pointed? If the greater forms of technologies are inevitable, what is next?

In the following chapters I show how innate tendencies in the technium converge upon recurring forms, just like biological evolution. This leads to inevitable inventions. And further, these self-generated biases also create a degree of autonomy, much like the autonomy earned by living creatures. And finally, this naturally emergent autonomy in technological systems also creates a suite of "wants." By following the long-term trends in evolution we can show what technology wants.

7

Convergence

In 2009, the world celebrated the 200th birthday of Charles Darwin and honored his theory's impact upon our science and culture. Overlooked in the celebrations was Alfred Russel Wallace, who came up with the same theory of evolution, at approximately the same time, 150 years ago. Weirdly, both Wallace and Darwin found the theory of natural selection after reading the same book on population growth by Thomas Malthus. Darwin did not publish his revelation until provoked by Wallace's parallel discovery. Had Darwin died at sea on his famous voyage (a not uncommon fate at that time) or been killed by one of his many ailments during his studious years in London, we would be celebrating the birthday of Wallace as the sole genius behind the theory. Wallace was a naturalist living in Southeast Asia, and he endured many serious illnesses as well. Indeed, he was suffering a debilitating jungle fever during the time he was reading Malthus. If poor Wallace, too, had succumbed to his Indonesian infection, and Darwin had died, it is clear from other naturalists' notebooks that someone else would have arrived at the theory of evolution by natural selection, even if they never read Malthus. Some think Malthus himself was close to recognizing the idea. None of them would have written up the theory in the same way, or used the same arguments, or cited the same evidence, but one way or another today we would be celebrating the 150th anniversary of the mechanics of natural evolution.

What seems to be an odd coincidence is repeated many times in technical invention as well as scientific discovery. Alexander Bell and Elisha Gray both applied to patent the telephone on the same day, February 14, 1876. This improbable simultaneity (Gray applied three hours before Bell) led to mutual accusations of espionage, plagiarism, bribery, and fraud. Gray was ill advised by his patent attorney to drop his claim for priority because the telephone "was not worth serious attention." But whether the winning inventor's dynasty became Ma Bell or Ma Gray, either way we would have telephone lines strung across our countryside, because while Bell got the master patent, at least three other tinkerers besides Gray had made working models of phones years earlier. In fact, Antonio Meucci had patented his "teletrofono" more than a decade earlier, in 1860, using the same principles as Bell and Gray, but because of his poor English, poverty, and lack of business acumen, he was unable to renew his patent in 1874. And not far behind them all was the inimitable Thomas Edison, who inexplicably didn't win the telephone race but did invent the microphone for it the next year.

Park Benjamin, author of *The Age of Electricity*, observed in 1901 that "not an electrical invention of any importance has been made but that the honor of its origin has been claimed by more than one person." Dig deep enough in the history of any type of discovery in any field and you'll find more than one claimant for the first priority. In fact, you are likely to find *many* parents for each novelty. Sunspots were first discovered not by two but by four separate observers, including Galileo, in the same year, 1611. We know of six different inventors of the thermometer, and three of the hypodermic needle. Edward Jenner was preceded by four other scientists who all independently discovered the efficacy of vaccinations. Adrenaline was "first" isolated four times. Three different geniuses discovered (or invented) decimal fractions. The electric telegraph was reinvented by Joseph Henry, Samuel Morse, William Cooke, Charles Wheatstone, and Karl Steinheil. The Frenchman Louis Daguerre is famous for inventing photography, but three others (Nicephore Niepce, Hercules Florence, and William Henry Fox Talbot) also independently came upon the same process. The invention of logarithms is usually credited to two mathematicians, John Napier and Henry Briggs, but actually a third one, Joost Burgi, invented them three years earlier. Several inventors in both England and America simultaneously came up with the typewriter. The existence of the eighth planet, Neptune, was independently predicted by two scientists in the same year, 1846. The liquefaction of oxygen, the electrolysis of aluminum, and the stereochemistry of carbon, for just three examples in chemistry, were each independently discovered by more than one person, and in each case the simultaneous discoveries occurred within a month or so.

Columbia University sociologists William Ogburn and Dorothy Thomas combed through scientists' biographies, correspondence, and notebooks to collect all the parallel discoveries and inventions they could find between 1420 and 1901. They write, "The steamboat is claimed as the 'exclusive' discovery of Fulton, Jouffroy, Rumsey, Stevens and Symmington. At least six different men, Davidson, Jacobi, Lilly, Davenport, Page and Hall, claim to have made independently the application of electricity to the railroad. Given the railroad and electric motors, is not the electric railroad inevitable?"

Inevitable! There is that word again. Common instances of equivalent inventions independently discovered at the same moment suggest that the evolution of technology converges in the same manner as biological evolution. If so, then if we could rewind and replay the tape of history, the very same sequence of inventions should roll out in a very similar sequence every time we reran it. Technologies would be inevitable. The appearance of morphological archetypes would further suggest that this technological invention has a direction, a tilt. A tilt that is independent to a certain extent of its human inventors.

Indeed, in all fields of technology we commonly find independent, equivalent, and simultaneous invention. If this convergence indicated that discoveries were inevitable, the inventors would appear as conduits filled by an invention that just had to happen. We would expect the people making them to be interchangeable, if not almost random.

That is exactly what psychologist Dean Simonton found. He took Ogburn and Thomas's catalog of simultaneous invention before 1900 and aggregated it with several other similar lists to map out the pattern of parallel discovery for 1,546 cases of invention. Simonton plotted the number of discoveries made by 2 individuals against the number of discoveries made by 3 people, or 4 people, or 5, or 6. The number of 6-person discoveries was of course lower, but the exact ratio between these multiples produced a pattern known in statistics as a Poisson distribution. This is the pattern you see in mutations on a DNA chromosome and in other rare chance events in a large pool of possible agents. The Poisson curve suggested that the system of "who found what" was essentially random.

Certainly talent is unequally distributed. Some innovators (like Edison, or Isaac Newton, or William Thomson Kelvin) are simply better than others. But if geniuses aren't able to jump far ahead of the inevitable, how do the better inventors become great? Simonton discovered that the higher the prominence of a scientist (as determined by the number of pages his biography occupies in encyclopedias), the greater the number of simultaneous discoveries he participated in. Kelvin was involved in 30 sets of simultaneous discoveries. Great discoverers not only contribute more than the average number of "next" steps, but they also take part in those steps that have the greatest impact, which are naturally the areas of investigation that attract many other players and so produce multiples. If discovery is a lottery, the greatest discoverers buy lots of tickets.

Simonton's set of historical cases reveals that the number of duplicated innovations has been increasing with time—simultaneous discovery is happening more often. Over the centuries the velocity of ideas has accelerated, speeding up codiscovery as well. The degree of synchronicity is also gaining. The gap between the first and last discovery in a concurrent multiple has been shrinking over the centuries. Long gone is the era when 10 years could elapse between the public announcement of an invention or discovery and the date the last researcher would hear about it.

Synchronicity is not just a phenomenon of the past, when communication was poor, but very much part of the present. Scientists at AT&T Bell Labs won a Nobel Prize for inventing the transistor in 1948, but two German physicists independently invented a transistor two months later at a Westinghouse laboratory in Paris. Popular accounts credit John von Neumann with the invention of a programmable binary computer during the last years of World War II, but the idea and a working punched-tape prototype were developed quite separately in Germany a few years earlier, in 1941, by Konrad Zuse. In a verifiable case of modern parallelism, Zuse's pioneering binary computer went completely unnoticed in the United States and the UK until many decades later. The ink-jet printer was invented twice: once in Japan in the labs of Canon and once in the United States at Hewlett-Packard, and the key patents were filed by the two companies within months of each other in 1977. "The whole history of inventions is one endless chain of parallel instances," writes anthropologist Alfred Kroeber. "There may be those who see in these pulsing events only a meaningless play of capricious fortuitousness; but there will be others to whom they reveal a glimpse of a great and inspiring inevitability which rises as far above the accidents of personality."

The strict wartime secrecy surrounding nuclear reactors during World War II created a model laboratory for retrospectively illuminating technological inevitability. Independent teams of nuclear scientists around the world raced against one another to harness atomic energy. Because of the obvious strategic military advantage of this power, the teams were isolated as enemies or kept ignorant as wary allies or separated by "need to know" secrecy within the same country. In other words, the history of discovery ran in parallel among seven teams. Each discrete team's highly collaborative work was well documented and progressed through multiple stages of technological development. Looking back, researchers can trace parallel paths as the same discoveries were made. In particular, physicist Spencer Weart examined how six of these teams each independently discovered an essential formula for making a nuclear bomb. This equation, called the four-factor formula, allows engineers to calculate the critical mass necessary for a chain reaction. Working in parallel but in isolation, teams in France, Germany, and the Soviet Union and three teams in the United States simultaneously discovered the formula. Japan came close but never quite reached it. This high degree of simultaneity—six simultaneous inventions—strongly suggests the formula was inevitable at this time.

However, when Weart examined each team's final formulation, he saw that the equations varied. Different countries used different mathematical notation to express it, emphasized different factors, varied in their assumptions and interpretation of results, and awarded the overall insight different status. In fact, the equation was chiefly ignored as merely theoretical by four teams. In only two teams was the equation integrated into experimental work—and one of those was the team that succeeded in making a bomb.

The formula in its abstract form was inevitable. Indisputably, if it had not been found by one, five others would have found it. But the specific expression of the formula was not at all inevitable, and that volitional expression can make a significant difference. (The political destiny of the country that put the formula to work, the United States, is vastly different from those that failed to exploit the discovery.)

Both Newton and Gottfried Leibniz are credited with inventing (or discovering) calculus, but in fact their figuring methods differed, and the two approaches were only harmonized over time. Joseph Priestley's method of generating oxygen differed from Carl Scheele's; using different logic they uncovered the same inevitable next stage. The two astronomers who both correctly predicted the existence of Neptune (John Couch Adams and Urbain Le Verrier) actually calculated different orbits for the planet. The two orbits just happened to coincide in 1846, so they found the same body by different means.

But aren't these kinds of anecdotes mere statistical coincidences? Given the millions of inventions in the annals of discovery, shouldn't we expect a few to happen simultaneously? The problem is that most multiples are unreported. Sociologist Robert Merton says, "All singleton discoveries are imminent multiples." By that he means that many potential multiples are abandoned when news of the firstborn is announced. A typical notebook entry goes like this one found in the records of mathematician Jacques Hadamard in 1949: "After having started a certain set of questions and seeing that several authors had begun to follow the same line, I happen to drop it and to investigate something else." Or a scientist will record their discoveries and inventions but never publish the work due to busyness, or their own dissatisfaction with the results. Only the notebooks of the great get a careful examination, so unless you are either Cavendish or Gauss (the notebooks of both reveal several unpublished multiples), your unreported ideas will never be counted. Further concurrent research is hidden by classified, corporate, or state-secret work. Much is not disseminated because of fear of competitors, and until very recently, many examples of duplicate discoveries and inventions remained obscure because they were published in obscure languages. A few coexistent inventions went unrecognized because they were described in impenetrable technical language. And occasionally a discovery is so contrarian or politically incorrect that it is ignored.

Furthermore, once a discovery has been revealed and entered into the repository of what is commonly known, all later investigations that arrive at the same results are reckoned as mere corroborations of the original—no matter how they are actually arrived at. A century ago the failure of communication was in its slow speed; a researcher in Moscow or Japan might not hear about an English invention for decades. Today the failure is due to volume. There is so much published, so fast, in so many areas, that it is very easy to miss what has already been done. Reinventions arise independently all the time, sometimes in full innocence centuries later. But because their independence can't be proven, these Johnny-come-latelies are counted as confirmations and not as evidence of inevitability.

By far the strongest bits of evidence for ubiquitous simultaneity of invention are scientists' own impressions. Most scientists consider getting scooped by another person working on the same ideas the unfortunate and painful norm. In 1974 sociologist Warren Hagstrom surveyed 1,718 U.S. academic research scientists and asked them if their research had ever been anticipated, or scooped, by others. He found that 46 percent believed that their work had been anticipated "once or twice" and 16 percent claimed they had been preempted three or more times. Jerry Gaston, another sociologist, surveyed 203 high-energy physicists in the UK and got similar results: 38 percent claimed to have been anticipated once and another 26 percent more than once.

Unlike scientific scholarship, which places a huge emphasis on previous work and proper credit, inventors tend to plunge ahead without methodically researching the past. This means reinvention is the norm from the patent office's viewpoint. When inventors file patents, they need to cite previous related inventions. One-third of inventors surveyed claimed they were unaware there were prior claims to their idea while developing their own invention. They did not learn about the competing patents until preparing their application with the required "prior art." More surprising, one-third claimed to be unaware of the prior inventions cited in their own patent until notified by the survey takers. (This is entirely possible, since patent citations can be added by the inventor's patent attorney or even the patent office examiner.) Patent law scholar Mark Lemley states that in patent law "a large percent of priority disputes involve near-simultaneous invention." One study of these near-simultaneous priority disputes, by Adam Jaffe of Brandeis University, showed that in 45 percent of cases both parties could prove they had a "working model" of the invention within six months of each other, and in 70 percent of cases within a year of each other. Jaffe writes, "These results provide some support for the idea that simultaneous or near-simultaneous invention is a regular feature of innovation."

There is the air of inevitability about these simultaneous discoveries. When the necessary web of supporting technology is established, then the next adjacent technological step seems to emerge as if on cue. If inventor X does not produce it, inventor Y will. But the step will come in the proper sequence.

This does not mean the iPod, with its perfect, milky case, was inevitable. We can say the invention of the microphone, the laser, the transistor, the steam turbine, and the waterwheel and the discovery of oxygen, DNA, and Boolean logic were all inevitable in roughly the era they appeared. However, the particular form of the microphone, its exact circuit, or the specific engineering of the laser, or the particular materials of the transistor, or the dimensions of the steam turbine, or the peculiar notation of the chemical formula, or the specifics of any invention are not inevitable. Rather, they will vary quite widely due to the personality of their finder, the resources at hand, the culture or society they are born into, the economics funding the discovery, and the influence of luck and chance. A light based on a coil of tungsten strung within an oval vacuum bulb is not inevitable, but the electric incandescent lightbulb is.

The general concept of the electric incandescent lightbulb can be abstracted from all the specific details allowed to vary (voltage, height, kind of bulb) while still producing the result—in this case, luminance from electricity. This general concept is similar to the archetype in biology, while the specific materialization of the concept is more like a species. The archetype is ordained by the technium's trajectory, while the species is contingent.

The electric incandescent lightbulb was invented, reinvented, coinvented, or "first invented" dozens of times. In their book *Edison's Electric Light: Biography of an Invention,* Robert Friedel, Paul Israel, and Bernard Finn list 23 inventors of incandescent bulbs prior to Edison. It might be fairer to say that Edison was the very *last* "first" inventor of the electric light. These 23 bulbs (each an original in its inventor's eyes) varied tremendously in how they fleshed out the abstraction of "electric lightbulb." Different inventors employed various shapes for the filament, different materials for the wires, different strengths of electricity, different plans for the bases. Yet they all seemed to be independently aiming for the same archetypal design. We can think of the prototypes as 23 different attempts to describe the inevitable generic lightbulb.

Quite a few scientists and inventors, and many outside science, are repulsed by the idea that the progress of technology is inevitable. It rubs

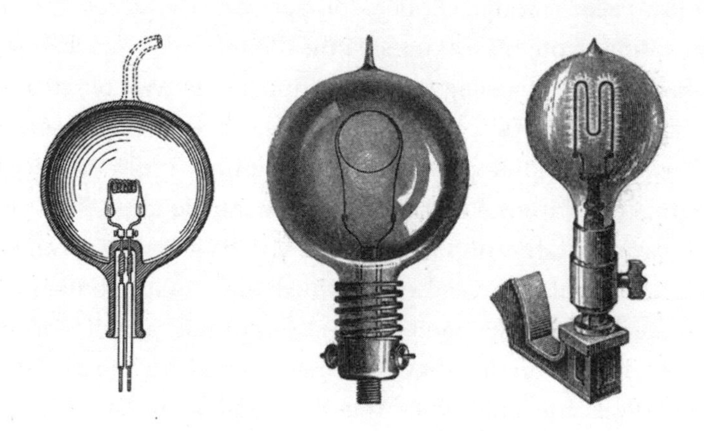

Varieties of the Lightbulb. Three independently invented electric lightbulbs: Edison's, Swan's, and Maxim's.

them the wrong way because it contradicts a deeply and widely held belief that human choice is central to our humanity and essential to a sustainable civilization. Admitting that anything is "inevitable" feels like a cop-out, a surrender to invisible, nonhuman forces beyond our reach. Such a false notion, the thinking goes, may lull us into abdicating our responsibility for shaping our own destiny.

On the other hand, if technologies really are inevitable, then we have only the illusion of choice, and we should smash all technologies to be free of this spell. I'll address these central concerns later, but I want to note one curious fact about this last belief. While many people claim to believe the notion of technological determinism is wrong (in either sense of that word), they don't act that way. No matter what they rationally think about inevitability, in my experience *all* inventors and creators act as if their own invention and discovery is imminently simultaneous. Every creator, inventor, and discoverer that I have known is rushing their ideas into distribution before someone else does, or they are in a mad hurry to patent before their competition does, or they are dashing to finish their masterpiece before something similar shows up. Has there ever been an inventor in the last two hundred years who felt that no one else would ever come up with his idea (and who was right)?

Nathan Myhrvold is a polymath and serial inventor who used to direct fast-paced research at Microsoft but wanted to accelerate the pace of innovation in other areas outside the digital realm—such as surgery, metallurgy, or archaeology—where innovation was often a second thought. Myhrvold came up with an idea factory called Intellectual Ventures. Myhrvold employs an interdisciplinary team of very bright innovators to sit around and dream up patentable ideas. These eclectic one- or two-day gatherings will generate 1,000 patents per year. In April 2009, author Malcolm Gladwell profiled Myhrvold's company in the *New Yorker* to make the point that it does not take a bunch of geniuses to invent the next great thing. Once an idea is "in the air" its many manifestations are inevitable. You just need a sufficient number of smart, prolific people to start catching them. And of course a lot of patent lawyers to patent what you generate in bulk. Gladwell observes, "The genius is not a unique source of insight; he is merely an efficient source of insight."

Gladwell never got around to asking Myhrvold how many of his own lab's inventions turn out to be ideas that others come up with, so I asked Myhrvold, and he replied: "Oh, about 20 percent—that we know about. We only file to patent one third of our ideas."

If parallel invention is the norm, then even Myhrvold's brilliant idea of creating a patent factory should have occurred to others at the same time. And of course it has. Years before the birth of Intellectual Ventures, internet entrepreneur Jay Walker launched Walker Digital Labs. Walker is famous for inventing Priceline, a name-your-own-price reservation system for hotels and airline flights. In his invention laboratory Walker set up an institutional process whereby interdisciplinary teams of brainy experts sit around thinking up ideas that would be useful in the next 20 years or so—the time horizon of patents. They winnow the thousands of ideas they come up with and refine a selection for eventual patenting. How many ideas do they abandon because they, or the patent office, find that the idea has been "anticipated" (the legal term meaning "scooped") by someone else? "It depends on the area," Walker says. "If it is a very crowded space where lots of innovation is happening, like e-commerce, and it is a 'tool,' probably 100 percent have been thought of before. We find the patent office rejects about two-thirds of challenged patents as 'anticipated.' Another space, say gaming inventions, about a third are either blocked by prior art or other inventors. But if the invention is a complex system, in an unusual space, there won't be many others. Look, most invention is a matter of time . . . of when, not if."

Danny Hillis, another polymath and serial inventor, is cofounder of an innovative prototype shop called Applied Minds, which is another idea factory. As you might guess from the name, they use smart people to invent stuff. Their corporate tagline is "the little Big Idea company." Like Myhrvold's Intellectual Ventures, they generate tons of ideas in interdisciplinary areas: bioengineering, toys, computer vision, amusement rides, military control rooms, cancer diagnostics, and mapping tools. Some ideas they sell as unadorned patents; others they complete as physical machines or operational software. I asked Hillis, "What percentage of your ideas do you find out later someone else had before you, or at the same time as you, or maybe even after you?" As a way of answering, Hillis offered a metaphor. He views the bias toward simultaneity as a funnel. He says, "There might be tens of thousands of people who conceive the possibility of the same invention at the same time. But less than one in ten of them imagines *how* it might be done. Of these who see how to do it, only one in ten will actually think through the practical details and specific solutions. Of these only one in ten will actually get the design to work for very long. And finally, usually only one of all those many thousands with the idea will get the invention to stick in the culture. At our lab we engage in all these levels of discovery, in the expected proportions." In other words, in the conceptual stage, simultaneity is ubiquitous and inevitable; your brilliant ideas will have lots of coparents. But there's less coparentage at each reducing stage. When you are trying to bring an idea to market, you may be alone, but by then you are a mere pinnacle of a large pyramid of others who all had the same idea.

INVENTORS	STAGE	TASK	EXAMPLE
10,000–1,000	Think of Possibility	Recognizing an opportunity for solutions	We should use electricity for lighting
1,000	Idea of How	Imagining the crucial elements of the solutions	An incadescent wire in a sealed bulb!
100	Details Specified	Selecting specific solutions	Welded tungsten, vacuum pump, solder exhaust port
10	Working Device	Proving your solutions work reliably	Prototypes by Swan, Latimer, Edison, Davy, etc.
1	Enabling Adoption	Convincing the world to adopt your solutions	Edison's bulb (and electric system)

The Inverted Pyramid of Invention. Time proceeds down, as the numbers involved at each level decrease.

Any reasonable person would look at that pyramid and say the likelihood of someone getting a lightbulb to stick is 100 percent, although the likelihood of Edison's being the inventor is, well, one in 10,000. Hillis also points out another consequence. Each stage of the incarnation can recruit new people. Those toiling in the later stages may not have been among the earliest pioneers of the idea. Given the magnitude of reduction, the numbers suggest that it is improbable that the first person to make an invention stick was also the first to think of the idea.

Another way to read this chart is to recognize that ideas start out abstract and become more specific over time. As universal ideas become more specific they become less inevitable, more conditional, and more responsive to human volition. Only the conceptual essence of an invention or discovery is inevitable. The specifics of how this essential core (the "chairness" of a chair) is manifested in practice (in plywood, or with a rounded back) are likely to vary widely depending on the resources available to the inventors at hand. The more abstract the new idea remains, the more universal and simultaneous it will be (shared by tens of thousands). As it steadily becomes embodied stage by stage into the constraints of a very particular material form, it is shared by fewer people and becomes less and less predictable. The final design of the first marketable lightbulb or transistor chip could not have been anticipated by anyone, even though the concept was inevitable.

What about great geniuses like Einstein? Doesn't he disprove the notion of inevitability? The conventional wisdom is that Einstein's wildly creative ideas about the nature of the universe, first announced to the world in 1905, were so out of the ordinary, so far ahead of his time, and so unique that if he had not been born we might not have his theories of relativity even today, a century later. Einstein was a unique genius, no doubt. But as always, others were working on the same problems. Hendrik Lorentz, a theoretical physicist who studied light waves, introduced a mathematical structure of space-time in July 1905, the same year as Einstein. In 1904 the French mathematician Henri Poincare pointed out that observers in different frames will have clocks that will "mark what one may call the local time" and that "as demanded by the relativity principle the observer cannot know whether he is at rest or in absolute motion." And the 1911 winner of the Nobel Prize in physics, Wilhelm Wien, proposed to the Swedish committee that Lorentz and Einstein be jointly awarded a Nobel Prize in 1912 for their work on special relativity. He told the committee, "While Lorentz must be considered as the first to have found the mathematical content of the relativity principle, Einstein succeeded in reducing it to a simple principle. One should therefore assess the merits of both investigators as being comparable." (Neither won that year.) However, according to Walter Isaacson, who wrote a brilliant biography of Einstein's ideas, *Einstein: His Life and Universe,* "Lorentz and Poincare never were able to make Einstein's leap even *after* they read his paper." But Isaacson, a celebrator of Einstein's special genius for the improbable insights of relativity, admits that "someone else would have come up with it, but not for at least ten years or more." So the greatest iconic genius of the human race is able to leap ahead of the inevitable by maybe 10 years. For the rest of humanity, the inevitable happens on schedule.

The technium's trajectory is more fixed in certain realms than in others. Based on the data, "mathematics has more apparent inevitability than the physical sciences," wrote Simonton, "and technological endeavors appear the most determined of all." The realm of artistic inventions—those engendered by the technologies of song, writing, media, and so on—is the home of idiosyncratic creativity, seemingly the very antithesis of the inevitable, but it also can't fully escape the currents of destiny.

Hollywood movies have an unnerving habit of arriving in pairs: two movies that arrive in theaters simultaneously featuring an apocalyptic hit by asteroids (*Deep Impact* and *Armageddon*), or an ant hero (*A Bug's Life* and *Antz*), or a hardened cop and his reluctant dog counterpart (*K-9* and *Turner & Hooch*). Is this similarity due to simultaneous genius or to greedy theft? One of the few reliable laws in the studio and publishing businesses is that the creator of a successful movie or novel will be immediately sued by someone who claims the winner stole their idea. Sometimes it *was* stolen, but just as often two authors, two singers, or two directors came up with similar works at the same time. Mark Dunn, a library clerk, wrote a play, *Frank's Life,* that was performed in 1992 in a small theater in New York City. *Frank's Life* is about a guy who is unaware that his life is a reality TV program. In his suit against the producers of the 1998 movie *The Truman Show,* Dunn lists 149 similarities between his story and theirs—which is a movie about a guy who is unaware that his life is a reality TV program. However, *The Truman Show*'s producers claim they have a copyrighted, dated script of the movie from 1991, a year before *Frank's Life* was staged. It is not too hard to believe that the idea of a movie about an unwitting reality TV hero was inevitable.

Writing in the *New Yorker*, Tad Friend tackled the issue of synchronistic cinematic expression by suggesting that "the giddiest aspect of copyright suits is how often the studios try to prove that their story was so derivative that they couldn't have stolen it from only one source." The studios essentially say: Every part of this movie is a cliche stolen from plots/stories/themes/jokes that are in the air. Friend continues,

> You might think that mankind's collective imagination could churn up dozens of fictional ways to track a tornado, but there seems to be only one. When Stephen Kessler sued Michael Crichton for "Twister," he was upset because his script about tornado chasers, "Catch the Wind," had placed a data-collection device called Toto II in the whirlwind's path, just like "Twister"'s data-collecting Dorothy. Not such a coincidence, the defense pointed out: years earlier two other writers had written a script called "Twister" involving a device called Toto.

Plots, themes, and puns may be inevitable once they are in the cultural atmosphere, but we yearn to encounter completely unexpected creations. Every now and then we believe a work of art must be truly original, not ordained. Its pattern, premise, and message originate with a distinctive human mind and shine as unique as they are. Say an original mind with an original story like J. K. Rowling, author of the highly imaginative Harry Potter series. After Rowling launched Harry Potter in 1997 to great success, she successfully rebuffed a lawsuit by an American author who published a series of children's books 13 years earlier about Larry Potter, an orphaned boy wizard wearing glasses and surrounded by Muggles. In 1990 Neil Gaiman wrote a comic book about a dark-haired English boy who finds out on his 12th birthday that he is a wizard and is given an owl by a magical visitor. Or keep in mind a 1991 story by Jane Yolen about Henry, a boy who attends a magical school for young wizards and must overthrow an evil wizard. Then there's *The Secret of Platform 13,* published in 1994, which features a gateway on a railway platform to a magical underworld. There are many good reasons to believe J. K. Rowling when she claims she read none of these (for instance, very few of the Muggle books were printed and almost none were sold; and Gaiman's teenage-boy comics don't usually appeal to single moms) and many more reasons to accept the fact that these ideas arose in simultaneous spontaneous creation. Multiple invention happens all the time in the arts as well as technology, but no one bothers to catalog similarities until a lot of money or fame is involved. Because a lot of money swirls around Harry Potter we have discovered that, strange as it sounds, stories of boy wizards in magical schools with pet owls who enter their otherworlds through railway station platforms are inevitable at this point in Western culture.

Just as in technology, the abstract core of an art form will crystallize into culture when the solvent is ready. It may appear more than once. But any particular species of creation will be flooded with irreplaceable texture and personality. If Rowling had not written Harry Potter, someone else would have written a similar story in broad outlines, because so many have already produced parallel parts. But the Harry Potter books, the ones that exist in their exquisite peculiar details, could not have been written by anyone other than Rowling. It is not the particular genius of human individuals like Rowling that is inevitable but the unfolding genius of the technium as a whole.

As in biological evolution, any claim of inevitability is difficult to prove. Convincing proof requires rerunning a progression more than once and showing that the outcome is the same each time. You must show a skeptic that no matter what perturbations are thrown at the system, it yields an identical result. To claim that the large-scale trajectory of the technium is inevitable would mean demonstrating that if we reran history, the same abstracted inventions would arise again and in roughly the same relative order. Without a reliable time machine, there'll be no indisputable proof, but we do have three types of evidence strongly suggesting that the paths of technologies are inevitable:

1. In all times we find that most inventions and discoveries have been made independently by more than one person.
2. In ancient times we find independent timelines of technology on different continents converging upon a set order.
3. In modern times we find sequences of improvement that are difficult to stop, derail, or alter.

In regard to the first point, we have a very clear modern record that simultaneous discovery is the norm in science and technology and not unknown in the arts. The second thread of evidence about ancient times is more difficult to produce because it entails tracking ideas during a period without writing. We must rely on the hints of buried artifacts in the archaeological record. Some of these suggest that independent discoveries converge in parallel to a uniform sequence of invention.

Until rapid communication networks wrapped the globe in stunning instantaneity, progress in civilization unrolled chiefly as independent strands on different continents. Earth's slippery landmasses, floating on tectonic plates, are giant islands. This geography produces a laboratory for testing parallelism. From 50,000 years ago, at the birth of Sapiens, until the year 1000 C.E. when sea travel and land communication ramped up, the sequence of inventions and discoveries on the four major continental landmasses—Europe, Africa, Asia, and the Americas—marched on as independent progressions.

In prehistory the diffusion of innovations might advance a few miles a year, consuming generations to traverse a mountain range and centuries to cross a country. An invention born in China might take a millennium to reach Europe, and it would never reach America. For thousands of years, discoveries in Africa trickled out very slowly to Asia and Europe. The American continents and Australia were cut off from the other continents by impassable oceans until the age of sailing ships. Any technology imported to America came over via a land bridge in a relatively short window between 20,000 and 10,000 B.C.E. and almost none thereafter. Any migration to Australia was also via a geologically temporary land bridge that closed 30,000 years ago, with only marginal flow afterward. Ideas primarily circulated within one landmass. The great cradle of societal discovery two millennia ago—Egypt, Greece, and the Levant—sat right between continents, making the common boundaries for that crossover spot meaningless. Yet despite ever-speedy conduits between adjacent areas, inventions still circulated slowly within one continental mass and rarely crossed oceans.

The enforced isolation back then gives us a way to rewind the tape of technology. According to archaeological evidence the blowgun was invented twice, once in the Americas and once in the islands of Southeast Asia. It was unknown anywhere else outside these two distant regions. This drastic separation makes the birth of the blowgun a prime example of convergent invention with two independent origins. The gun as devised by these two separate cultures is expectedly similar—a hollow tube, often carved in two halves bound together. In essence it is a bamboo or cane pipe, so it couldn't be much simpler. What's remarkable is the nearly identical set of inventions supporting the air pipe. Tribes in both the Americas and Asia use a similar kind of dart padded by a fibrous piston,

Parallels in Blow Gun Culture. Shooting position for a blowgun in the Amazon (left) compared to the position in Borneo (right).

both coat the ends with a poison that is deadly to animals but does not taint the meat, both carry the darts in a quill to prevent the poisoned tip from accidentally pricking the skin, and both employ a similarly peculiar stance when shooting. The longer the pipe, the more accurate the trajectory, but the longer the pipe the more it wavers during aiming. So in both America and Asia the hunters hold the pipe in a nonintuitive stance, with both hands near the mouth, elbows out, and gyrate the shooting end of the pipe in small circles. On each small revolution the tip will briefly cover the target. Accuracy, then, is a matter of the exquisite timing of when to blow. All this invention arose twice, like the same crystals found on two worlds.

In prehistory, parallel paths were played out again and again. From the archaeological record we know technicians in West Africa developed steel centuries before the Chinese did. In fact, bronze and steel were discovered independently on four continents. Native Americans and Asians independently domesticated ruminants such as llamas and cattle. Archaeologist John Rowe compiled a list of 60 cultural innovations common to two civilizations separated by 12,000 kilometers: the ancient Mediterranean and the high Andean cultures. Included on his list of parallel inventions are slingshots, boats made of bundled reeds, circular bronze mirrors with handles, pointed plumb bobs, and pebble-counting boards, or what we call abacus. Between societies, recurring inventions are the norm. Anthropologists Laurie Godfrey and John Cole conclude that "cultural evolution followed similar trajectories in various parts of the world."

But perhaps there was far more communication between civilizations in the ancient world than we sophisticated moderns think. Trade in prehistoric times was very robust, but trade between continents was still rare. Nonetheless, with little evidence, a few minority theories (called the Shang-Olmec hypothesis) claim Mesoamerican civilizations maintained substantial transoceanic trade with China. Other speculations suggest extended cultural exchange between the Maya and west Africa, or between the Aztecs and Egypt (those pyramids in the jungle!), or even between the Maya and the Vikings. Most historians discount these possibilities and similar theories about deep, ongoing relations between

Australia and South America or Africa and China before 1400. Beyond some superficial similarities in a few art forms, there is no empirical archaeological or recorded evidence of sustained transoceanic contact in the ancient world. Even if a few isolated ships from China or Africa might have reached, say, the shores of the pre-Columbian new world, these occasional landings would not have been sufficient to kindle the many parallels we find. It is highly improbable that the sewed-and-pitched bark canoe of the northern Australian aborigines came from the same source as the sewed-and-pitched bark canoe of the American Algonquin. It is much more likely that they are examples of convergent invention and arose independently on parallel tracks.

When viewed along continental tracks, a familiar sequence of inventions plays out. Each technological progression around the world follows a remarkably similar approximate order. Stone flakes yield to control of fire, then to cleavers and ball weapons. Next come ocher pigments, human burials, fishing gear, light projectiles, holes in stones, sewing, and figurine sculptures. The sequence is fairly uniform. Knifepoints always follow fire, human burials always follow knifepoints, and the arch precedes welding. A lot of the ordering is "natural" mechanics. You obviously need to be able to master blades before you make an ax. And textiles always follow sewing, since threads are needed for any kind of fabric. But many other sequences don't have a simple causal logic. There is no obvious reason that we are currently aware of why the first rock art always precedes the first sewing technology, yet it does each time. Metalwork does not have to follow claywork (pottery), but it always does.

Geographer Neil Roberts examined the parallel paths of domestication of crops and animals on four continents. Because the potential biological raw material on each continent varies so greatly (a theme explored in full by Jared Diamond in *Guns, Germs, and Steel*), only a few native species of crops or animals are first tamed on more than one landmass. Contrary to earlier assumptions, agriculture and animal husbandry were not invented once and then diffused around the world. Rather, as Roberts states, "Bio-archeological evidence taken overall indicates that global diffusion of domesticates was rare prior to the last 500 years.

Farming systems based on the three great grain crops—wheat, rice, and maze—have independent centers of origin." The current consensus is that agriculture was (re)invented six times. And this "invention" is a series of inventions, a string of domestications and tools. The order of these inventions and tamings is similar across regions. For instance, on more than one continent humans domesticated dogs before camels and grains before root crops.

Archaeologist John Troeng cataloged 53 prehistoric innovations beyond agriculture that independently originated not just twice but three times in three distinct separate regions of the globe: Africa, western Eurasia, and east Asia/Australia. Twenty-two of the inventions were also discovered by inhabitants of the Americas, meaning these innovations spontaneously erupted on four continents. The four regions are sufficiently separated that Troeng reasonably accepts that any invention in them is an independent parallel discovery. As technology invariably does, one invention prepares the ground for the next, and every corner of the technium evolves in a seemingly predetermined sequence.

With the help of a statistician, I analyzed the degree to which the four sequences of these 53 inventions paralleled one another. I found they correlated to an identical sequence by a coefficiency of 0.93 for the three regions and 0.85 for all four regions. In layman's terms, a coefficiency above 0.50 is better than random, while a coefficiency of 1.00 is a perfect match; a coefficiency of 0.93 indicates that the sequences of discoveries were nearly the same, and 0.85 slightly less so. That degree of overlap in the sequence is significant given the incomplete records and the loose dating inherent in prehistory. In essence, the direction of technological development is the same anytime it happens.

To confirm this direction, research librarian Michele McGinnis and I also compiled a list of the dates when preindustrial inventions, such as the loom, sundial, vault, and magnet, first appeared on each of the five major continents: Africa, the Americas, Europe, Asia, and Australia. Some of these discoveries occurred during eras when communication and travel were more frequent than in prehistoric times, so the independence of each invention is less certain. We found historical evidence for 83 innovations that were invented on more than one continent. And again, when matched up, the sequence of technology's unfolding in Asia is similar to that in the Americas and Europe to a significant degree.

We can conclude that in historic times as well as in prehistory, technologies with globally distinct origins converge along the same developmental path. Independent of the different cultures that host it, or the diverse political systems that rule it, or the different reserves of natural resources that feed it, the technium develops along a universal path. The large-scale outlines of technology's course are predetermined.

Anthropologist Kroeber warns, "Inventions are culturally determined. Such a statement must not be given a mystical connotation. It does not mean, for instance, that it was predetermined from the beginning of time that type printing would be discovered in Germany about 1450, or the telephone in the United States in 1876." It means only that when all the required conditions generated by previous technologies are in place, the next technology can arise. "Discoveries become virtually inevitable when prerequisite kinds of knowledge and tools accumulate," says sociologist Robert Merton, who studied simultaneous inventions in history. The ever-thickening mix of existing technologies in a society creates a supersaturated matrix charged with restless potential. When the right idea is seeded within, the inevitable invention practically explodes into existence, like an ice crystal freezing out of water. Yet as science has shown, even though water is destined to become ice crystals when it is cold enough, no two snowflakes are the same. The path of freezing water is predetermined, but there is great leeway, freedom, and beauty in the individual expression of its predestined state. The actual pattern of each snowflake is unpredictable, although its archetypal six-sided form is determined. For such a simple molecule, its variations upon an expected theme are endless. That's even truer for extremely complex inventions today. The crystalline form of the incandescent lightbulb or the telephone or the steam engine is ordained, while its unpredictable expression will vary in a million possible formations, depending on the conditions in which it evolved.

It is not much different from the natural world. The birth of any species depends on an ecosystem of other species in place to support, divert, and goad its metamorphosis. We call it coevolution because of the reciprocal influence of one species upon another. In the technium many discoveries await the invention of another technological species: the proper tool or platform. The moons of Jupiter were discovered by a number of folks only a year after the telescope was invented. But the instruments by themselves didn't make the discovery. Celestial bodies were expected by astronomers. Because no one expected germs, it took 200 years after the microscope was invented before Antonie van Leeuwenhoek spied microbes. In addition to instruments and tools, a discovery needs the proper beliefs, expectations, vocabulary, explanation, know-how, resources, funds, and appreciation to appear. But these, too, are fueled by new technologies.

An invention or discovery that is too far ahead of its time is worthless; no one can follow. Ideally, an innovation opens up only the next adjacent step from what is known and invites the culture to move forward one hop. An overly futuristic, unconventional, or visionary invention can fail initially (it may lack essential not-yet-invented materials or a critical market or proper understanding) yet succeed later, when the ecology of supporting ideas catches up. Gregor Mendel's 1865 theories of genetic heredity were correct but ignored for 35 years. His keen insights were not embraced because they did not explain the problems biologists had at the time, nor did his explanation operate by known mechanisms, so his discoveries were out of reach even for the early adopters. Decades later science faced the urgent questions that Mendel's discoveries could answer. Now his insights were only one step away. Within a few years of one another, three different scientists (Hugo de Vries, Karl Erich Correns, and Erich Tschermak) each independently rediscovered Mendel's forgotten work, which of course had been there all along. Kroeber claims that if you had prevented those three from rediscovery and waited another year, six scientists, not just three, would had made the then-obvious next step.

The technium's inherent sequence makes leapfrogging ahead very difficult. It would be wonderful if a society that lacks all technology infrastructure could jump to 100 percent clean, lightweight digital technology and simply skip over the heavy, dirty industrial stage. The fact that billions of poor in the developing world have purchased cheap cell phones and bypassed long waits for industrial-age landline telephones has given hope that other technologies could also leapfrog into the future. But my close examination of cell-phone adoption in China, India, Brazil, and Africa shows that the boom in cell phones around the world is accompanied by a parallel boom in copper-wire landlines. Cell phones don't cancel landlines. Instead, where cell phones go, copper follows. Cell phones train newly educated customers to need higher-bandwidth internet connections and higher-quality voice connections, which then follow in copper wires. Cell phones and solar panels and other potential leapfrog technologies are not skipping over the industrial age as much as sprinting ahead to accelerate industry's overdue arrival.

To a degree that is invisible to us, new tech sits on a foundation of old tech. Despite the vital layer of electrons that constitutes our modern economy, a huge portion of what goes on each day is fairly industrial in scope: moving atoms, rearranging atoms, mining atoms, burning atoms, refining atoms, stacking atoms. Cell phones, web pages, solar panels all rest upon heavy industry, and industry rests upon agriculture.

It is no different with our brains. Most of our brain's activity is spent on primitive processes—like walking—that we can't even perceive consciously. Instead, we are aware of only a thin, newly evolved layer of cognition that sits on and depends upon the reliable workings of older processes. You can't do calculus unless you do counting. Likewise, you can't do cell phones unless you do wires. You can't do digital infrastructure unless you do industrial. For example, a recent high-profile effort to computerize every hospital in Ethiopia was abandoned because the hospitals did not have reliable electricity. According to a study by the World Bank, a fancy technology introduced in developing countries typically reaches only 5 percent penetration before it stalls. It doesn't disseminate further until older foundational technologies catch up. Wisely, low-income countries are still rapidly inhaling industrial technologies. Big-budget infrastructure—roads, waterworks, airports, machine factories, electrical systems, power plants—are needed to make the high-tech stuff work. In a report on technological leapfrogging the *Economist* concluded: "Countries that failed to adopt old technologies are at a disadvantage when it comes to new ones."

Does this mean that if we were to try to colonize an uninhabited Earth-like planet we would be required to recapitulate history and start with sharp sticks, smoke signals, and mud-brick buildings and then work our way through each era? Would we not try to create a society from scratch using the most sophisticated technology we had?

I think we would try but that it would not work. If we were civilizing Mars, a bulldozer would be as valuable as a radio. Just like the predominance of lower functions in our brains, industrial processes predominate in the technium, even though they are gilded with informational veneers. The demassification of high technology is at times an illusion. Although the technium really does advance by using fewer atoms to do more work, information technology is not an abstract virtual world. Atoms still count. As the technium progresses, it embeds information in materials, in the same way that information and order is embedded in the atoms of a DNA molecule. Advanced high technology is the seamless fusion of bits *and* atoms. It is adding intelligence to industry, rather than removing industry and leaving only information.

Technologies are like organisms that require a sequence of developments to reach a particular stage. Inventions follow this uniform developmental sequence in every civilization and society, independent of human genius. You can't effectively jump ahead when you want to. But when the web of supporting technological species are in place, an invention will erupt with such urgency that it will occur to many people at once. The progression of inventions is in many ways the march toward forms dictated by physics and chemistry in a sequence determined by the rules of complexity. We might call this technology's imperative.

8

Listen to the Technology

In the early 1950s, the same thought occurred to many people at once: Things are improving so fast and so regularly that there might be a pattern to the improvements. Maybe we could plot technological progress to date, then extrapolate the curves and see what the future holds. Among the first to do this systemically was the U.S. Air Force. They needed a long-term schedule of what kinds of planes they should be funding, but aerospace was one of the fastest-moving frontiers in technology. Obviously, they would build the fastest planes possible, but since it took decades to design, approve, and then deliver a new type of plane, the generals thought it prudent to glimpse what futuristic technologies they should be funding.

So in 1953 the Air Force Office of Scientific Research plotted out the history of the fastest air vehicles. The Wright Brothers' first flight reached 6.8 kilometers per hour in 1903, and they jumped to 60 kilometers per hour two years later. The airspeed record kept increasing a bit each year, and in 1947 the fastest flight passed 1,000 kilometers per hour in a Lockheed Shoot Star flown by Colonel Albert Boyd. The record was broken four times in 1953, ending with the F-100 Super Sabre doing 1,215 kilometers per hour. Things were moving fast. And everything was pointed toward space. According to Damien Broderick, the author of *The Spike,* the Air Force

> charted the curves and metacurves of speed. It told them something preposterous. They could not believe their eyes.

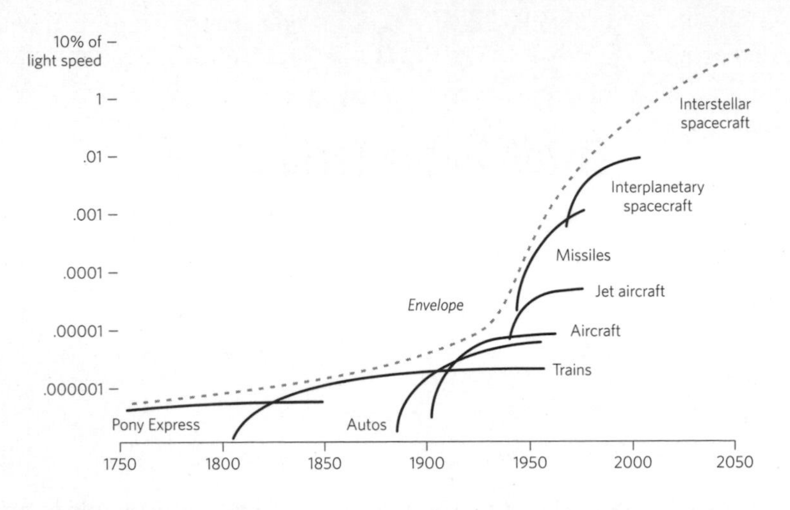

Speed Trend Curve. The U.S. Air Force's plot of historical speed records up to the 1950s and their expectations of the fastest speeds in the near future.

> The curve said they could have machines that attained orbital speed . . . within four years. And they could get their payload right out of Earth's immediate gravity well just a little later. They could have satellites almost at once, the curve insinuated, and if they wished—if they wanted to spend the money, and do the research and the engineering—they could go to the Moon quite soon after that.

It is important to remember that in 1953 none of the technology for these futuristic journeys existed. No one knew how to go that fast and survive. Even the most optimistic, die-hard visionaries did not expect a lunar landing any sooner than the proverbial "year 2000." The only voice telling them they could do it sooner was a curve on a piece of paper. But the curve turned out to be correct. Except not politically correct. In 1957 the Soviet Union (not America!) launched Sputnik, right on schedule. Then U.S. rockets zipped to the moon 12 years later. As Broderick notes, humans arrived on the moon "close to a third of a century sooner than loony space travel buffs like Arthur C. Clarke had expected it to occur."

What did the curve know that Arthur C. Clarke did not? How did it account for the secretive efforts of the Russians as well as dozens of teams around the world? Was the curve a self-fulfilling prophecy or a revelation of an inevitable trend rooted deep in the nature of the technium? The answer may lie in the many other trends plotted since then. The most famous of them all is the trend known as Moore's Law. In brief, Moore's Law predicts that computing chips will shrink by half in size and cost every 18 to 24 months. For the past 50 years it has been astoundingly correct.

It has been steady and true, but does Moore's Law reveal an imperative in the technium? In other words is Moore's Law in some way inevitable? The answer is pivotal for civilization for several reasons. First, Moore's Law represents the acceleration in computer technology, which is accelerating everything else. Faster jet engines don't lead to higher corn yields, nor do better lasers lead to faster drug discoveries, but faster computer chips lead to all of these. These days all technology follows computer technology. Second, finding inevitability in one key area of technology suggests invariance and directionality may be found in the rest of the technium.

This seminal trend of steadily increasing computing power was first noticed in 1960 by Doug Engelbart, a researcher at Stanford Research Institute (now SRI International) in Palo Alto, California, who would later go on to invent the "windows and mouse" computer interface that is now ubiquitous. When he first started as an engineer, Engelbart worked in the aerospace industry testing airplane models in wind tunnels, where he learned how systematic scaling down led to all kinds of benefits and unexpected consequences. The smaller the model, the better it flew. Engelbart imagined how the benefits of scaling down, or as he called it, "similitude," might transfer to a new invention SRI was tracking—multiple transistors on one integrated silicon chip. Perhaps as they were made smaller, circuits might deliver a similar kind of magical similitude: the smaller a chip, the better. Engelbart presented his ideas on similitude to an audience of engineers at the 1960 Solid State Circuits Conference that included Gordon Moore, a researcher at Fairchild Semiconductor, a start-up making the integrated chips.

In the following years Moore began tracking the actual statistics of the earliest prototype chips. By 1964 he had enough data points to extrapolate the slope of the curve so far. Moore kept adding data points as the semiconductor industry grew. He was tracking all kinds of parameters—number of transistors made, cost per transistor, number of pins, logic speed, and components per wafer. But one of them was cohering into a nice curve. The trends were saying something no one else was: that the chips would keep getting smaller at a predictable rate. But how far would the trend really go?

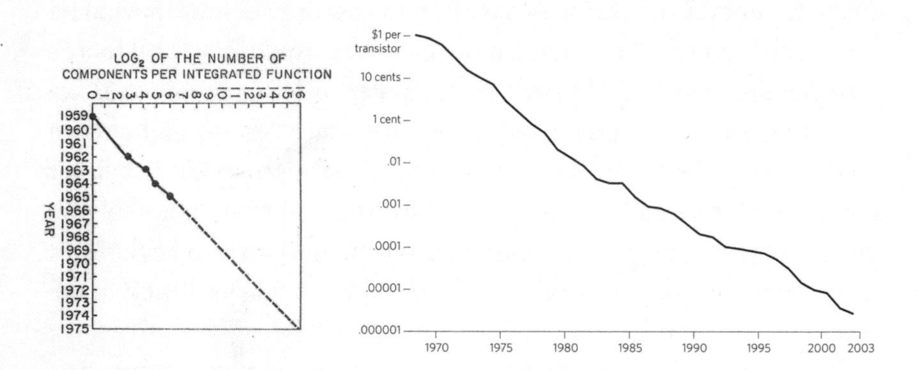

Plotting Moore's Law. The original chart of Moore's Law contained only five data points and a bold extrapolation for the next 10 years (left). The continuation of Moore's Law since 1968 (right).

Moore hooked up with Carver Mead, a fellow Caltech alumnus. Mead was an electrical engineer and early transistor expert. In 1967 Moore asked Mead what kind of theoretical limits were in store for microelectronic miniaturization. Mead had no idea, but as he did his calculations he made an amazing discovery: The efficiency of the chip would increase by the cube of the scale's reduction. The benefits from shrinking were exponential. Microelectronics would not only become cheaper, but they would also become better. As Moore puts it, "By making things smaller, everything gets better simultaneously. There is little need for tradeoffs. The speed of our products goes up, the power consumption goes down, system reliability improves by leaps and bounds, but especially the cost of doing things drops as a result of the technology."

Today when we stare at the plot of Moore's Law we can spot several striking characteristics of its 50-year run. First, this is a picture of *acceleration*. The straight line marks not just an increase, but a 10-time increase for each point on the line (because the horizontal axis is an exponential scale). Silicon computation is not simply getting better, but getting better faster and faster. Relentless acceleration for five decades is rare in biology and unknown in the technium before this century. So this graph is as much about the phenomenon of cultural acceleration as about silicon chips. In fact, Moore's Law has come to represent the principle of an accelerating future that underpins our expectations of the technium.

Second, even a cursory glance reveals the astounding regularity of Moore's line. From the earliest points its progress has been eerily mechanical. Without interruption for 50 years, chips improve exponentially at the same speed of acceleration, neither more nor less. It could not be more straight if it had been engineered by a technological tyrant. Is it really possible that this strict, unwavering trajectory came about via the chaos of the global marketplace and uncoordinated, ruthless scientific competition? Is Moore's Law a direction pushed forward by the nature of matter and computation, or is this steady growth an artifact of economic ambition?

Moore and Mead themselves believe the latter. Writing in 2005, on the 40th anniversary of his law, Moore says, "Moore's Law is really about economics." Carver Mead made it clearer yet: Moore's Law, he says, "is really about people's belief system, it's not a law of physics, it's about human belief, and when people believe in something, they'll put energy behind it to make it come to pass." In case that was not clear enough, he spells it out further:

> After [it] happened long enough, people begin to talk about it in retrospect, and in retrospect it's really a curve that goes through some points and so it looks like a physical law and people talk about it that way. But actually if you're living it, which I am, then it doesn't feel like a physical law. It's really a thing about human activity, it's about vision, it's about what you're allowed to believe.

Finally, in another reference, Carver Mead adds: "Permission to believe that [the law] will keep going" is what keeps the law going. Gordon Moore agreed in a 1996 article: "More than anything, once something like this gets established, it becomes more or less a self-fulfilling prophecy. The Semiconductor Industry Association puts out a technology road map, which continues this [generational improvement] every three years. Everyone in the industry recognizes that if you don't stay on essentially that curve they will fall behind. So it sort of drives itself."

Clearly, expectations of future progress guide current investments, not just in semiconductors but in all aspects of technology. The invariant curve of Moore's Law helps focus money and intelligence on very specific goals—keeping up with the law. The only problem with accepting self-constructed goals as the source of such regular progress is that other technologies that might benefit from the same belief do not show the same zooming rise. Why don't we see Moore's Law type of growth in the performance of jet engines or steel alloys or corn hybrids if this is simply a matter of believing in a self-fulfilling prophecy? Surely such a fantastic faith-based acceleration would be ideal for consumers and generate billions of dollars for investors. It would be easy to find entrepreneurs eager to believe in such prophecies.

So what is the curve of Moore's Law telling us that expert insiders don't see? That this steady acceleration is more than an agreement. It originates within the technology. There are other technologies, also solid-state materials, that exhibit a steady curve of progress, just as in Moore's Law. They, too, seem to obey a rough law of remarkably steady exponential improvement. Consider the cost performance of communication bandwidth and digital storage in the past two decades. The picture of their exponential growth parallels the integrated circuit's. Except for the slope, these graphs are so similar, in fact, that it is fair to ask whether these curves are just reflections of Moore's Law. Telephones are heavily computerized, and storage disks are organs of computers. Since progress in speed and cheapness of bandwidth and storage capacity rely directly and indirectly on accelerating computing power, it may be impossible to untangle the destiny of bandwidth and storage from computer chips. Perhaps the curves of bandwidth and storage are simply derivatives of the one uberlaw? Without Moore's Law ticking beneath them, would they even remain solvent?

In the inner circle of the tech industry the fast-paced drop in prices for magnetic storage is called Kryder's Law. It's the Moore's Law for computer storage and is named after Mark Kryder, the former chief technical officer of Seagate, a major manufacturer of hard disks. Kryder's Law says that the cost per performance of hard disks is decreasing exponentially at a steady rate of 40 percent per year. Kryder says that if computers stopped getting better and cheaper every year, storage would still continue to improve. In Kryder's own words: "There is no direct relationship between Moore's Law and Kryder's Law. The physics and fabrication processes are different for the semiconductor devices and magnetic storage. Hence, it is quite possible that semiconductor scaling could stop while scaling of disk drives continues."

Larry Roberts, the principal architect of the ARPANET, the earliest version of the internet, keeps detailed stats on communication improvements. He has noticed that communication technology in general also exhibits a Moore's Law–like rise in quality. Roberts's curve shows a steady, exponential fall in communication costs. Might progress in wires also be correlated to progress in chips? Roberts says that the performance of communication technology "is strongly influenced by and very similar to Moore's Law but not identical as might be expected."

Consider another encapsulation of accelerating progress. For a decade or so biophysicist Rob Carlson has been tabulating progress in DNA sequencing and synthesis. Graphed similarly to Moore's Law in cost performance per base pair, this technology, too, displays a steady drop when plotted on a log axis. If computers did not get better, faster, cheaper each year, would DNA sequencing and synthesis continue to accelerate? Carlson says: "If Moore's Law stopped, I don't think it would have much effect. The one area it might affect is processing the raw sequence information into something comprehensible by humans. Crunching the data of DNA is at least as expensive as getting the sequence of the physical DNA."

The same kind of steady exponential progress that drives computer chips also drives three information industries, and the keenest observ-

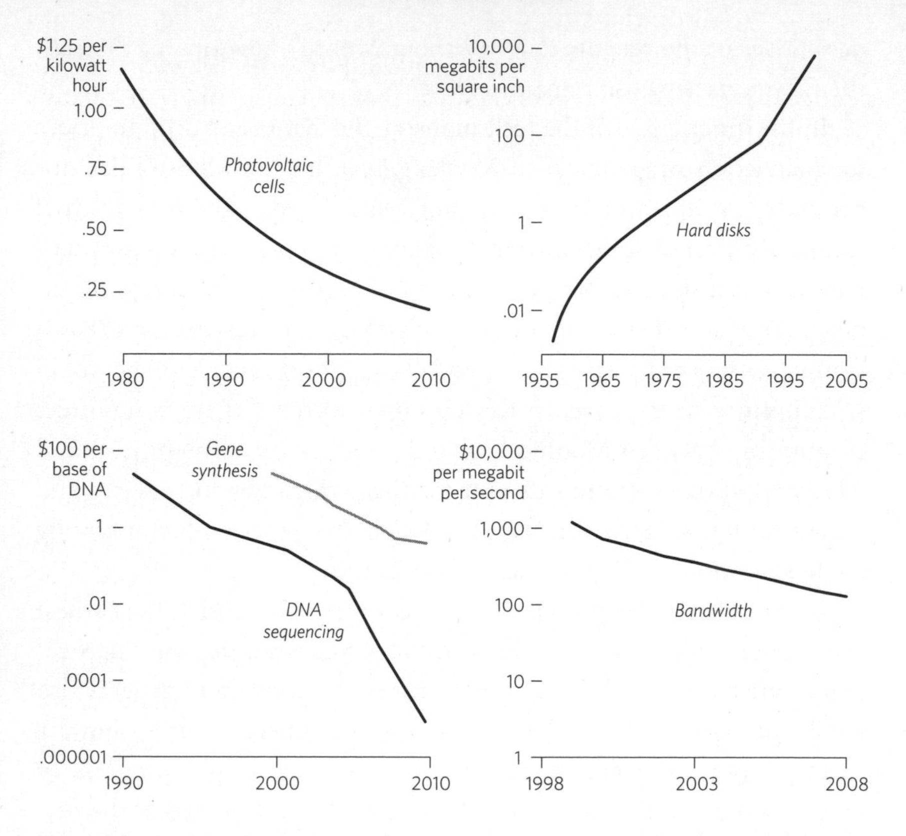

Four Other Laws. Photovoltaic cells: the cost of solar electricity drops (dollars per kilowatt) and is expected to continue in a linear fashion. Hard disks: the maximum density of storage available per year. DNA sequencing: The cost per base pair of DNA sequenced (dark line) or synthesized (light line) drops exponentially. Bandwidth: The cost per megabit per second drops exponentially.

ers of these trajectories—the very founders of their respective "laws"—all believe that these trajectories of improvement are independent lines of acceleration and are not derivative of the overarching progress of computer chips.

Consistent, lawlike improvement must be more than self-fulfilling prophecy for another reason: This obedience to a curve often begins long before anyone notices there is a law, and way before anyone would be able to influence it. The exponential growth of magnetic storage began in 1956, almost a whole decade before Moore formulated his law for semiconductors and 50 years before Kryder formulized the existence of its slope. Rob Carlson says, "When I first published the DNA exponential curves, I got reviewers claiming that they were unaware of any evidence that sequencing costs were falling exponentially. In this way the trends were operative even when people disbelieved it."

Inventor and author Ray Kurzweil dug into the archives to show that something like Moore's Law had its origins as far back as 1900, long before electronic computers existed, and of course long before the path could have been constructed by self-fulfillment. Kurzweil estimated the number of calculations per second per $1,000 performed by turn-of-the-century analog machines, by mechanical calculators, and later by the first vacuum-tube computers and extended the same calculation to modern semiconductor chips. He established that this ratio increased exponentially for the past 109 years. More important, the curve (let's call it Kurzweil's Law) transects five different technological species of computation: electromechanical, relay, vacuum tube, transistors, and integrated circuits. An unobserved constant operating in five distinct paradigms of technology for over a century must be more than an industry road map. It suggests that the nature of these ratios is baked deep into the fabric of the technium.

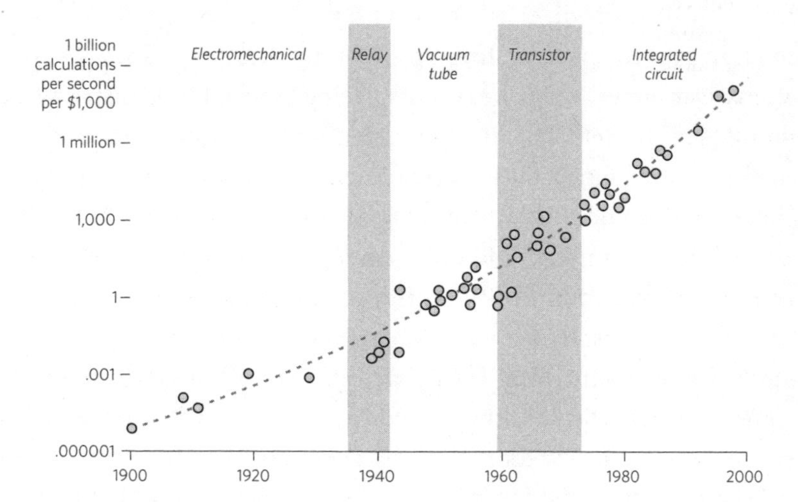

Kurzweil's Law. Ray Kurzweil translated earlier calculating methods into a uniform metric of computation to yield a steady foreshadowing of Moore's Law.

Technology's imperative can be seen in the rigid acceleration of progress in DNA sequencing, magnetic storage, semiconductors, bandwidth, and pixel density. Once a fixed curve is revealed, scientists, investors, marketers, and journalists all grab hold of this trajectory and use it to guide experiments, investments, schedules, and publicity. The map becomes the territory. At the same time, since these curves begin and advance independent of our awareness and do not waver very much from a straight line under enormous competition and investment pressures, their course must in some way be bound to the materials.

To see how far this type of imperative extended into the technium I gathered as many examples of current exponential progress as I could find. I was not seeking examples where the total quantity produced (watts, kilometers, bits, base pairs, traffic, etc.) was rising exponentially, because these quantities are skewed by our rising population. More people use more stuff, even if it is not improving. Rather, I looked for examples that showed performance ratios (such as pounds per inch and illumination per dollar) steadily increasing, if not accelerating. On the opposite page is a set of quickly found examples, and the rate at which their performance is doubling. The shorter the time period, the faster the acceleration.

The first thing to notice is that all these examples demonstrate the effects of scaling down, or working with the small. We don't find exponential improvement in scaling up, as in making skyscrapers or space stations ever larger. Airplanes aren't getting bigger, flying faster, or becoming more fuel efficient at an exponential rate. Gordon Moore jokes that if the technology of air travel experienced the same kind of progress as Intel chips, a modern-day commercial aircraft would cost $500, circle the Earth in 20 minutes, and only use five gallons of fuel for the trip. However, the plane would only be the size of a shoebox!

In this microcosmic realm, unlike the macroworld we live in, energy is not very important. That is why we don't see a Moore's Law type of progress at work when scaling up: energy requirements scale up just as fast, and energy is a major limiting constraint, unlike information, which can be duplicated freely. This is also why we don't see exponential progress in the performance of solar panels (only linear progress) or batteries—because they generate or store lots of energy. So our entire

TECHNOLOGY	METRIC	MONTHS
Fiber-optic throughput	Wavelengths per fiber	9
Optical network	Dollars per bit	9
Wireless	Bits per second	10
Communication	Bits per dollar	12
Magnetic areal storage	Gigabits per square inch	12
Digital cameras	Pixels per dollar	12
Microprocessor	Dollars per cycle	13
Supercomputer power	FLOPS	14
RAM	Mebibytes per dollar	16
Transistor	Dollars per transistor	18
PCU power consumption	Watts per square centimeter	18
Pixels	Per array	19
Hard-drive storage	Gigabytes per dollar	20
Chip	MIPS	21
DNA sequencing	Dollars per base pair	22
Trunk-line data speed	Bits per second	22
Microprocessor	Transistors per chip	24
Chip processor	Megahertz per dollar	27
Bandwidth	Kilobits per second per dollar	30
Microprocessor	Hertz	36

Doubling Times. Performance ratios of various technologies measured as the number of months required to double their performance.

new economy is built around technologies that need little energy and scale down well—photons, electrons, bits, pixels, frequencies, and genes. As these inventions miniaturize, they reach closer to bare atoms, raw bits, and the essence of the immaterial. And so the fixed and inevitable path of their progress derives from this elemental essence.

The second thing to notice about this set of examples is the narrow range of slopes, or doubling time (in months). The particular power being optimized in these technologies is doubling in between 8 and 30 months. (Moore's Law calls for doubling every 18 months.) Every one of these parameters is getting twice as better every year or two. What's up with that? Engineer Mark Kryder's explanation is that "twice as better every two years" is an artifact of corporate structure, where most of these inventions happen. It just takes one to two years of calendar time to conceive, design, prototype, test, manufacture, and market a new improvement, and while a five- or 10-fold increase is very difficult to achieve, almost any engineer can deliver a factor of two. Voila! Twice as better every two years. If true, this suggests that while the steady trajectory of progress stems directly from the technium, the actual angle of incline is not a supernatural number (doubling every 18 months) but is simply dependent on human work cycles.

At the moment there is no end in sight for any of these curves, but at some point in the future, each curve will plateau. Moore's Law will not continue forever. That's just life. Any specific exponential growth will inevitably smooth out into a typical *S*-shaped curve. This is the archetypal pattern of growth: After a slow ramp-up, gains take off straight up like a rocket, and then after a long run level out slowly. Back in 1830 only 37 kilometers of railroad track had been laid in the United States. That count doubled in the next ten years, and then doubled in the decade after that, and kept doubling every decade for 60 years. In 1890 any reasonable railroad buff would have predicted that the United States would have hundreds of millions of kilometers of railroad by a hundred years later. There would be railroad to everyone's house. Instead, there were fewer than 400,000 kilometers. However, Americans did not cease to be mobile. We merely shifted our mobility and transportation to other species of invention. We built automobile highways and airports. The miles we travel keep expanding, but the exponential growth of that particular technology peaked and plateaued.

Much of the churn in the technium is due to our tendency to shift what we care about. Mastering one technology engenders new technological desires. A recent example: The first digital cameras had very rough picture resolution. Then scientists began cramming more and more pixels onto one sensor to increase photo quality. Before they knew it, the number of pixels possible per array was on an exponential curve, heading into megapixel territory and beyond. The rising megapixel count became the chief selling point for new cameras. But after a decade of acceleration, consumers shrugged off the increasing number of pixels because the current resolution was sufficient. Their concern instead shifted to the speed of the pixel sensors or the response in low light—things no one had cared about before. So a new metric is born, and a new curve started, and the exponential curve of ever more pixels per array will gradually abate.

Moore's Law is headed to a similar fate. When, no one knows. Decades ago Gordon Moore himself predicted his law would end when it reached 250-nanometer manufacturing, which it passed in 1997. Today the industry is aiming for 20 nanometers. Whether Moore's Law—as the count of transistor density—has one, two, or three decades left to zoom and drive our economy, we can be sure it will peter out as other past trends have by being sublimated into another rising trend. As the old Moore's Law abates, we'll find alternative solutions to making a million times more transistors. In fact, we may already have enough transistors per chip to do what we want, if only we knew how.

Moore began by measuring the number of "components" per square inch, then switched to transistors; now we measure transistors per dollar. Just as happened in pixel counts, once one exponential trend in computer chips (say, the density of transistors) decelerates, we begin caring about a new parameter (say, speed of operations or number of connections), and so we begin measuring a new metric and plotting a new graph. Suddenly, another "law" is revealed. As the character of this new technique is studied, exploited, and optimized, its natural pace is revealed, and when this trajectory is extrapolated, it becomes the

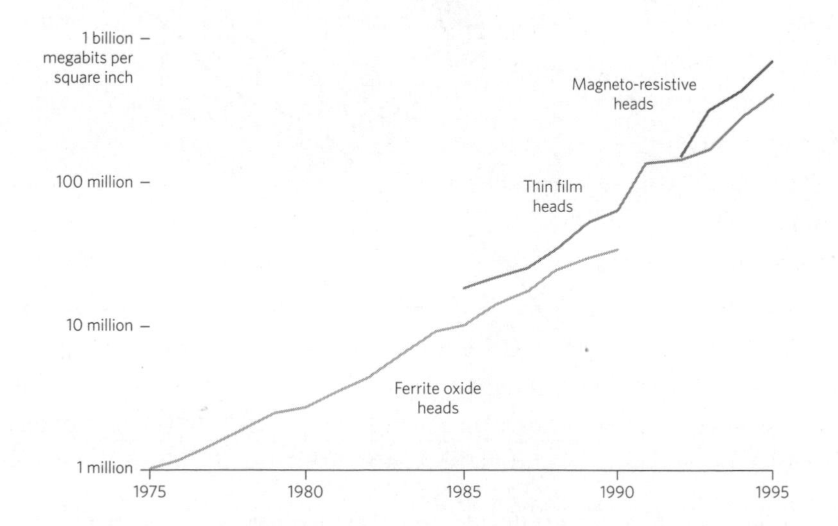

The Continuum of Kryder's Law. Improvements in the recording density of magnetic technologies continue uninterrupted across different technological platforms.

creators' goal. In the case of computing, this newly realized attribute of microprocessors will become, over time, the new Moore's Law.

Like the Air Force's 1953 graph of top speed, the curve is one way the technium speaks to us. Carver Mead, who barnstormed the country waving plots of Moore's Law, believes we need to "listen to the technology." The curves speak in concert. As one curve inevitably flattens out, its momentum is taken up by another *S* curve. If we inspect any enduring curve closely, we can see how definitions and metrics shift over time to accommodate new substitute technologies.

For instance, close scrutiny of Kryder's Law in hard-disk densities shows that it is composed of a sequence of overlapping smaller trend lines. The first hard-disk technology, ferrite oxide, ran from 1975 to 1990. The second technology, thin film, had a slightly better performance and slightly faster acceleration and overlapped ferrite oxide, running from 1985 to 1995. The third technological innovation, magneto resistance, began in 1993 and improved at a still faster rate. Their slightly uneven slopes combine to yield an unwavering trajectory.

The graph below dissects what is happening for a generic technology.

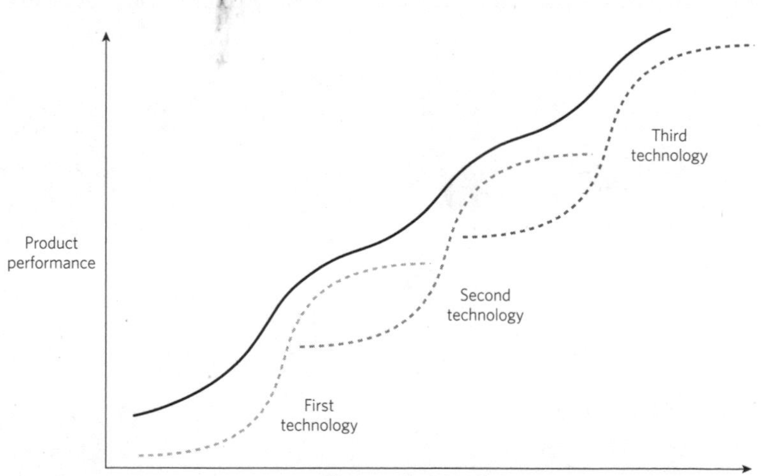

Compound *S* Curves. On this idealized chart, technological performance is measured on the vertical axis and time or engineering effort captured on the horizontal. A series of sub-*S* curves create an emergent larger-scale invariant slope.

A stack of *S* curves, each one containing its own limited run of exponential growth, overlap to produce a long-run emergent exponential growth line. The megatrend bridges more than one technology, giving it a transcendent power. As one exponential boom is subsumed into the next, an established technology relays its momentum to the next paradigm and carries forward an unrelenting growth. The exact unit of what is being measured can also morph from one subcurve to the next. We may start out counting pixel size, then shift to pixel density, then to pixel speed. The final performance trait may not be evident in the initial technologies and reveal itself only over the long term, perhaps as a macrotrend that continues indefinitely. In the case of computers, as the performance metric of chips is constantly recalibrated from one technological stage to the next, Moore's Law—redefined—will never end.

The slow demise of the more-transistors-per-chip trend is inevitable. But on average, digital technologies will roughly double in performance every two years for the foreseeable future. That means our most culturally important devices and systems will get faster, cheaper, better by 50 percent every year. Imagine if you got half again smarter every year or could remember 50 percent more this year than last. Embedded deep in the technium (as we now know it) is the remarkable capacity of half-again annual improvement. The optimism of our age rests on the reliable advance of Moore's promise: that stuff will get significantly, seriously, desirably better and cheaper tomorrow. If the things we make will get better the next time, that means that the golden age is ahead of us, and not in the past. But if Moore's Law ceased, would our optimism end, too?

Even if it we wanted to, what on Earth could derail the long version of Moore's Law? Suppose we were part of a vast conspiracy to halt Moore's Law. Maybe we believed it elevates undue optimism and encourages misguided expectations of a super artificial intelligence that will bring us immortality. What could we do? How would you stop it? Those who believe its power rests primarily in its self-reinforced expectations would say: Simply announce that Moore's Law will end. If enough smart believers declare Moore's Law over, then it will be over. The loop of self-fulfilling prophecy would be broken. But all it takes is one maverick to push ahead and make further progress, and the spell would be broken. The race would resume until the physics of scaling down gave out.

More clever folk might reason that since the economic regime as a whole determines the doubling time of Moore's Law, you could keep decreasing the quality of the economy until it stopped. Perhaps through armed revolution you could install an authoritative command-style policy (like an old state communism) whose lackadaisical economic growth would kill the infrastructure for exponential increases in computing power. I find that possibility intriguing, but I have my doubts. If in a counterfactual history, communism had won the cold war and microelectronics had been invented in a global Soviet-style society, my guess is that even that alternative policy apparatus could not stifle Moore's Law. Progress might roll out slower at a lower slope, maybe with a doubling time of five years, but I don't doubt that Stalinist scientists would tap into the law of the microcosm and soon marvel at the same technical wonder we do: chips improving exponentially as constant linear effort is applied.

I suspect Moore's Law is something we don't have much sway over, other than its doubling period. Moore's Law is the Moirae of our age. In Greek mythology the Moirae were the three Fates, usually depicted as dour spinsters. One Moira spun the thread of a newborn's life. The other Moira counted out the thread's length. And the third Moira cut the thread at death. A person's beginning and end were predetermined. But what happened in between was not inevitable. Humans and gods could work within the confines of one's ultimate destiny.

The unbending trajectories uncovered by Moore, Kryder, Roberts, Carlson, and Kurzweil spin through the technium, forming a long thread. The direction of the thread is inevitable, destined by the nature of matter and discovery. But its meander is open, left for us to finish.

Listen to the technology, Carver Mead says. What do the curves say? Imagine it is 1965. You've seen the curves Gordon Moore discovered. What if you believed the story they were trying to tell us: that each year, as sure as winter follows summer and as day follows night, computers would get half again better, and half again smaller, and half again cheaper, year after year, and that in 5 decades they would be 30 million times more powerful than they were then. (This is what happened.) If you were sure of that in 1965, or even mostly persuaded, what good fortune you could have harvested! You would have needed no other prophecies, no other predictions, no other details to optimize the coming benefits. As a society, if we just believed that single trajectory of Moore's, and none other, we would have educated differently, invested differently, prepared more wisely to grasp the amazing powers it would sprout.

The invariant growth ratios found in transistors, bandwidth, storage, pixels, and DNA sequencing are some of the first few Moira threads we've teased out in our short history in the accelerated technium. There must be others still to be uncovered by tools not yet invented. These "laws" are reflexes of the technium that kick in regardless of the social climate. They, too, will spawn progress and inspire new powers and new desires as they unroll in ordered sequence. Perhaps these self-governing dynamics will appear in genetics, or in pharmaceuticals, or in cognition. Once a growth dynamic is launched and made visible, the fuels of finance, competition, and markets will push the law to its limits and keep it riding along that curve until it has consumed its potential.

Our choice, and it is significant, is to prepare for the gift—and the problems it will also bring. We can choose to get better at anticipating these inevitable surges. We can choose to educate ourselves and our children to become intelligently literate and wise in their employment. And we can choose to modify our legal and political and economic assumptions to meet the ordained trajectories ahead. But we cannot escape from them.

When we spy our technological fate in the distance, we should not reel back in horror of its inevitability; rather, we should lurch forward in preparation.

9

Choosing the Inevitable

I once saw our future technological fate myself. In 1964 I visited the New York World's Fair as a wide-eyed, slack-jawed kid. The inevitable future was on display, and I swallowed it up in great gulps. At the AT&T pavilion they had a working picture phone. The idea of a videophone had been circulating in science fiction for a hundred years, in a clear case of prophetic foreshadowing. Now here was one that actually worked. Although I was able to see it, I didn't get to use it, but photos of how it would enliven our suburban lives ran in the pages of *Popular Science* and other magazines. We all expected it to appear in our lives any day. Well, the other day, 45 years later, I was using a picture phone just like the one predicted way back in 1964. As my wife and I gathered in our California den to lean toward a curved white screen displaying the moving image of our daughter in Shanghai, we mirrored the old magazine's illustration of a family crowded around a picture phone. While our daughter watched us on her screen in China, we chatted leisurely about unimportant family matters. Our picture phone was exactly what everyone imagined it to be, except in three significant ways: the device was not exactly a phone, it was our iMac and her laptop; the call was free (via Skype, not AT&T); and despite being perfectly useable, and free, picture-phoning has not become common—even for us. So unlike the earlier futuristic vision, the inevitable picture phone has not become the standard modern way of communicating.

First Glimpse of the Picture Phone. From Bell Telephone's pavilion at the 1964 New York World's Fair.

So was the picture phone inevitable? There are two senses of "inevitable" when used with regard to technology. In the first case, an invention merely has to exist once. In that sense, every realizable technology is inevitable because sooner or later some mad tinkerer will cobble together almost anything that can be cobbled together. Jetpacks, underwater homes, glow-in-the-dark cats, forgetting pills—in the goodness of time every invention will inevitably be conjured up as a prototype or demo. And since simultaneous invention is the rule, not the exception, any invention that can be invented will be invented more than once. But few will be widely adopted. Most won't work very well. Or more commonly they will work but be unwanted. So in this trivial sense, all technology is inevitable. Rewind the tape of time and it will be reinvented.

The second, more substantial sense of "inevitable" demands a level of common acceptance and viability. A technology's use must come to dominate the technium or at least its corner of the technosphere. But more than ubiquity, the inevitable must contain a large-scale momentum and proceed on its own determination beyond the free choices of several billion humans. It can't be diverted by mere social whims.

The picture phone was imagined in sufficient detail a number of times, in different eras and different economic regimes. It really wanted to happen. One artist sketched out a fantasy of it in 1878, only two years after the telephone was patented. A series of working prototypes were demoed by the German post office in 1938. Commercial versions, called Picturephones, were installed in public phone booths on the streets in New York City after the 1964 World's Fair, but AT&T canceled the product ten years later due to lack of interest. At its peak the Picturephone had only 500 or so paid subscribers, even though nearly everyone recognized the vision. One could argue that rather than being inevitable progress, this was an invention battling its own inevitable bypass.

Yet today it is back. Perhaps it is more inevitable over a 50-year span. Maybe it was too early back then, and the necessary supporting technology absent and social dynamics not ripe. In this respect the repeated earlier tries can be taken as evidence of its inevitability, its relentless urge to be born. And perhaps it is still being born. There may be other innovations yet to be invented that could make the videophone more common. Such needed innovations as ways to direct the gaze of the speaker into your eyes instead of toward the off-center camera or methods for the screen to switch gazes among multiple parties in the conversation.

The hesitant arrival of the picture phone is evidence for both arguments: (a) that it clearly had to happen and (b) that it clearly does not have to happen. That brings up the question: Does any technology lurch forward on its own inertia as "a self-propelling, self-sustaining, ineluctable flow," in the words of technology critic Langdon Winner, or do we have clear free-will choice in the sequence of technological change, a stance that makes us (individually or corporately) responsible for each step?

I'd like to suggest an analogy.

Who you are is determined in part by your genes. Every single day scientists identify new genes that code for a particular trait in humans, revealing the ways in which inherited "software" drives your body and brain. We now know that behaviors such as addiction, ambition, risk-taking, shyness, and many others have strong genetic components. At the same time, "who you are" is clearly determined by your environment and upbringing. Every day science uncovers more evidence of the ways in which our family, peers, and cultural background shape our being. The strength of what others believe about us is enormous. And more recently we have increasing proof that environmental factors can influence genes, so that these two factors are cofactors in the strongest sense of the word—they determine each other. Your environment (like what you eat) can affect your genetic code, and your code will steer you into certain environments—making untangling the two influences a conundrum.

Last, who you are in the richest sense of the word—your character, your spirit, what you do with your life—is determined by what you choose. An awful lot of the shape of your life is given to you and is beyond your control, but your freedom to choose within those givens is huge and significant. The course of your life within the constraints of your genes and environment is up to you. You decide whether to speak the truth at any trial, even if you have a genetic or familial propensity to lie. You decide whether or not to risk befriending a stranger, no matter your genetic or cultural bias toward shyness. You decide beyond your inherent tendencies or conditioning. Your freedom is far from total. It is not your choice alone whether to be the fastest runner in the world (your genetics and upbringing play a large role), but you can choose to be faster than you have been. Your inheritance and education at home and school set the outer boundaries of how smart or generous or sneaky you can be, but you choose whether you will be smarter, more generous, or sneakier today than yesterday. You may inhabit a body and brain that wants to be lazy or sloppy or imaginative, but you choose to what degree those qualities progress (even if you aren't inherently decisive).

Curiously, this freely chosen aspect of ourselves is what other people remember about us. How we handle life's cascade of real choices within the larger cages of our birth and background is what makes us who we are. It is what people talk about when we are gone. Not the given, but the choices we made.

It is the same with technology. The technium is in some part preordained by its inherent nature—which is the larger theme of this book. Just as our genes drive the inevitable unfolding of human development, starting from a fertilized egg, proceeding to an embryo, then to a fetus, an infant, a toddler, a kid, and a teenager, so, too, the largest trends of technology unroll in developmental stages.

In our lives we have no choice about becoming teenagers. The strange hormones will flow, and our bodies and minds must morph. Civilizations follow a similar developmental pathway, although their outlines are less certain because we have witnessed fewer of them. But we can discern a necessary ordering: A society must control fire first, then metalworking before electricity, and electricity before global communications. We might disagree on what exactly is sequenced, but a sequence there is.

At the same time, history matters. Technological systems gain their own momentum and become so complex and self-aggregating that they form a reciprocal environment for other technologies. The infrastructure built to support the gasoline automobile is so extensive that after a century of expansion it now affects technologies outside of transportation. For instance, the invention of air-conditioning in concert with the highway system encouraged subtropical suburbs. The invention of cheap refrigerated air altered the landscape of the American South and Southwest. If air-conditioning had been implemented in a nonauto society, its pattern of consequences would have been different, even though air-cooling systems contain their own technological momentum and inherencies. So every new development in the technium is contingent upon the historical antecedents of previous technologies. In biology this effect is called coevolution, and it means that the "environment" of one species is the ecosystem of all the other species it interacts with, all of them in flux. For example, prey and predator evolve together, and evolve each other, in a never-ending arms race. Host and parasite become one duet as they try to outdo each other, and an ecosystem will adapt to the moving target of a new species adapting to it.

Within the borders laid out by inevitable forces, our choices unleash consequences that gain momentum over time until these contingencies harden into technological necessities and become nearly unchangeable in future generations. There's an old story about the long reach of early choices that is basically true: Ordinary Roman carts were constructed to match the width of imperial Roman war chariots because it was easier to follow the ruts in the road left by the war chariots. The chariots were sized to accommodate the width of two large warhorses, which translates into our English measurement of 4' 8.5". Roads throughout the vast Roman Empire were built to this specification. When the legions of Rome marched into Britain, they constructed long-distance imperial roads 4' 8.5" wide. When the English started building tramways, they used the same width so the same horse carriages could be used. And when they started building railways with horseless carriages, naturally the rails were 4' 8.5" wide. Imported laborers from the British Isles built the first railways in the Americas using the same tools and jigs they were used to. Fast-forward to the U.S. space shuttle, which is built in parts around the country and assembled in Florida. Because the two large solid-fuel rocket engines on the side of the launch shuttle were sent by railroad from Utah, and that line traversed a tunnel not much wider than the standard track, the rockets themselves could not be much wider in diameter than 4' 8.5". As one wag concluded: "So, a major design feature of what is arguably the world's most advanced transportation system was determined over two thousand years ago by the width of two horses' arse." More or less, this is how technology constrains itself over time.

The past 10,000 years of technology influence the preordained march of technology in each new era. The initial conditions of an embryonic electrical system, for example, can guide the character of its eventual network in several ways. The engineers might choose alternating current (AC), favoring centralization, or direct current (DC), favoring decentralization. Or the system could be installed in 12 volts (by amateurs) or 250 volts (by professionals). The legal regime could favor patent protection or not, and the business models could be built around profits or charitable nonprofit. These initial specifications affected how the internet developed on top of the electric network. All these variables bend the unrolling system in different cultural directions. Yet electrification in some form was a necessary, unavoidable phase for the technium. The internet that followed it, too, was inevitable, but the character of its incarnation is contingent on the tenor of the technologies that preceded it. Telephones were inevitable, but the iPhone wasn't. We accept the biological analog: Human adolescence is inevitable, but delinquency is not. The exact pattern that the inevitable teenagehood manifests in any individual will depend in part on his or her biology, which depends in part on his or her past health and environment but also hinges on his or her free-will choices.

Like personality, technology is shaped by a triad of forces. The primary driver is preordained development—what technology wants. The second driver is the influence of technological history, the gravity of the past, as in the way the size of a horse's yoke determines the size of a space rocket. The third force is society's collective free will in shaping the technium, or our choices. Under the first force of inevitability, the path of technological evolution is steered both by the laws of physics and by self-organizing tendencies within its large, complex, adaptive system. The technium will tend toward certain macroforms, even if you rerun the tape of time. What happens next is contingent on the second force, or what has already happened, and so the momentum of history constrains our choices forward. These two forces channel the technium along a limited path and severely restrict our choices. We like to think that "anything is possible next," but in fact anything is not possible in technology.

In contrast to these two, the third force is our free will to make individual choices of use and collective policy decisions. Compared to all possibilities that we can imagine, we have a very narrow range of choices. But compared to 10,000 years ago, or even 1,000 years ago, or even last year, our possibilities are expanding. Although restricted in the cosmic sense, we have more choice than we know what to do with. And via the engine of the technium, these real choices will keep expanding (even though the larger path is preordained).

This paradox is recognized not just by historians of technology but by ordinary historians as well. In the view of cultural historian David Apter, "Human freedom actually exists within the limits set by the historical process. While not everything is possible, there is much that can still be chosen." Historian of technology Langdon Winner sums up this convergence of free will and the ordained in these terms: "Technology moves steadily onward as if by cause and effect. This does not deny human creativity, intelligence, idiosyncrasy, chance, or the willful desire to head in one direction rather than another. All of these are absorbed into the process and become moments in the progressions."

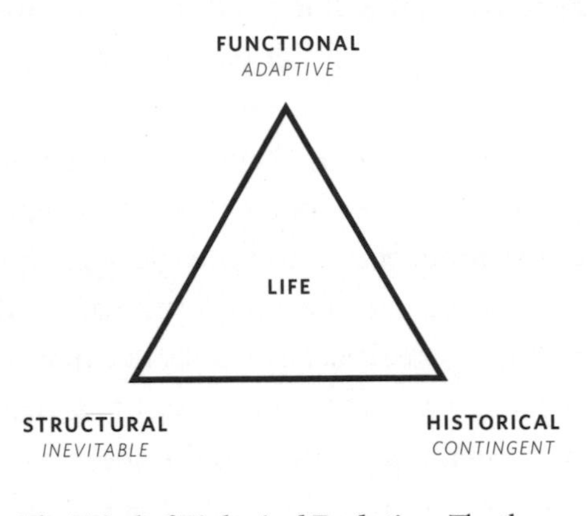

The Triad of Biological Evolution. The three evolutionary vectors in life.

It is no coincidence that the triadic nature of the technium is the same as the triadic nature of biological evolution. If the technium is indeed the extended acceleration of the evolution of life, it should be governed by the same three forces.

One vector is the inevitable. The basic laws of physics and emergent self-organization drive evolution toward certain forms. Specific species (either biological or technological) are unpredictable in their microdetails, but the macropatterns (electrical motors, binary computing) are ordained by the physics of matter and self-organization. This inescapable force can be thought of as the structural inevitability of biological and technological evolution (shown as the lower left corner in the diagram above).

The second corner of the triad is the historical/contingent aspect of evolutionary change (lower right). Accidents and circumstantial opportunities bend the course of evolution this way and that, and those contingencies add up over time to create ecosystems with their own internal momentum. The past matters.

The third force working within evolution is the adaptive function—the relentless engine of optimization and creative innovation that con-

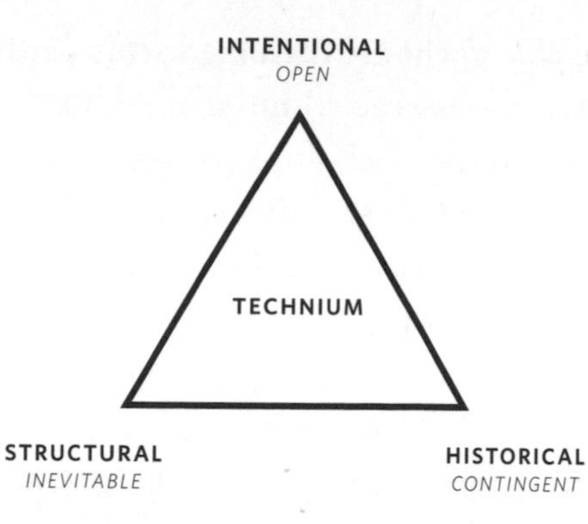

The Triad of Technological Evolution. In the technium, the functional vector is occupied by an equivalent force: the intentional.

tinually solves the problems of survival. In biology this is the incredible force of unconscious, blind natural selection (shown as the top corner).

But in the technium the adaptive function is not unconscious, as it is in natural selection. Instead it is open to human free will and choice. This intentional domain consists of the many decisions we make in the political expressions of inevitable inventions and of the billions of personal decisions individuals make about whether (and how) they use or avoid certain inventions. In biological evolution there is no designer, but in the technium there is an intelligent designer—Sapiens. And of course, this conscious open design (shown as the top corner) is why the technium has become the most powerful force in the world.

The other two legs of technological evolution are identical to the other two legs of biological evolution. The basic laws of physics and emergent self-organization drive technological evolution through an inevitable series of structural forms—four-wheeled vehicles, hemispherical boats, books of pages, etc. At the same time, the historical contingency of past inventions forms an inertia that bends evolution this way or that—within the bounds of the inevitable developments. It is the third leg, the collective choice of free-willed individuals, that provides the character of the technium. And just as our free-will choices in our individual lives create the kind of person we are (our ineffable "person"), our choices, too, shape the technium.

We may not be able to choose the macroscale outlines of an industrial automation system—assembly-line factories, fossil-fuel power, mass education, allegiance to the clock—but we can choose the character of those parts. We have latitude in selecting the defaults of our mass education, so that we can nudge the system to maximize equality or to favor excellence or to foster innovation. We can bias the invention of the industrial assembly line either toward optimization of output or toward optimization of worker skills; those two paths yield different cultures. Every technological system can be set with alternative defaults that will change the character and personality of that technology.

The consequences of choice can easily be seen from space. Satellites sweeping over the skies record city lights at night. From orbit each lighted town on the Earth below acts like a pixel in a night portrait of the technium. An evenly distributed coat of light indicates technological development. In Asia the steady scatter of light is interrupted by a large, dark, unlit blob. The dark outline follows the exact contour of the renegade country North Korea.

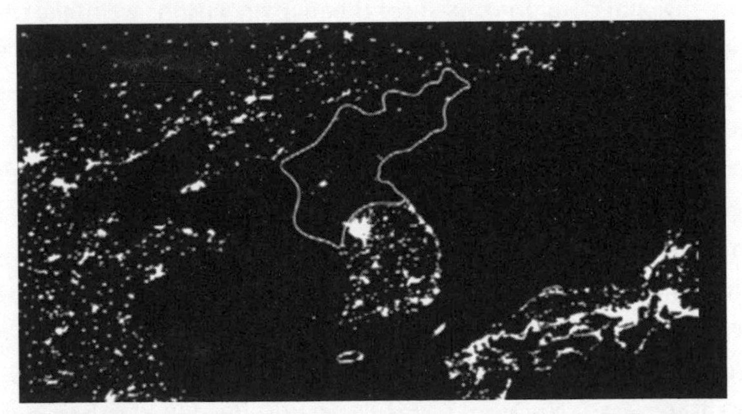

North Korea at Night. The absence of modern technological abundance displayed by night satellite photography over east Asia. The outline of North Korea is drawn in white.

Paul Romer, an economist at Stanford, points out that this remarkable negative space is the result of political policy. All the technological ingredients for nighttime light exist for North Korea, as evidenced by the brightly lit areas surrounding it, but as a country North Korea has adopted an expression of an electrical system that is sparse and minimal. The result is a stunning map of technological choice.

In *Nonzero,* author Robert Wright offers a wonderful analogy for understanding the role of the inevitable as applied to technology, which I paraphrase here. It's appropriate, Wright says, to claim that the destiny of a tiny seed, say, a poppy seed, is to grow into a plant. Flower yields seed, seed sprouts plant, according to an eternal fixed routine burned in by a billion years of flowers. Sprouting is what seeds do. In that fundamental sense, it is inevitable that a poppy seed becomes a plant, even though a fair number of poppy seeds wind up on bagels. We don't require that 100 percent of seeds arrive at their next stage to acknowledge the inexorable direction of the poppy's growth because we know that inside the poppy seed is a DNA program. The seed "wants" to be a plant. More precisely, the poppy seed is designed to grow stems, leaves, and flowers of a precise type. We regard the destiny of the seed less as the statistical probability of how many complete the journey, and more as a matter of what it is designed for.

To claim that the technium pushes itself through certain inevitable technological forms is not to say that every technology was a mathematical certainty. Rather, it indicates a direction more than a destiny. More precisely, the technium's long-term trends reveal the design of the technium; this design indicates what the technium is built to do.

Inevitability is not a flaw. Inevitability makes prediction easier. The better we can forecast, the better we can be prepared for what comes. If we can discern the large outlines of persistent forces, we can better educate our children in the appropriate skills and literacies need for thriving in that world. We can shift the defaults in our laws and public institutions to reflect that coming reality. If, for instance, we realize that everyone's full DNA will be sequenced from birth or before (that is inevitable), then instructing everyone in genetic literacy becomes essential. Each must know the limits to what can and cannot be gleaned from this code, how it varies or does not among related individuals, what might impact its integrity, what information about it can be shared, what concepts such as "race" and "ethnicity" mean in this context, and how to use this knowledge to get therapeutics tailored to it. There's a whole world to open up, and it will take time, but we can begin to sort these choices out now because its arrival, in alignment with the exotropic principles, is pretty inevitable.

As the technium progresses, better tools for forecasting and prediction help us spot the inevitable. To return to the adolescence analogy, because we can anticipate the inevitable onset of human adolescence, we are better able to thrive in it. Teenagers are biologically compelled to take risks as a means of establishing their independence. Evolution "wants" risky teenagers. Knowing that risky behavior is expected in adolescence is both reassuring to teenagers (you are normal, not a freak) and to society (they will grow out of it) and an invitation to harness that normal riskiness for improvement and gain. If we ascertain that a global web of continuous connection is an inevitable phase in a growing civilization, then we can both be reassured by this inevitability and take it as an invitation to make the best global web we can.

As technology advances, we gain both more possibilities and, if we are smart and wise, better ways to anticipate these ordained trends. Our real choices in technology matter. Although constrained by predetermined forms of development, the particular specifics of a technological phase matter to us greatly.

Inventions and discoveries are crystals inherent in the technium, waiting to be manifested. There is nothing magical about these patterns, nothing mystical about technology having a direction. All complex, adaptive systems that maintain a stable self-organization—from galaxies to starfish to human minds—will exhibit emergent forms and inherent directions. We call these forms inevitable because, like a spiral whirlpool in draining water or snowflakes in a winter storm, they will manifest themselves whenever the conditions are right. But of course, they never render themselves in exactly the same details.

The vortex of the technium has grown its own agenda, its own imperative, its own direction. It is no longer under the full control and mastery of its parent and creator, humanity. We worry, like all parents, particularly as the technium's power and independence increase.

But its autonomy also brings us great benefits. The long-term rise in real progress is due to its growth as a lifelike system. And the most attractive aspects of technology are also due to these self-augmenting long-term trends.

The urge for self-preservation, self-extension, and self-growth is the natural state of any living thing. We don't begrudge the selfish nature of a lion, or a grasshopper, or ourselves. But there comes a moment in the childhood of our biological offspring when their childish selfish nature confronts us, and we have to acknowledge that they have their own agenda. Even though their very own life is an unambiguous continuation of our life (all their cells derive uninterrupted from our cells), our children have their own life. No matter how many babies we have seen, we are unsettled each time these moments of independence arrive.

Collectively we are at one of these moments with the technium. We encounter this natural life cycle every day in biology, but this is the first time we have met it in technology, and it is unnerving us. Our shock at meeting selfishness in technology has to do with the fact that, by definition, we are, and will always remain, part of the technium itself. In the words of psychologist Sherry Turkle, technology is our "second self." It is both "other" and "us." Unlike our biological children, who grow up to have minds completely separate from us, the technium's autonomy includes us and our collective minds. We are part of its selfish nature.

The ongoing dilemma of technology, then, will never leave us. It is an ever-elaborate tool that we wield and continually update to improve our world; and it is an ever-ripening superorganism, of which we are but a part, that is following a direction beyond our own making. Humans are both master and slave to the technium, and our fate is to remain in this uncomfortable dual role. Therefore, we will always be conflicted about technology and find making our choices difficult.

But our concern should not be about whether to embrace it. We are beyond embrace; we are already symbiotic with it. At a macroscale, the technium is following its inevitable progression. Yet at the microscale, volition rules. Our choice is to align ourselves with this direction, to expand choice and possibilities for everyone and everything, and to play out the details with grace and beauty. Or we can choose (unwisely, I believe) to resist our second self.

The conflict that the technium triggers in our hearts is due to our refusal to accept our nature—the truth is that we are continuous with the machines we create. We are self-made humans, our own best invention. When we reject technology as a whole, it is a brand of self-hatred.

"We trust in nature, but we hope in technology," says Brian Arthur. That hope lies in embracing our own natures. By aligning ourselves with the imperative of the technium, we can be more prepared to steer it where we can and more aware of where we are going. By following what technology wants, we can be more ready to capture its full gifts.

PART THREE

CHOICES

10

The Unabomber Was Right

In 1917 Orville Wright predicted that "the aeroplane will help peace in more ways than one—in particular I think it will have a tendency to make war impossible." He was echoing earlier sentiments from American journalist John Walker, who declared in 1904, "As a peace machine, the value [of the aeroplane] to the world will be beyond computation." This wasn't the first grand promise of technology. In that same year Jules Verne announced, "The submarine may be the cause of bringing battle to a stoppage altogether, for fleets will become useless, and as other war material continues to improve, war will become impossible."

Alfred Nobel, the Swedish inventor of dynamite and founder of the Nobel Prize, sincerely believed his explosives would be a war deterrent: "My dynamite will sooner lead to peace than a thousand world conventions." In the same vein, when Hiram Maxim, inventor of the machine gun, was asked in 1893, "Will this gun not make war more terrible?" he answered, "No, it will make war impossible." Guglielmo Marconi, inventor of the radio, told the world in 1912, "The coming of the wireless era will make war impossible, because it will make war ridiculous." General James Harbord, chairman of the board of RCA in 1925, believed, "Radio will serve to make the concept of Peace on Earth, Good Will Toward Men a reality."

Not long after the telephone was commercialized in the 1890s, John J.

Carty, AT&T's chief engineer, prophesied, "Someday we will build up a world telephone system, making necessary to all peoples the use of a common language or common understanding of languages, which will join all the people of the Earth into one brotherhood. There will be heard throughout the Earth a great voice coming out of the ether which will proclaim, 'Peace on Earth, good will towards men.'"

Nikola Tesla claimed that his invention of "the economic transmission of power without wire . . . will bring peace and harmony on Earth." That was back in 1905; since we still don't have economic transmission of power without wires, there is still hope for world peace.

David Nye, a historian of technology, adds to the list of inventions envisioned as abolishing war once and for all and ushering in universal peace the torpedo, the hot-air balloon, poison gas, land mines, missiles, and laser guns. Nye says, "Each new form of communication, from the telegraph and telephone to radio, film, television and the internet, has been heralded as the guarantor of free speech and the unfettered movement of ideas."

George Gent, writing in a 1971 *New York Times* article about interactive cable television, said, "Supporters have hailed the program as . . . a major step toward the political philosopher's dream of participatory democracy." Today promises about the democratizing and peaceful effects of the internet overshadow any similar claims about television. Yet as futurist Joel Garreau marvels, "Given what we know happened with television, I am astonished that computer technology is now seen as a sacrament."

It is not that all these inventions are without benefits—even benefits toward democracy. Rather, it's the case that each new technology creates more problems than it solves. "Problems are the answers to solutions," says Brian Arthur.

Most of the new problems in the world are problems created by previous technology. These technogenic problems are nearly invisible to us. Every year 1.2 million people die in automobile accidents. The dominant technological transportation system kills more people than cancer. Global warming, environmental toxins, obesity, nuclear terrorism, propaganda, species loss, and substance abuse are only a few of the many other serious technogenic problems troubling the technium. As techno-critic Theodore Roszak says, "How much of what we readily identify as 'progress' in the urban-industrial society is really the undoing of evils inherited from the last round of technological innovation?"

If we embrace technology we need to confront its costs. Thousands of traditional livelihoods have been sidetracked by progress, and the lifestyles around those occupations eliminated. Hundreds of millions of humans today toil at jobs they hate, producing things they have no love for. Sometimes these jobs cause physical pain, disability, or chronic disease. Technology creates many new occupations that are indisputably dangerous (coal mining, for one). At the same time, mass education and media train humans to shy away from low-tech manual work, to seek jobs working for the digital technium. The divorce of the hands from the head puts a strain on the human psyche. Indeed, the sedentary nature of the best-paying jobs is a health hazard—for body and mind.

Technology swells till it fills all holes and spaces between us. We monitor not only our neighbors' affairs but those of anyone we care to spy on. We have 5,000 "friends" on our list but space in our heart for only 50. Our ability to impact has expanded beyond our ability to care. By turning our lives inside out with technological mediation, we are open to manipulation by mobs, clever advertisers, governments, and the inadvertent biases of the system.

Time spent with machines must come from somewhere. The flood of newly invented consumer gadgets suck time from the use of other gadgets or from other human activities. One hundred thousand years ago, Sapiens' foraging day was predominantly clear of technology. Ten thousand years ago a farming human might spend a few hours a day with a tool in one hand. Only 1,000 years ago, medieval technology was ubiquitous on the periphery of human relationships, but not central. Today technology places itself in the middle of everything we do, see, hear, and make. Technology has permeated eating, romance, sex, child rearing, education, death. Our lives run on clock time.

As the most powerful force in the world, technology tends to dominate our thinking. Because of its ubiquity, it monopolizes any activity and questions any nontechnological solution as unreliable or impotent.

Because of its power to augment us, we give precedence to the made over the born. Which do we expect to more be effective, a wild herb or an engineered drug? Even our cultural compliments for excellence have drifted to the mechanical: "smooth as glass," "bright and shiny," "sterling," "watertight," and "like clockwork"—all suggesting the superiority of the man-made. We have become imprisoned in the technological framework of what the poet William Blake called "the mind-forg'd manacles."

Simply the fact that a machine is *able* to perform a task often becomes sufficient reason to have it *do* the task, even if it does it poorly at first. The first machine-made versions of things, such as garments, china bowls, writing paper, baskets, and canned soups, were not very good, just very cheap. Often we will invent a machine for a particular and limited purpose, and then, in what Neil Postman calls the Frankenstein syndrome, the invention's own agenda blossoms. "Once the machine is built," Postman writes, "we discover, always to our surprise—that it has ideas of its own; that it is quite capable not only of changing our habits but . . . of changing our habits of mind." In this way, humans have become an adjunct to or, in Karl Marx's phrase, appendages of the machine.

There is a widespread belief that the technium grows only by consuming irreplaceable resources, ancient habitats, and myriad wild creatures and yet returns to the biosphere only pollution, pavement, and myriad obsolete junk. And worse, this same technology takes from the least in the world—the nations with the most natural resources and least economic power—to enrich the most powerful. So as progress fattens the lives of the lucky few, it starves the unfortunate poor. Many people who acknowledge the technium's progress are held back from a full embrace of the technological imperative because of its adverse effect on the natural environment.

This encroachment is real. Often technological progress has been produced at the expense of ecological habitats. The technium's steel is mined from the Earth; its lumber is taken by cutting down forests; its plastics and energy are sucked from oil and burned into the air. Its factories displace wetlands or meadows. One third of the Earth's land surface is already altered by agriculture and human habitation. You could compile a very long list of mountains leveled, lakes poisoned, rivers dammed, jungles flattened, air dirtied, and diversity slashed. More damning, civilization is responsible for the permanent extinction of many unique species of life. Over geological time the normal, or background, rate of species loss is one species every four years. Today, at the minimum, it is four times that; probably we are now eliminating species at thousands of times that rate.

(I happen to know a little bit about this decimation because for a decade I chaired an initiative to catalog all the life on Earth. We have historical evidence for the extinction of about 2,000 species in the last 2,000 years, or one per year, or four times the natural rate. The bulk of those extinctions, however, are in the last 200 years, so the known annual average today is significantly higher. Since we have identified about 5 percent of all species on Earth, and many of those yet-to-be-named species are in the same vanishing habitats as the documented extinctions, we can extrapolate what the total number of species going extinct might be. These estimates run at the higher end of 50,000 per year. In truth, no one has any idea how many species are actually on Earth, or what percentage we have identified, even to the nearest magnitude, so all we can say for sure is that we are eliminating species faster than before, which is criminal enough.)

And yet there is nothing inherent in the technium that insists on species loss. For every technological method we currently use that causes loss of habitat, we can imagine an alternative solution that does not. In fact, for every technology X that we can invent there is—or could be—a corresponding technology Y that is potentially greener. There will always be ways to increase energy and material efficiency, to better mimic biological processes, or to ease the pressure on ecosystems. "I cannot imagine a technology that cannot be made orders of magnitude greener," says Paul Hawken, a renowned advocate for environmentally sound technology. "But in my opinion we have not even stepped inside the realm of green technology yet." A greener improvement, it is true, may adversely affect the environment in a new unknown way, but that only means yet another innovation needs to remedy that shortfall. In this way, we will never exhaust the potential for greener technology. Since we can detect no limit to how biophilic technology can become, this open-ended horizon indicates to us that the nature of technology is inherently prolife. The technium is, at its most fundamental level, potentially compatible with life. It just needs to grow into that potential.

In the apt phrase of futurist Paul Saffo, we often confuse a clear view of the future with a short distance. But in reality, technology creates a worrisome dissonance between what we can imagine and what we can do. I can't think of a better explanation of this than filmmaker George Lucas's rendition of technology's eternal dilemma. In 1997, I interviewed Lucas about the new, high-tech method of filmmaking that he had devised for his prequel *Star Wars* films. It entailed stitching computers, cameras, animation, and live action together into one seamless cinematic world, building up layers of images, almost like painting in film. It has since been adopted by other avant-garde directors of action films, including James Cameron in *Avatar.* At the time, Lucas's radical new process was the apogee of advanced technology. But while his innovative technique was futuristic, many viewers claimed it didn't make his newer films any better. I asked him, "Do you think technology is making the world better or worse?" Lucas's answer:

> If you watch the curve of science and everything we know, it shoots up like a rocket. We're on this rocket and we're going perfectly vertical into the stars. But the emotional intelligence of humankind is equally if not more important than our intellectual intelligence. We're just as emotionally illiterate as we were 5,000 years ago; so emotionally our line is completely horizontal. The problem is the horizontal and the vertical are getting farther and farther apart. And as these things grow apart, there's going to be some kind of consequence of that.

I think we underestimate the strain of that gap. In the long term, the erosion of the traditional self may prove to be a larger part of the technium's cost than its erosion of the biosphere. Langdon Winner suggests there is a kind of conservation of life force: "Insofar as men pour their own life into the apparatus, their own vitality is that much diminished. The transference of human energy and character leaves men empty, although they may never acknowledge the void."

That transference is not inevitable, but it does happen. As machines take over more of what humans once did, we tend to do less of the familiar. We don't walk as much, letting our autos do our walking. We don't dig anymore, except with backhoes. We don't hunt for food, we don't gather. We don't hammer or sew. We don't read if we don't have to. We don't calculate. We are in the process of offloading our remembering to Google, and we are eager to stop cleaning as soon as those cleaning bots get cheap enough. Eric Brende, an engineering student who spent two years living like an Amish, says, "Duplicating vital human capacities can have one of only two consequences: atrophying the capacities or creating competition between *Homo sapiens* and machine. Neither of these is savory to self-respecting members of the former." Technology chips away at our human dignity, calling into question our role in the world and our own nature.

This can make us crazy. The technium is a global force beyond human control that appears to have no boundaries. Popular wisdom perceives no counterforce to prevent technology from usurping all available surfaces of the planet, creating an extreme ecumenopolis—planet-sized city—like the fictional Trantor in Isaac Asimov's sci-fi stories or the planet Coruscant in Lucas's *Star Wars*. Pragmatic ecologists would argue that long before an ecumenopolis could form, the technium would outstrip the capacity of Earth's natural systems and thus would either stall or collapse. The cornucopians, who believe the technium capable of infinite substitutions, see no hurdle to endless growth of civilization's imprint and welcome the ecumenopolis. Either prospect is unsettling.

About 10,000 years ago, humans passed a tipping point where our ability to modify the biosphere exceeded the planet's ability to modify us. That threshold was the beginning of the technium. We are at a second tipping point where the technium's ability to alter us exceeds our ability to alter the technium. Some people call this the Singularity, but I don't think we have a good name for it yet. Langdon Winner claims that "technical artifice as an aggregate phenomenon [or what I call the technium] dwarfs human consciousness and makes unintelligible the systems that people supposedly manipulate and control; by this tendency to exceed human grasp and yet to operate successfully according to its own internal makeup, technology is a total phenomenon which constitutes a 'second nature' far exceeding any desires or expectations for the particular components."

Ted Kaczynski, the convicted bomber who blew up dozens of technophilic professionals, killing three of them, was right about one thing: Technology has its own agenda. It is selfish. The technium is not, as most people think, a series of individual artifacts and gadgets for sale. Rather, Kaczynski, speaking as the Unabomber, echoes the arguments of Winner and many of the points I am making in this book, claiming that technology is a dynamic, holistic system. It is not mere hardware; rather, it is more akin to an organism. It is not inert, nor passive; rather, the technium seeks and grabs resources for its own expansion. It is not merely the sum of human action, but in fact it transcends human actions and desires. I think Kaczynski was right about these claims. In his sprawling, infamous 35,000-word manifesto, the Unabomber wrote:

> The system does not and cannot exist to satisfy human needs. Instead, it is human behavior that has to be modified to fit the needs of the system. This has nothing to do with the political or social ideology that may pretend to guide the technological system. It is the fault of technology, because the system is guided not by ideology but by technical necessity.

I, too, argue that the technium is guided by "technical necessity." That is, baked into the nature of this vast complex of technological systems are self-serving aspects—technologies that enable more technology, and systems that preserve themselves—as well as inherent biases that lead the technium in certain directions, outside human desire. Kaczynski writes, "Modern technology is a unified system in which all parts are dependent on one another. You can't get rid of the 'bad' parts of technology and retain only the 'good' parts."

The truth of Kaczynski's observations does not absolve him of his murders or justify his insane hatred. Kaczynski saw something in technology that caused him to lash out with violence. But despite his mental imbalance and moral sins, he was able to articulate that view with surprising clarity. Kaczynski set off 16 bombs and murdered 3 people (and injured 23 more) in order to get his manifesto published. His desperation and despicable crimes hide a critique that has gained a minority following among other Luddites. Here, in meticulous, scholarly precision, Kaczynski makes his primary claim that "freedom and technological progress are incompatible" and that therefore technological progress must be undone. The center section of his argument is clear, remarkably so, given the cranky personal grievances against leftists that bookend his rant.

I have read almost every book on the philosophy and theory of technology and interviewed many of the wisest people pondering the nature of this force. So I was utterly dismayed to discover that one of the most astute analyses of the technium was written by a mentally ill mass murderer and terrorist. What to do? A few friends and colleagues counseled me to not even mention the Unabomber in this book. Some are deeply upset that I have.

I quote at length from the Unabomber's manifesto for three reasons. First, it succinctly states, often better than I can, the case for autonomy in the technium. Second, I have not found a better example of the view held by many skeptics of technology (a view shared by many ordinary citizens less strongly) that the greatest problems in the world are due not to individual inventions but to the entire self-supporting system of technology itself. Third, I think it is important to convey the fact that the emergent autonomy of the technium is recognized not only by supporters of technology like myself, but also by those who despise it.

The Unabomber was right about the self-aggrandizing nature of the technium. But I disagree with many other of Kacyznski's points, especially his conclusions. Kacyznski was misled because he followed logic divorced from ethics, but as befits a mathematician, his logic was insightful.

As best I understand, the Unabomber's argument goes like this:

- Personal freedoms are constrained by society, as they must be in any civilization for the sake of order.
- The stronger that technology makes the society, the less individual freedom there is.
- Technology destroys nature, strengthening itself further.
- But because it is destroying nature, the technium will ultimately collapse.
- In the meantime, the ratchet of technological self-amplification is stronger than politics.
- Using technology to try to tame the system only strengthens the technium.
- Because it cannot be tamed, technological civilization must be destroyed rather than reformed.
- Since it cannot be destroyed by technology or politics, humans must push the technium toward its inevitable self-collapse.
- Then we should pounce on it when it is down and kill it before it rises again.

In short, Kaczynski claims that civilization is the source of our problems and not the cure for them. He wasn't the first to make this claim. Rants against the machine of civilization go back as far as Freud and beyond. But the assaults against industrial society sped up as industry sped up. Edward Abbey, the legendary wilderness activist, considered industrial civilization to be a "destroying juggernaut" wrecking both the planet and humans. Abbey did all he could personally do to stop the juggernaut with monkey-wrenching maneuvers—sabotaging logging equipment and so forth. Abbey was the iconic Earth First warrior who inspired many fire-throwing followers. The Luddite theorist Kirkpatrick Sale, who, unlike Abbey, railed against the machine while living in a brownstone in Manhattan, refined the idea of "civilization as disease." (In 1995, at my instigation, Sale bet me $1,000 in the pages of *Wired* magazine that civilization would collapse by 2020.) Recently, the call to undo civilization and return to a purer, more humane, primitive state has accelerated in pace with the rapidly thickening mesh of global connections and always-on technology. A rash of armchair revolutionaries has issued books and websites announcing the end-times. In 1999 John Zerzan published an anthology of contemporary readings focused on the theme called *Against Civilization.* And in 2006 Derrick Jensen penned a 1,500-page treatise on how and why to topple technological civilization, with hands-on suggestions of the ideal places to start—for instance, power and gas lines and the information infrastructure.

Kaczynski had read earlier jeremiads against industrial society and arrived at his hatred of civilization in the same way many other nature lovers, mountain men, and back-to-the-landers have. He was driven there in a retreat from the rest of us. Kaczynski buckled under the many rules and expectations society put up for him as an aspiring professor of mathematics. He said, "Rules and regulations are by nature oppressive. Even 'good' rules are reductions in freedom." He was deeply frustrated at not being able to integrate into professional society (he resigned from his position as an assistant professor), which he and society had groomed him for. His frustration is expressed in these words from his manifesto:

> Modern man is strapped down by a network of rules and regulations. . . . Most of these regulations cannot be disposed with, because they are necessary for the functioning of industrial society. When one does not have adequate opportunity . . . the consequences are boredom, demoralization, low self-esteem, inferiority feelings, defeatism, depression, anxiety, guilt, frustration, hostility, spouse or child abuse, insatiable hedonism, abnormal sexual behavior, sleep disorders, eating disorders, etc. [The rules of industrial society] have made life unfulfilling, have subjected human beings to indignities, have led to widespread psychological suffering. By "feelings of inferiority" we mean not only inferiority feelings in the strictest sense but a whole spectrum of related traits:

> low self-esteem, feelings of powerlessness, depressive tendencies, defeatism, guilt, self-hatred, etc.

Kaczynski suffered these indignities, which he blamed on society, and escaped to the hills, where he perceived he could enjoy more freedoms. In Montana he built a cabin without running water or electricity. Here he lived a fairly self-sustained life—away from the rules and the reach of technological civilization. (But just as Thoreau did at Walden, he came into town to restock his supplies.) However, his escape from technology was disturbed around 1983. One of the wilderness oases Kaczynski loved to visit was what he describes as a "plateau that dated from the Tertiary Age," a two-day hike from his cabin. The spot was sort of a secret retreat for him. As Kaczynski later told a reporter from *Earth First! Journal*, "It's kind of rolling country, not flat, and when you get to the edge of it you find these ravines that cut very steeply into cliff-like drop-offs. There was even a waterfall there." The area around his own cabin was getting too much traffic from hikers and hunters, so in the summer of 1983 he retreated to his secret spot on the plateau. As he told another interviewer later in prison,

> When I got there I found they had put a road right through the middle of it. [His voice trails off; he pauses, then continues.] You just can't imagine how upset I was. It was from that point on I decided that, rather than trying to acquire further wilderness skills, I would work on getting back at the system. Revenge. That wasn't the first time I ever did any monkey wrenching, but at that point, that sort of thing became a priority for me.

It is easy to sympathize with Kaczynski's plight as a dissenter. You politely try to escape the squeeze of technological civilization by retreating to its furthest reaches, where you establish a relatively techno-free lifestyle—and then the beast of civilization/development/industrial technology stalks you and destroys your paradise. Is there no escape? The machine is ubiquitous! It is relentless! It must be stopped!

Ted Kaczynski, of course, is not the only wilderness lover to suffer the encroachment of civilization. Entire tribes of indigenous Americans were driven to remote areas by the advance of European culture. They were not running from technology per se (they happily picked up the latest guns when they could), but the effect was the same—to distance themselves from industrial society.

Kaczynski argues that it is impossible to escape the ratcheting clutches of industrial technology for several reasons: one, because if you use *any* part of the technium, the system demands servitude; two, because technology does not "reverse" itself, never releasing what is in its hold; and three, because we don't have a choice of what technology to use in the long run. In his words, from the manifesto:

> The system HAS TO regulate human behavior closely in order to function. At work, people have to do what they are told to do, otherwise production would be thrown into chaos. Bureaucracies HAVE TO be run according to rigid rules. To allow any substantial personal discretion to lower-level bureaucrats would disrupt the system and lead to charges of unfairness due to differences in the way individual bureaucrats exercised their discretion. It is true that some restrictions on our freedom could be eliminated, but GENERALLY SPEAKING the regulation of our lives by large organizations is necessary for the functioning of industrial-technological society. The result is a sense of powerlessness on the part of the average person.
>
> Another reason why technology is such a powerful social force is that, within the context of a given society, technological progress marches in only one direction; it can never be reversed. Once a technical innovation has been introduced, people usually become dependent on it, unless it is replaced by some still more advanced innovation. Not only do people become dependent as individuals on a new item of technology, but, even more, the system as a whole becomes dependent on it.

> When a new item of technology is introduced as an option that an individual can accept or not as he chooses, it does not necessarily REMAIN optional. In many cases the new technology changes society in such a way that people eventually find themselves FORCED to use it.

Kaczynski felt so strongly about the last point that he repeated it once more in a different section of his treatise. It is an important criticism. Once you accept the fact that individuals surrender freedom and dignity to "the machine" and that they increasingly have no choice but to do so, then the rest of Kaczynski's argument flows fairly logically:

> But we are suggesting neither that the human race would voluntarily turn power over to the machines nor that the machines would willfully seize power. What we do suggest is that the human race might easily permit itself to drift into a position of such dependence on the machines that it would have no practical choice but to accept all of the machines decisions. As society and the problems that face it become more and more complex and machines become more and more intelligent, people will let machines make more of their decision for them, simply because machine-made decisions will bring better result than man-made ones. Eventually a stage may be reached at which the decisions necessary to keep the system running will be so complex that human beings will be incapable of making them intelligently. At that stage the machines will be in effective control. People won't be able to just turn the machines off, because they will be so dependent on them that turning them off would amount to suicide. . . . Technology will eventually acquire something approaching complete control over human behavior.
>
> Will public resistance prevent the introduction of technological control of human behavior? It certainly would if an attempt were made to introduce such control all at once. But

> since technological control will be introduced through a long sequence of small advances, there will be no rational and effective public resistance.

I find it hard to argue against this last section. It is true that as the complexity of our built world increases we will necessarily need to rely on mechanical (computerized) means to manage this complexity. We already do. Autopilots fly our very complex flying machines. Algorithms control our very complex communications and electrical grids. And for better or worse, computers control our very complex economy. Certainly as we construct yet more complex infrastructure (location-based mobile communications, genetic engineering, fusion generators, autopiloted cars) we will rely further on machines to run it and make decisions. For those services, turning off the switch is not an option. In fact, if we wanted to turn off the internet right now, it would not be easy to do, particularly if others wanted to keep it on. In many ways the internet is designed to never turn off. Ever.

Finally, if the triumph of a technological takeover is the disaster that Kaczynski outlines—robbing souls of freedom, initiative, and sanity and robbing the environment of its sustainability—and if this prison is inescapable, then the system must be destroyed. Not reformed, because that will merely extend it, but eliminated. From his manifesto:

> Until the industrial system has been thoroughly wrecked, the destruction of that system must be the revolutionaries' ONLY goal. Other goals would distract attention and energy from the main goal. More importantly, if the revolutionaries permit themselves to have any other goal than the destruction of technology, they will be tempted to use technology as a tool for reaching that other goal. If they give in to that temptation, they will fall right back into the technological trap, because modern technology is a unified, tightly organized system, so that, in order to retain SOME technology, one finds oneself obliged to retain MOST technology, hence one ends up sacrificing only token amounts of technology.

> Success can be hoped for only by fighting the technological system as a whole; but that is revolution not reform. . . . While the industrial system is sick we must destroy it. If we compromise with it and let it recover from its sickness, it will eventually wipe out all of our freedom.

For these reasons Ted Kaczynski went to the mountains to escape the clutches of civilization and then later to plot his destruction of it. His plan was to make his own tools (anything he could hand fashion) while avoiding technology (stuff it takes a system to make). His small one-room shed was so well constructed that the feds later moved it off his property as a single intact unit, like a piece of plastic, and put it in storage (it now sits reconstructed in the Newseum in Washington, D.C.). His place was way off the road; he used a mountain bike to get into town. He dried hunted meat in his tiny attic and spent his evenings in the yellow light of a kerosene lamp crafting intricate bomb mechanisms. The bombs were strikes at the professionals running the civilization he hated. While his bombs were deadly, they were ineffective in achieving his goal, because no one knew what their purpose was. He needed a billboard to announce why civilization needed to be destroyed. He needed a manifesto published in the major papers and magazines of the world. Once they read it, a special few would see how imprisoned they were and would join his cause. Perhaps others would also start bombing the choke points in civilization. Then his imaginary Freedom Club ("FC" is how he signed his manifesto, written with the plural "we") would be a club of more than himself.

The attacks on civilization did not materialize in bulk once his manifesto was published (although it did help authorities arrest him). Occasionally an Earth Firster would burn a building in an encroaching development or pour sugar into a bulldozer's gas tank. During the otherwise peaceful protests against the G7, some anticivilization anarchists (who call themselves anarcho-primitivists) broke fast-food storefront windows and smashed property. But the mass assault on civilization never happened.

The problem is that Kaczynski's most basic premise, the first axiom in his argument, is not true. The Unabomber claims that technology robs people of freedom. But most people of the world find the opposite. They gravitate toward technology because they recognize that they have more freedoms when they are empowered with it. They (that is, we) realistically weigh the fact that yes, indeed, some options are closed off when adopting new technology, but many others are opened, so that the net gain is an increase in freedom, choices, and possibilities.

Consider Kaczynski himself. For 25 years he lived in a type of self-enforced solitary confinement in a dirty, smoky shack without electricity, running water, or a toilet. He cut a hole in the floor for late-night peeing. In terms of material standards, the cell he now occupies in the Colorado supermax prison is a four-star upgrade: His new place is larger, cleaner, and warmer, with the running water, electricity, and the toilet he did not have, plus free food and a much better library. In his Montana hermitage he was free to move about as much as the snow and weather permitted him. He could freely choose among a limited set of choices of what to do in the evenings. He may have personally been

Inside the Unabomber's Shack. Ted Kaczynski's library and workbench where he made bombs.

content with his limited world, but overall his choices were very constrained, although he had unshackled freedom within those limited choices—sort of like, "You are free to hoe the potatoes any hour of the day you want." Kaczynski confused latitude with freedom. He enjoyed great liberty within limited choices, but he erroneously believed this parochial freedom was superior to an expanding number of alternative choices that may offer less latitude within each choice. An exploding circle of choices encompasses much more actual freedom than simply increasing the latitude within limited choices.

I can only compare his constraints in his cabin to mine, or perhaps anyone else's reading this today. I am plugged into the belly of the machine. Yet technology allows me to work at home, so I hike in the mountains, where cougars and coyotes roam, most afternoons. I can hear a mathematician give a talk on the latest theory of numbers one day and the next day be lost in the wilderness of Death Valley with as little survivor gear as possible. My choices in how I spend my day are vast. They are not infinite, and some options are not available, but in comparison to the degree of choices and freedoms available to Ted Kaczynski in his shack, my freedoms are overwhelmingly greater.

This is the chief reason billions of people migrate from mountain shacks—very much like Kaczynski's—all around the world. A smart kid living in a smoky one-room hut in the hills of Laos or Cameroon or Bolivia will do all he can to make his way against all odds to the city, where there are—so obvious to the migrant—vastly more freedom and choices. He would find Kaczynski's argument that there is more freedom back in the stifling prison he just escaped from plain crazy.

The young are not under some kind of technological spell that warps their minds into believing civilization is better. Sitting in the mountains, they are under no spell but poverty's. They clearly know what they give up when they leave. They understand the comfort and support of family, the priceless value of community acquired in a small village, the blessings of clean air, and the soothing wholeness of the natural world. They feel the loss of immediate access to these, but they leave their shacks anyway because in the end, the tally favors the freedoms created by civilization. They can (and will) return to the hills to be rejuvenated.

My family doesn't have TV, and while we have a car, I have plenty of city friends who do not. Avoiding particular technologies is certainly possible. The Amish do it well. Many individuals do it well. However, the Unabomber is right that choices that begin as optional can over time become less so. First, there are certain technologies (say, sewage treatment, vaccinations, traffic lights) that were once matters of choice but that are now mandated and enforced by the system. Then there are other systematic technologies, such as automobiles, that are self-reinforcing. The success and ease of cars shift money away from public transport, making it less desirable and encouraging the purchase of a car. Thousands of other technologies follow the same dynamic: The more people who participate, the more essential it becomes. Living without these embedded technologies requires more effort, or at least more deliberate alternatives. This web of self-reinforcing technologies would be a type of noose if the total gains in choices, possibilities, and freedoms brought about by them did not exceed the losses.

Anticivilizationists would argue that we embrace more because we are brainwashed by the system itself and we have no choice but to say yes to more. We can't, say, resist more than a few individual technologies, so we are imprisoned in this elaborate artificial lie.

It is possible that the technium has brainwashed us all, except for a few clear-eyed anarcho-primitivists who like to blow up stuff. I would be inclined to believe in breaking this spell if the Unabomber's alternative to civilization was more clear. After we destroy civilization, then what?

I've been reading the literature of the anticivilization collapsitarians to find out what they have in mind after the collapse of the technium. Anticivilization dreamers spend a lot of time devising ways to bring down civilization (befriend hackers, unbolt power towers, blow up dams) but not so much on what replaces it. They do have a notion of what the world looked like before civilization. According to them it looks like this (from the *Green Anarchy Primer*):

> Prior to civilization there generally existed ample leisure time, considerable gender autonomy and equality, a non-destructive approach to the natural world, the absence of

organized violence, no mediating or formal institutions, and strong health and robusticity.

Then came civilization and all the ills (literally) of the Earth:

> Civilization inaugurated warfare, the subjugation of women, population growth, drudge work, concepts of property, entrenched hierarchies, and virtually every known disease, to name a few of its devastating derivatives.

Among the green anarchists there's talk of recovering your soul, of making fire by rubbing sticks together, discussions of whether vegetarianism is a good idea for hunters, but there is no outline of how groups of people go beyond survival mode, or whether they do. We are supposed to aim for "rewilding" but the rewilders are shy about describing what life is like in this rewild state. One prolific green anarchy author whom I spoke to, Derrick Jensen, dismisses the lack of alternatives to civilization and told me simply, "I do not provide alternatives because there is no need. The alternatives already exist, and they have existed—and worked—for thousands and tens of thousands of years." He means, of course, tribal life, but not modern tribal; he means tribal as in no agriculture, no antibiotics, no nothing beyond wood, fur, and stone.

The great difficulty of the anticivilizationists is that a sustainable, desirable alternative to civilization is unimaginable. We cannot picture it. We cannot see how it would be a place we'd like to move to. We can't imagine how this primitive arrangement of stone and fur would satisfy each of our individual talents. And because we cannot imagine it, it will never happen, because nothing has ever been created without being imagined first.

Despite their inability to imagine a desirable, coherent alternative, the anarcho-primitivists all agree that some combination of being in tune with nature, eating low-calorie diets, owning very little, and using only things you make yourself will bring on a level of contentment, happiness, and meaning we have not seen for 10,000 years.

But if this state of happy poverty is so desirable and good for the soul, why do none of the anticivilizationists live like this? As far as I can tell from my research and personal interviews with them, all self-identifying anarcho-primitivists live in modernity. They are living in the trap identified by the Unabomber. They compose their rants against the machine on very fast desktop machines. While they sip coffee. Their routines are only marginally different from mine. They have not relinquished the conveniences of civilization for the better shores of nomadic hunter-gathering.

Except, perhaps, one purist: the Unabomber. Kaczynski went further than other critics in living the story he believed in. At first glance his story seems promising, but on second look, it collapses into the familiar conclusion: He was living off the fat of civilization. The Unabomber's shack was crammed with stuff he purchased from the machine: snowshoes, boots, sweatshirts, food, explosives, mattresses, plastic jugs and buckets, etc.—all things that he could have made himself but did not. After 25 years on the job, why did he not make his own tools separate from the system? Based on photographs of his cabin's untidy interior, it looks like he shopped at Wal-Mart. The food he scavenged from the wild was minimal. Instead he regularly rode his bike to town and there rented an old car to drive to the big city to restock his food and supplies from supermarkets. He was unwilling to support himself without civilization.

Besides the lack of a desirable alternative, the final problem with destroying civilization as we know it is that the alternative, such as it has been imagined by the self-described "haters of civilization," would not support but a fraction of the people alive today. In other words, the collapse of civilization would kill billions. Ironically, the poorest rural inhabitants would fare the best, as they could retreat to hunting and gathering with the least trouble, but billions of urbanites would die within months or even weeks, once food ran out and disease took over. The anarcho-primitives are rather sanguine about this catastrophe, arguing that accelerating the collapse early might save lives in total.

Again the exception seems to be Ted Kaczynski, who reckoned with the die-off with very clear eyes in a postarrest interview:

> For those who realize the need to do away with the techno-industrial system, if you work for its collapse, in effect you are killing a lot of people. If it collapses, there is going to be social disorder, there is going to be starvation, there aren't going to be any more spare parts or fuel for farm equipment, there won't be any more pesticide or fertilizer on which modern agriculture is dependent. So there isn't going to be enough food to go around, so then what happens? This is something that, as far as I've read, I haven't seen any radicals facing up to.

Presumably Kaczynski personally "faced up to" the logical conclusion of taking down civilization; it would kill billions of people. He must have decided that murdering a few more people up front in the process would not matter. After all, the techno-industrial complex had snuffed out the humanity from him, so if he had to snuff out a few dozen humans on the way to snuffing out the system that enslaves billions, that would be worth it. The death of billions would also be justified because all those unfortunate people in the grasp of technology were now soulless, like he was. Once civilization was gone, the next generation would be really free. They would all be in his Freedom Club.

The ultimate problem is that the paradise that Kaczynski is offering, the solution to civilization, so to speak, the alternative to the emerging autonomous technium, is the tiny, smoky, dingy, smelly wooden shack that absolutely nobody else wants to dwell in. It is a "paradise" billions are fleeing from. Civilization has its problems, but in almost every way it is better than the Unabomber's shack.

The Unabomber is right that technology is a holistic, self-perpetuating machine. He is also right that the selfish nature of this system causes specific harms. Certain aspects of the technium are detrimental to the human self, because they defuse our identity. The technium also contains power to harm itself; because it is no longer regulated by either nature or humans, it could accelerate so fast as to extinguish itself. Finally, the technium can harm nature if not redirected.

But despite the reality of technology's faults, the Unabomber is wrong to want to exterminate it, for many reasons, not the least of which is that the machine of civilization offers us more actual freedoms than the alternative. There is a cost to running this machine, a cost we are only beginning to reckon with, but so far the gains from this ever-enlarging technium outweigh the alternative of no machine at all.

A lot of people don't believe this. Not for a second. I know from many conversations that a certain percentage of readers of this book will reject this conclusion and side with Kaczynski. My arguments that the positive aspects of technology slightly exceed the negative don't persuade them.

Instead they believe—very strongly—that the expanding technium robs us of our humanity, and it steals our children's future. Therefore, the so-called benefits of technology that I outline in these chapters must be an illusion, a sleight-of-hand trick we perform upon ourselves to permit our addiction to the new.

They point to the vices that I cannot deny. We seem to be less content, less wise, less happy the "more" we have. They rightly point out that this unease is captured in many polls and surveys. The most cynical believe that progress simply extends our lives so that we can be unsatisfied for decades longer. Some year in the future, science will enable us to live forever, so we'll be unhappy forever.

My question is this: If technology is so rotten, why do we keep grabbing it, even after Ted Kaczynski has exposed its true nature? Why do really smart, committed ecowarriors not give it all up, as the Unabomber tried to do?

One theory: The technium's rampant materialism outlaws greater meaning in life by focusing our spirits on stuff. In a blind fury to find some kind of meaning in life, we consume technology madly, energetically, ceaselessly, obsessively buying the only answer that seems for sale—more technology. We end up needing more and more technology to feel less and less satisfied. "Needing more to be satisfied less" is one definition of addiction. According to this logic, technology is, therefore, an addiction. Instead of a compulsive obsession with television or the internet or texting, we have a compulsive obsession with the technium as a whole. Perhaps we are addicted to the dopamine rush of the new.

That might explain why even those who intellectually despise technology still buy stuff. In other words, we are aware of how bad it is for us, and even of how it enslaves us (we scanned the Unabomber's tract), yet we continue to amass vast doses of gadgets and things (perhaps guiltily) because we can't help it. We are powerless to resist technology.

If that was true, the remedy is a bit unsettling. All addictions are fixed by effecting change not in the offending pleasure but in the person addicted. Whether it is via a 12-step program or medication, the problem is resolved in the heads of the addicted. In the end they are liberated not by changing the nature of television, the internet, gambling machines, or alcohol but by changing their relation to it. Those who overcome addictions do so by assuming power over their powerlessness. If the technium is an addiction, we can't solve this addiction by trying to change the technium.

A variant of this explanation is that we are addicted but unaware of our addiction. We are bewitched. Hypnotized by glitter. Technology, by some black magic, has impaired our discernment. In this account the technology of media disguises the true colors of the technium behind the front of utopia. Its shiny new benefits instantly blind us to its powerful new vices. We operate under some kind of spell.

But this global spell must be a consensual hallucination, because we all want the same new stuff: the best medicines, the coolest vehicle, the smallest cell phone. It must be a most powerful spell, because it affects all members of our species without regard to race, age, geography, or wealth. This means that everyone reading this text is under this hex. The hip college-campus theory is that we are duped and cast under this curse by corporations peddling technology and presumably by the executives running corporations. But that would mean that the CEOs are aware of, or above, the hoax themselves. In my experience, they are in the same boat as the rest of us. Believe me, having consulted with many of them, I know they are not capable of such a conspiracy.

The unhip theory is that technology is duping us itself of its own accord. It uses technological media to brainwash us into thinking that it is wholly benevolent and then removes its downsides from our minds. As one who believes the technium has its own agenda, I find this theory plausible. Its anthropomorphism doesn't bother me at all. But by this logic we should expect the least technologically cultured people to be the least duped and to be the most aware of the plainly visible dangers. They should be like the children who see the emperor without clothes. Or with wolf's clothing. But in fact, those disenfranchised people not under media's spell are often the most eager to trade in the old for the new. They look the juggernaut of the technium in the eye and say to it: Give me it all, right now. Or if they consider themselves wise, they say: Give me only your good stuff, none of that addictive crap.

On the other hand, it is often the most technologically mediated people, those experts driving Priuses, blogging, and twittering, who "see" or believe in the presence of the technium's spell. This reversal does not add up for me.

That leaves one remaining theory: We willingly choose technology, with its great defects and obvious detriments, because we unconsciously calculate its virtues. In an entirely wordless calculus, we note the addictions in others, the degradations in the environment, the distractions in our own lives, the confusion about character that various technologies generate, and then we sum these up against the benefits. I don't believe this is a wholly rational procedure; I think we also tell each other stories about technology, and these are added in with as much weight as the pluses and minuses. But in a real way we do a risk-benefit analysis. Even the most primitive shaman trying to decide whether to trade a wild skin for a machete will make such a calculation. He's seen what happens when others get a steel blade. We do the same with unknown technologies, too, just not as well. And most of the time, after we've weighed downsides and upsides in the balance of our experience, we find that technology offers a greater benefit, but not by much. In other words, we freely choose to embrace it—and pay the price.

But as irrational humans we sometimes don't make the best possible choice for several reasons. The costs of technology are not easily visible, and the expectations of virtue often hyped. To improve our chances of making better decisions, we need—I almost hate to say it—more technology. The way to reveal the full costs of technology and deflate its hype is with better information tools and processes. We require technologies such as real-time self-monitoring of our use, transparent sharing of problems, deep analysis of testing results, relentless retesting, accurate recording of the chain of sources in manufacturing, and honest accounting of negative externalities such as pollution. Technology can help us reveal the costs of technology and help us make better choices about how we adopt it.

Better technological tools for illuminating the downsides of technology would, paradoxically, boost the reputation of technology. They would bring the calculation out of the unconscious and rationalize it. With proper tools, the trade-offs could be brought into science.

Finally, a true articulation of each particular technology's vices will allow us to see that our embrace of the technium is done willingly, and is neither an addiction nor a spell.

11

Lessons of Amish Hackers

In any discussion about the merits of avoiding the addictive grip of technology, the Amish stand out as offering an honorable alternative. The Amish have the reputation of being Luddites, people who refuse to employ faddish new technology. It's well known that the strictest of them don't use electricity or automobiles, but rather farm with manual tools and drive a horse and buggy. They favor technology they can either build or repair themselves, and they are, on the whole, thrifty and relatively self-reliant. They work outside in the fresh air with their hands, which endears them to the average Dilbert working inside at a computer screen in a cubicle. Plus, their minimal lifestyle is prospering (Amish population grows at 4 percent annually) while middle-class white-collar and factory workers are increasingly unemployed and withering.

The Unabomber was not Amish, and the Amish are no collapsitarians. They have created a civilization of sorts that seems to offer valuable lessons on how to balance the blessings and ills of technology.

Yet Amish lives are anything but antitechnological. In fact, on my several visits with them, I have found them to be ingenious hackers and tinkerers, the ultimate makers and do-it-yourselfers. They are often, surprisingly, protechnology.

First, a few caveats. The Amish are not a monolithic group. Their practices vary parish by parish. What one group does in Ohio another church in New York may not do or a parish in Iowa may do more so. Also, their relationship to technology is uneven. Most Amish use a mixture of old and very new stuff, like the rest of us. It's important to note that Amish practices are ultimately driven by religious belief: The technological consequences are secondary. They often don't have logical reasons for their policies. Last, Amish practices change over time and are, at this moment, adapting to the world by embracing new technologies at their own rate. In many ways the view of the Amish as old-fashioned Luddites is an urban myth.

Like all legends, the Amish myth is based on some facts. The Amish, particularly the Old Order Amish—the stereotypical Amish on postcards—really *are* reluctant to adopt new things. In contemporary society our default is set to say yes to new things, and in Old Order Amish communities the default is set to "not yet." When new things come around, the Old Order Amish automatically react by ignoring them. Thus, many Old Order Amish never said yes when automobiles were new. Instead, they travel around in a buggy hauled by a horse, as they always have. Some orders require the buggy to be an open carriage (so riders—teenagers, say—are not tempted by a private place to fool around); others will permit closed carriages. Some orders allow tractors on the farm, if the tractors have steel wheels; that way a tractor can't be "cheated" to drive on the road like a car. Some groups allow farmers to power their combines or threshers with diesel engines, as long as the engine only spins the threshers and does not propel the vehicle—which means the whole smoking, noisy contraption is pulled by horses. Some sects allow cars, but only if they are painted entirely black (no chrome) to ease the temptation to upgrade to the latest model.

Behind all of these variations is the Amish motivation to strengthen their communities. When cars first appeared at the turn of the last century, the Amish noticed that drivers would leave the community to go picnicking or sightseeing in other towns, instead of visiting family or the sick on Sundays or patronizing local shops on Saturday. Therefore, the ban on unbridled mobility was intended to make it hard to travel far and to keep energy focused in the local community. Some parishes did this with more strictness than others.

A similar communal motivation lies behind the Old Order Amish practice of living without electricity. The Amish noticed that when their homes were electrified with wires from a generator in town, they became more tied to the rhythms, policies, and concerns of the town. Amish religious belief is founded on the principle that they should remain "in the world, not of it" and so should remain separate in as many ways as possible. Being tied to electricity tied them into the world, so they forfeited electrical benefits in order to stay outside the world. Visiting many Amish households even today, you'll see no power lines weaving toward their homes. They live off the grid. To live without electricity or cars eliminates most of what we expect from modernity. No electricity means no internet, TV, or phones, either, so suddenly the Amish life stands in stark contrast to our complex modern lives.

But when you visit an Amish farm, that simplicity vanishes. Indeed, the simplicity vanishes even before you get to the farm. Cruising down the road you may see an Amish kid in a straw hat and suspenders zipping by on Rollerblades. In front of one schoolhouse I spied a flock of parked push-scooters, which is how the kids had arrived there. But on the same street a constant stream of grimy minivans paraded past the school. Each was packed with full-bearded Amish men sitting in the back. What was that about?

Turns out the Amish make a distinction between using something and owning it. The Old Order won't own a pickup truck, but they will ride in one. They won't get a license, purchase an automobile, pay insurance, and become dependent on the automobile and the industrial-car complex, but they will call a taxi. Since there are more Amish men than farms, many men work at small factories, and these guys will hire vans driven by outsiders to take them to and from work. So even the horse-and-buggy folk will use cars—on their own terms. (Very thrifty, too.)

The Amish also make a distinction between technology they have at work and technology they have at home. I remember an early visit to an Amish man who ran a woodworking shop near Lancaster, Pennsylvania. Let's call him Amos, although Amos was not his real name: The Amish prefer not to call attention to themselves, thus their reluctance to be photographed or have their names in the press. I followed Amos into a grubby concrete building. Most of the interior was dimly lit naturally from windows, but hanging over the wooden meeting table in a very cluttered room was a single electrical lightbulb. The host saw me staring at it, and when I looked at him, he just shrugged and said that it was for the benefit of visitors like me.

While the rest of his large workshop lacked electricity beyond that naked bulb, it did not lack power machines. The place was vibrating with an ear-cracking racket of power sanders, power saws, power planers, power drills, and so on. Everywhere I turned there were bearded men covered in sawdust pushing wood through screaming machines. This was not a circle of Renaissance craftsman hand-tooling masterpieces. This was a small-time factory cranking out wooden furniture with machine power. But where was the power coming from? Not from windmills.

Amos took me around to the back where a huge SUV-sized diesel generator sat. It was massive. In addition to a gas engine there was a very large tank, which, I learned, stored compressed air. The diesel engine burned petroleum fuel to drive the compressor that filled the reservoir with pressure. From the tank, a series of high-pressure pipes snaked off toward every corner of the factory. A hard rubber flexible hose connected each tool to a pipe. The entire shop ran on compressed air. Every piece of machinery was running on pneumatic power. Amos even showed me a pneumatic switch, which he could flick like a light switch to turn on some paint-drying fans running on air.

The Amish call this pneumatic system "Amish electricity." At first, pneumatics were devised for Amish workshops, but air power was seen as so useful that it migrated to Amish households. In fact, there is an entire cottage industry in retrofitting tools and appliances to run on Amish electricity. The retrofitters buy a heavy-duty blender, say, and yank out the electrical motor. They then substitute an air-powered motor of appropriate size, add pneumatic connectors, and bingo, your Amish mom now has a blender in her electricity-less kitchen. You can get a pneumatic sewing machine and a pneumatic washer/dryer (with propane heat). In a display of pure steam-punk (air-punk?) nerdiness, Amish hackers try to outdo one another in building pneumatic versions of electrified contraptions. Their mechanical skill is quite impressive, particularly since none went to school beyond the eighth grade. They love to show off their geekiest hacks. And every tinkerer I met claimed that pneumatics were superior to electrical devices because air was more powerful and durable, outlasting motors that burned out after a few years of hard labor. I don't know if this claim of superiority is true or merely a justification, but it was a constant refrain.

I visited one retrofitted workshop run by a strict Mennonite. Marlin was a short, beardless man (no beards for the Mennonites). He used a horse and buggy and had no phone, but electricity ran in the shop behind his home. They used electricity to make pneumatic parts. As was the case in most of his community, his kids worked alongside him. A few of his boys, in Plain Folk clothes, used a propane-powered forklift with metal wheels (no rubber so you can't drive it on the road) to cart around stacks of heavy metal as they manufactured very precise milled metal parts for pneumatic motors and for kerosene cooking stoves, an Amish favorite. The tolerances needed are a thousandth of an inch. So a few years ago they installed a $400,000 computer-controlled milling machine in his backyard, behind the horse stable. This massive tool was about the size of a delivery truck. It was operated by Marlin's 14-year-old daughter, in a bonnet and long dress. With this computer-controlled machine she made parts for grid-free horse-and-buggy living.

I say "grid-free" rather than "electricity-free" because I kept finding electricity in Amish homes. Once you have a huge diesel generator running behind your barn to power the refrigeration units that store the milk (the main cash crop for the Amish), it's a small thing to stick on a small electrical generator. For recharging batteries, say. You can find battery-powered calculators, flashlights, and electric fences and generator-powered electric welders on Amish farms. The Amish also use batteries to run a radio or phone (outside in the barn or shop), or to power the required headlights and turn signals on their horse buggies. One clever Amish fellow spent a half hour explaining to me the ingenious way he had hacked up a mechanism to make a buggy turn signal automatically shut off when the turn was finished, just as it does in your car.

Nowadays solar panels are becoming popular among the Amish. With these they can get electricity without being tied to the grid, which was their main worry. Solar is used primarily for utilitarian chores like pumping water, but it will slowly leak into the household. As do most innovations.

The Amish use disposable diapers (why not?), chemical fertilizers, and pesticides, and they are big boosters of genetically modified corn. In Europe this corn is called Frankenfood. I asked a few of the Amish elders about that last one. Why do they plant GMOs? Well, they reply, corn is susceptible to the corn borer, which nibbles away at the bottom of the stem and occasionally topples the stalk. Modern 500-horsepower harvesters don't notice this fall; they just suck up all the material and spit out the corn into a bin. The Amish harvest their corn semimanually. It's cut by a chopper device and then pitched into a thresher. But if there are a lot of stalks that are broken, they have to be pitched by hand. That is a lot of very hard, sweaty work. So they plant Bt corn. This genetic mutant carries the genes of the corn borer's enemy, *Bacillus thuringiensis,* which produces a toxin deadly to the corn borer. Fewer stalks are broken and the harvest can be aided with machines, so yields are up. One elder Amish man whose sons run his farm said he was too old to be pitching heavy, broken cornstalks, and he told his sons that he'd only help them with the harvest if they planted Bt corn. The alternative was to purchase expensive, modern harvesting equipment, which none of them wanted. So the technology of genetically modified crops allowed the Amish to continue using old, well-proven, debt-free equipment, which accomplished their main goal of keeping the family farm together. They did not use these words, but they made it clear that they considered genetically modified crops appropriate technology for family farms.

Artificial insemination, solar power, and the web are technologies that Amish are still debating. They use the web at libraries (using but not owning). In fact, from cubicles in public libraries Amish sometimes set up a website for their business. So while an "Amish website" sounds like the punch line to a joke, there are actually quite a few of them. What about postmodern innovations like credit cards? A few Amish did get them, presumably for their businesses at first. But over time local Amish bishops noticed problems of overspending and the resultant crippling interest rates. Farmers got into debt, which impacted not only them but also their community, since their families had to help them recover (that's what community and families are for). So after a trial period, the elders ruled against credit cards.

One Amish man told me that the problem with phones, pagers, BlackBerrys, and iPhones (yes, he knew about them) was that "you got messages rather than conversations." That's about as accurate a summary of our times as any. Henry, his long white beard contrasting with his young bright eyes, told me, "If I had a TV, I'd watch it." What could be simpler?

No looming decision is riveting the Amish themselves as much as the question of whether they should accept cell phones. Previously, the Amish would build a shanty at the end of their driveway that housed an answering machine and phone to be shared by neighbors. The shanty sheltered the caller from rain and cold and kept the grid away from the house, and the long walk outside reduced phone use to essential calls rather than gossip and chatting. Cell phones are a new twist. You get a phone without wires, off the grid. As one Amish guy told me, "What is the difference if I stand in my phone booth with a wireless phone or stand outside with a cell phone? There's no difference." Further, cell phones have been embraced by women, who can keep in touch with their far-flung families, since they don't drive. And the bishops have noticed that the cell phone is so small it can be kept hidden, which is a concern for a people dedicated to discouraging individualism. The Amish have still not decided on the cell phone. Or perhaps it is more accurate to say they have decided "maybe."

For people who live off the grid, without TV, internet, or books beyond one Bible, the Amish are perplexingly well informed. There's not much I could tell them that they didn't know about and already have an opinion on. And surprisingly, there's not much new that at least one person in their church has not tried to use. In fact, the Amish rely on the enthusiasm of those early adopters to try stuff out until it proves harmful.

The typical adoption pattern for a new technology goes like this: Ivan is an Amish alpha geek. He is always the first to try a new gadget or technique. He gets in his head that the new flowbitzmodulator would be really useful. He comes up with a justification of how it fits into the Amish orientation. So he goes to his bishop with this proposal: "I'd like to try this out." The bishop says to Ivan, "Okay, Ivan, do whatever you want with this. But you have to be ready to give it up if we decide it is not helping you or is hurting others." So Ivan acquires the tech and ramps it up, while his neighbors, family, and bishops watch intently. They weigh the benefits and drawbacks. What is it doing to the community? To Ivan? Cell-phone use among the Amish began that way. According to anecdote, the first Amish alpha geeks to request permission to use cell phones were two ministers who were also contractors. The bishops were reluctant to give permission but suggested a compromise: Keep the cell phones in the vans of the drivers. The van would be a mobile phone shanty. Then the community would watch the contractors. It seemed to work, so other early adopters picked it up. But still, at any time, even years later, the bishops can say no.

I visited a shop that built the Amish's famous buggies. From the outside, the carts look simple and old-fashioned. But when I inspected the process in the shop, I could see that they are quite high-tech and surprisingly complicated rigs. Made of lightweight fiberglass, they are hand cast and outfitted with stainless steel hardware and cool LED lights. The owner's teenage son, David, also worked at the shop. Like a lot of Amish, who work alongside their parents from an early age, he was incredibly poised and mature. I asked him what he thought the Amish would do about cell phones. He snuck his hand into his overalls and pulled one out. "They'll probably accept them," he said and smiled. He then quickly added that he worked for the local volunteer fire department, which was why he had one. (Sure!) But, his dad chimed in, if cell phones are accepted, "there won't be wires running down the street to our homes."

In pursuit of their goal to remain off the grid yet modernize, some Amish have installed inverters on their diesel generators linked to batteries to provide them with 110 off-grid volts. They power specialty appliances at first, such as an electric coffeepot. I saw one home with an electric copier in the home office part of the living room. Will the slow acceptance of modern appliances creep along until, 100 years hence, the Amish have what we have now (but will by then have left behind)? What about cars? Will the Old Order ever drive old-fashioned internal combustion clunkers, say, when the rest of the world is using personal jet packs? Or will they embrace electric cars? I asked David, the 18-year-old Amish, what he expects to use in the future. Much to my surprise, he had a ready teenage answer. "If the bishops allow the church to leave behind buggies, I know exactly what I will get: a black Ford 460 V8." That's a 500-horsepower muscle car. Some Mennonite orders permit generic cars if they are black—no chrome or fanciness. So a black hot rod is okay! His dad, the carriage maker, again chimed in, "Even if that happens, there will always be some horse-and-carriage Amish."

David then admitted, "When I was deciding whether to join the church or not, I thought of my future children and whether they would be brought up without restrictions. I could not imagine it." A common phrase among the Amish is "holding the line." They all recognize the line keeps moving, but a line must remain.

The book *Living Without Electricity* charts out how many years later the Amish adopt a technology after it has been adopted by the rest of America. My impression is that the Amish are living about 50 years behind us. Half of the inventions they use now were invented within the last 100 years. They don't adopt everything new, but when they do embrace it, it's half a century after everyone else does. By that time, the benefits and costs are clear, the technology stable, and it is cheap. The Amish are steadily adopting technology—at their pace. They are slow geeks. As one Amish man said, "We don't want to stop progress, we just want to slow it down." But their manner of slow adoption is instructive:

1. They are selective. They know how to say no and are not afraid to refuse new things. They ignore more than they adopt.

2. They evaluate new things by experience instead of by theory. They let the early adopters get their jollies by pioneering new stuff under watchful eyes.

3. They have criteria by which to make choices: Technologies must enhance family and community and distance themselves from the outside world.

4. The choices are not individual but communal. The community shapes and enforces technological direction.

This method works for the Amish, but can it work for the rest of us? I don't know. It has not really been tried yet anywhere else. And if the Amish hackers and early adopters teach us anything, it's that you have to try things first. Their motto is "try first and relinquish later, if need be." We are good at trying first, not good at relinquishing. To fulfill the Amish model we'd have to get better at relinquishing as a group—which is very difficult for a pluralistic society. Social relinquishing relies on mutual support. I have not seen any evidence of that happening outside of Amish communities, but it would be a telling sign if it did appear.

The Amish have become very good at managing technologies. But what do they gain by this discipline? Are their lives really any better for this effort? We can see what they give up, but have they earned anything we would want?

Recently an Amish guy rode his bicycle out to our home along the foggy coast of the Pacific, and I had a chance to ask this question in depth. He appeared at our door sweaty and out of breath from the long uphill climb to our house under the redwoods. Parked a few feet away was his ingenious Dahon fold-up bike, which he had pedaled from the train station. Like most Amish, he did not fly, so he had stored his bike on the three-day cross-country train ride from Pennsylvania. This was not his first trip to San Francisco. He had previously ridden his bike along the entire coast of California and had in fact seen a lot of the world by train, bike, and boat.

For the next week, our Amish visitor couch-surfed in our spare bedroom, and at dinner he regaled us with tales of his life growing up in a horse-and-buggy, Old Order, Plain Folk community. I'll call our friend Leon. He is an unusual Amish in many ways. I met Leon online. Online, is of course, the last place you'd ever expect to meet an Amish man. But Leon had read some things I had posted about the Amish on my website and wrote to me. While he never went to high school (Amish formal education ceases after eighth grade) he is among the few Plain Folk to go to college, where he is currently an older student. (He is in his 30s.) He hopes to study medicine and perhaps become the first Amish doctor. Many former Amish have gone on to college or become doctors, but none have done that while remaining in the Old Order church. Leon is unusual in that he is a Plain Folk church member yet relishes his ability to live in the "outside" world as well.

The Amish practice a remarkable tradition called *rumspringa*, wherein their teenagers are allowed to ditch their homemade uniforms—suspenders and hats for boys, long dresses and bonnets for girls—and don baggy pants and short skirts, buy a car, listen to music, and party for a few years before they decide to forever give up these modern amenities and join the Old Order church. This intimate, real exposure to the technological universe means that they are fully cognizant of what that world has to offer and what exactly they are denying themselves. Leon is on a sort of permanent *rumspringa*—although he doesn't party but works very hard. His father runs a machine shop (a common Amish occupation), so Leon is a genius with tools. I was in the middle of a bathroom plumbing job on the afternoon when Leon first showed up, and he quickly took over the chore. I was impressed by his complete mastery of hardware store parts. I've heard of Amish auto mechanics who don't drive cars but can fix any model you bring them.

As Leon spoke of what his boyhood was like with only a horse and buggy for transportation, and what he learned in his multigrade, one-room schoolhouse, a fervent wistfulness played over his face. He missed the comfort of Old Order life now that he was away from it. We outsiders think of life without electricity, central heat, or cars as hard punishment. But curiously, Amish life offers more leisure than contemporary urbanity does. In Leon's account, they always had time for a game of baseball, reading, visiting neighbors, and hobbies.

Many observers of the Amish have remarked on how hardworking they are. So it was a complete surprise to someone like Eric Brende, an MIT grad student who gave up an engineering degree and instead dropped out to live alongside an Old Order Amish/Mennonite community, to find out how much leisure this lifestyle generated. Brende, who is not Amish, eliminated as much gear as he could from his home with his wife and tried to live as Plain as possible, a tale he recounts in his book, *Better Off*. For over two years Brende gradually adopted what he calls a minimite lifestyle. A minimite uses "the least amount of technology needed to accomplish something." Like his Old Order Amish/ Mennonite neighbors, he employed a minimum of technology: no power tools or electric appliances. Brende found that the absence of electronic entertainment, of long auto commutes, and of chores aimed at simply maintaining existing complex technology resulted in more time of real leisure. In fact, the constraints of cutting wood by hand, hauling manure with horses, and doing dishes by lamplight liberated the first genuine leisure time he had ever had. At the same time, the hard, strenuous manual work was satisfying and rewarding. He told me he found not only more leisure but more fulfillment as well.

Wendell Berry is a thinker and farmer who works his farm in an old-fashioned way using horses instead of tractors, very much like the Amish. Like Eric Brende, Berry finds tremendous satisfaction in the visible arrangement of bodily labor and agricultural results. Berry is a master wordsmith as well, and no one has been able to convey the "gift" that minimalism can deliver as well as he. One particular story from his collection *The Gift of Good Land* captures the almost ecstatic sense of fulfillment won with minimal technology.

> Last summer we put up our second cutting of alfalfa on an extremely hot, humid afternoon. . . . There was no breeze at all. The hot, bright, moist air seemed to wrap around us and stick to us while we loaded the wagons. It was worse in the barn, where the tin roof raised the temperature and held the air even closer and stiller. We worked more quietly than we usually do, not having breath for talk. It was miserable, no doubt about it. And there was not a push button anywhere in reach.
>
> But we stayed there and did the work, were even glad to do it, and experienced no futurological fits. When we were

> done we told stories and laughed and talked a long time, sitting on a post pile in the shade of a big elm. It was a pleasing day.
>
> Why was it pleasing? Nobody will ever figure that out by a "logical projection." The matter is too complex and too profound for logic. It was pleasing, for one thing, because we got done. That does not make logic, but it makes sense. For another thing, it was good hay, and we got it up in good shape. For another, we like each other and we work together because we want to.
>
> And so, six months after we shed all that sweat, there comes a bitter cold January evening when I go up to the horse barn to feed. It is nearly nightfall, and snowing hard. The north wind is driving the snow through the cracks in the barn wall. I bed the stalls, put corn in the troughs, climb into the loft and drop the rations of fragrant hay into the mangers. I go to the back door and open it; the horses come in and file along the driveway to their stalls, the snow piled white on their backs. The barn fills with the sounds of their eating. It is time to go home. I have my comfort ahead of me: talk, supper, fire in the stove, something to read. But I know too that all my animals are well fed and comfortable, and my comfort is enlarged in theirs. . . . And when I go out and shut the door, I am satisfied.

Our Amish friend Leon spoke of the same equation: fewer distractions, more satisfaction. The ever-ready embrace of his community was palpable. Imagine it: Neighbors would pay your medical bill if needed, or build your house in a few weeks without pay, and, more important, allow you to do the same for them. Minimal technology, unburdened by cultural innovations such as insurance or credit cards, forces a daily reliance on neighbors and friends. Hospital stays are paid by church members, who also visit the sick regularly. Barns destroyed by fire or storm are rebuilt in a barn raising and not by insurance money. Financial, marital, and behavioral counseling are done by peers. The community is as self-reliant as it can make itself and only as self-reliant as it is because it is a community. I began to understand the strong attraction the Amish lifestyle exerts on its young adults and why, even today, only a very few leave after their *rumspringa*. Leon observed that of the 300 or so friends his age in his church, only 2 or 3 have abandoned this very technologically constrained life, and they joined a church slightly less strict but still not mainstream.

But the cost of this closeness and dependency is limited choice. No education beyond eighth grade. Few career options for guys, none besides homemaker for girls. For the Amish and minimites, one's fulfillment must blossom inside the traditional confines of a farmer, tradesman, or housewife. But not everyone is born to be a farmer. Not every human is ideally matched to the rhythms of horse and corn and seasons and the eternal close inspection of village conformity. Where in the Amish scheme of things is the support for a mathematical genius or a person who might spend all day composing new music?

I asked Leon whether all the goodness of the Amish life—all that comforting mutual aid, satisfying hands-on work, reliable community infrastructure—could still issue forth if, say, all kids attended school up to 10th grade instead of eighth, as they now do? Just for starters. Well, you know, he said, "hormones kick in around the ninth grade, and boys, and even some girls, just don't want to sit at desks and do paperwork. They need to use their hands as well as their heads, and they ache to be useful. Kids learn more doing real things at that age." Fair enough. When I was a teen I wished I had been "doing real stuff" instead of being holed up in a stuffy high-school classroom.

The Amish are a little sensitive about this, but their self-reliant lifestyle as it is currently practiced is heavily dependent on the greater technium that surrounds their enclaves. They do not mine the metal they build their mowers from. They do not drill or process the kerosene they use. They don't manufacture the solar panels on their roofs. They don't grow or weave the cotton in their clothes. They don't educate or train their own doctors. They also famously do not enroll in armed forces of any kind. (But in compensation for that, the Amish are world-class volunteers in the outside world. Few people volunteer more often, or with more expertise and passion, than the Amish/Mennonites. They travel by bus or boat to distant lands to build homes and schools for the needy.) If the Amish had to generate all their own energy, grow all their clothing fibers, mine all metal, harvest and mill all lumber, they would not be Amish at all because they would be running large machines, dangerous factories, and other types of industry that would not sit well in their backyards (one of the criteria they use to decide whether a craft is appropriate for them). But without someone manufacturing this stuff, they could not maintain their lifestyle or prosperity. In short, the Amish depend on the outside world for the way they currently live. Their choice of minimal technology adoption is a choice—but a choice enabled by the technium. Their lifestyle is within the technium, not outside it.

For a long time I had been perplexed as to why Amish-like dissenters were primarily found only in North America. (The related Mennonites have a few satellite settlements in South America.) I looked long and hard to find Japanese "Amish," Chinese Amish, Indian Amish, even Islamic Amish but discovered none. I found some ultraorthodox Jews in Israel who reject computers, and likewise one or two small Islamic sects that prohibit TV and internet and some Jain monks in India who refuse to ride in automobiles or trains. As far as I can tell, there are no other ongoing large-scale communities based outside North America that have built a lifestyle around minimal technology. That's because outside technological America the idea seems crazy. This opt-out option makes sense only when there is something to opt out of. The original Amish protesters (or Protestants) were indistinguishable from neighboring European peasants. Fiercely persecuted by the state church, the Amish maintained their separation from the "worldly" mainstream by not upgrading their technology. No longer persecuted, the Amish today are a counterpoint to the incredibly technological aspect of American society. Their alternative thrives in opposition to the unrelenting thrust of individual personal reinvention and progress that is the hallmark of America. The Amish lifestyle is too familiar to poor peasants in China or India to have any meaning there. Such elegant rejection can only exist in, and because of, a modern technium.

The overabundance of the technium in North America has sprouted other dropouts as well. In the late 1960s and early 1970s tens of thousands of self-described hippies stampeded to small farms and makeshift communes to live simply, not too differently from the Amish. I was part of that movement. Wendell Berry was one of the clear-thinking gurus we listened to. In small experiments in rural America, we jettisoned the technology of the modern world (because it seemed to crush individualism) and tried to rebuild a new world while digging wells by hand, grinding our own flour, keeping bees, erecting homes from sun-dried clay, and even getting windmills and water generators to occasionally work. Some found religion, too. Our discoveries paralleled what the Amish knew—that this simplicity worked best in community, that the solution wasn't no technology but some technology, and that what seem to work best were the low-tech solutions we called "appropriate technology." This tie-dyed, deliberate, conscious engagement with appropriate technology was deeply satisfying for a while.

But only for a while. The *Whole Earth Catalog,* which I edited at one point, was the field manual for those millions of simple technology experiments. We ran pages and pages of information on how to build chicken coops, grow your own veggies, curdle your own cheese, school your children, and start a home business in a house made from bales of straw. And so I got to witness close up how the early enthusiasm for restricted technology would inevitably give way to unease and restlessness. Slowly the hippies drifted away from their deliberately low-tech world. One by one they left their domes for suburban garages and lofts where, much to our collective astonishment, many of them transformed their small-is-beautiful skills into small-is-start-up entrepreneurship. The origins of the *Wired* generation and the long-hair computer culture (think open-source UNIX) lay in the counterculture dropouts of the 70s. As Stewart Brand, hippie founder of the *Whole Earth Catalog,* remembers, "'Do your own thing' easily translated into 'Start your own business.'" I've lost count of the hundreds of individuals I personally know who left communes to eventually start high-tech companies in Silicon Valley. It's almost a cliche by now—barefoot to billionaire, just like Steve Jobs.

The hippies of the previous generation did not remain in their

Amish-like mode because as satisfying and attractive as the work in those communities was, the siren call of choices was more attractive. The hippies left the farm for the same reason the young have always left: The possibilities leveraged by technology beckon all night and day. In retrospect we might say the hippies left for the same reason Thoreau left his Walden; they both came and left to experience life to its fullest. Voluntary simplicity is a possibility, an option, a choice that one should experience for at least part of one's life. I highly recommend elective poverty and minimalism as a fantastic education, not least because it will help you sort out your technology priorities. But I have observed that simplicity's fullest potential requires that one consider minimalism one phase of many (even if a recurring phase, as is meditation or the Sabbath). In the past decade, a new generation of minimites has arisen, and they are now urban homesteading—living lightly in cities, supported by ad hoc communities of like-minded homesteaders. They are trying to have both—the Amish satisfaction of intense mutual aid and hand labor and the ever-cascading choices of a city.

Because of my own personal journey from low tech to high choice, I admire Leon and Berry and Brende and the Old Order Plain Folk communities. I am convinced that the Amish and minimites are more content and satisfied as people than the rest of us fast-forward urban technophiles. In their deliberate constraint of technology they have figured out how to optimize an alluring combination of leisure, comfort, and certainty over the optimization of uncertain possibilities. The honest truth is that as the technium explodes with new self-made options, we find it harder to find fulfillment. How can we be fulfilled when we don't know what is being filled?

So why not steer everyone in this direction? Why don't we all give up more choices and become Amish? After all, Wendell Berry and the Amish see our multimillion choices as illusory and meaningless, or as choices that are really entrapments.

I believe these two different routes for technological lifestyle—either optimizing contentment or optimizing choices—come down to very different ideas of what humans are to be.

It is only possible to optimize human satisfaction if you believe human nature is fixed. Needs cannot be maximally satisfied if they are in flux. Minimal technologists maintain that human nature is unchanging. If they refer to evolution at all, they claim that millions of years surviving on the savannah shaped our social natures in such a way they are not easily satiated with new gizmos. Instead, our enduring souls crave timeless goods.

If the nature of humans is indeed invariant, then it is possible to achieve a peak technological solution to support it. For example, Wendell Berry believes that a solid cast-iron hand pump is far superior to hauling water in buckets on a yoke. And he says that domesticated horses are better than pulling a plow yourself, as many an ancient farmer before him has done. But for Berry, who uses horses to drive his farm gear, anything beyond the innovation of hand pump and horsepower works against the satisfaction of human nature and natural systems. When tractors were introduced in the 1940s, "the speed of work could be increased, but not the quality." He writes:

> Consider, for example, the International High Gear No. 9 mowing machine. This is a horse-drawn mower that certainly improved on everything that came before it, from the scythe to previous machines in the International line. . . . I own one of these mowers. I have used it in my hayfield at the same time that a neighbor mowed there with a tractor mower; I have gone from my own freshly cut hayfield into others just mowed by tractors; and I can say unhesitatingly that, though the tractors do faster work, they do not do it better. The same is substantially true, I think, of other tools: plows, cultivators, harrows, grain drills, seeders, spreaders, etc. . . . The coming of the tractor made it possible for a farmer to do more work, but not better.

For Berry, technology peaked in 1940, about the moment when all these farm implements were as good as they could get. In his eyes, and in those of the Amish, too, the elaborate circular solution of a small, mixed family farm, where the farmer produces plant feed for the animals, who produce manure (power and food to grow more plants), is the perfect pattern for the health and satisfaction of a human being, human society, and the environment. After thousands of years of tinkering, humans found a way to optimize human work and leisure. But now found, additional choices overshoot this peak and only make things worse.

I could be wrong, of course, but it seems pure foolishness, if not the height of conceit and hubris, to believe that in the long course of human history, and by that I mean the next 10,000 years in addition to the past 10,000 years, the peak of human invention and satisfaction should turn out to be 1940. It is no coincidence that this date also happens to be the time when Wendell Berry was a young boy growing up on a farm with horses. Berry seems to follow Alan Kay's definition of technology. Kay, a brilliant polymath who has worked at Atari, Xerox, Apple, and Disney, came up with as good a definition of technology as I've heard: "Technology," Kay says, "is anything that was invented after you were born." The year 1940 cannot be the end of technological perfection for human fulfillment simply because human nature is not at its end.

We have domesticated our humanity as much as we have domesticated our horses. Our human nature itself is a malleable crop that we planted 50,000 years ago and continue to garden even today. The field of our nature has never been static. We know that genetically our bodies are changing faster now than at any time in the past million years. Our minds are being rewired by our culture. With no exaggeration and no metaphor, we are not the same people who first started to plow 10,000 years ago. The snug interlocking system of horse and buggy, wood-fire cooking, compost gardening, and minimal industry may be perfectly fit for a human nature—of an ancient agrarian epoch. But this devotion to a traditional way of being ignores the way in which our nature—our wants, desires, fears, primeval instincts, and loftiest aspirations—is being recast by ourselves and by our inventions, and it excludes the needs of our new natures. We need new jobs in part because we are new people at our core.

We are different physical beings from our ancestors. We think differently. Our educated and literate brains work differently. More than our hunter-gatherer ancestors, we are shaped by the accumulating wisdom, practices, traditions, and culture of all those who've lived before us and live with us. We are cramming our lives with ubiquitous messages, science, pervasive entertainment, travel, surplus food, abundant nutrition, and new possibilities every day. At the same time, our genes are racing to keep up with culture. And we are speeding the acceleration of those genes by several means, including medical interventions such as gene therapy. In fact, every trend of the technium—especially its increasing evolvability—points to a much more rapid change of human nature in the future.

Curiously, many of the same traditionalists who deny we are changing insist that we had better not.

I wish I had been an Amish boy in high school, making things, far from a classroom, sure of who I was. But reading books in high school opened up my mind to possibilities I had never imagined in grade school. My world began expanding in those years and has never stopped. Chief among those expanding possibilities were new ways to be human. Writing in 1950, sociologist David Riesman observed: "The more advanced the technology, on the whole, the more possible it is for a considerable number of human beings to imagine being somebody else." We expand technology to find out who we are and who we can be.

I know the Amish and Wendell Berry and Eric Brende and the minimites well enough to know that they believe we don't need exploding technology to expand ourselves. They are, after all, minimalists. The Amish find incredible contentment in their enactment of a fixed human nature. This deep human fulfillment is real, visceral, renewable, and so attractive that Amish numbers are doubling every generation. But I believe the Amish and minimites have traded contentment for revelation. They have not discovered, and cannot discover, who they can become.

That's their choice, which is fine as far as it goes. And because it is a choice, we should celebrate their development of it.

I may not tweet, watch TV, or use a laptop, but I certainly benefit from the effect of others who do. In that way I am not that different from the Amish, who benefit from the outsiders around them fully engaged with electricity, phones, and cars. But unlike individuals who opt out of individual technologies, Amish society indirectly constrains others as well as themselves. If we apply the ubiquity test—what happens if everyone does it?—to the Amish way, the optimization of choice collapses. By constraining the suite of acceptable occupations and narrowing education, the Amish are holding back possibilities not just for their children but indirectly for all.

If you are a web designer today, it is only because many tens of thousands of other people around you and before you have been expanding the realm of possibilities. They have gone beyond farms and home shops to invent a complex ecology of electronic devices that require new expertise and new ways of thinking. If you are an accountant, untold numbers of creative people in the past devised the logic and tools of accounting for you. If you do science, your instruments and field of study have been created by others. If you are a photographer, or an extreme sports athlete, or a baker, or an auto mechanic, or a nurse—*then your potential has been given an opportunity by the work of others.* You are being expanded as others expand themselves.

Unlike the Amish and minimites, the tens of millions of migrants headed into cities each year may invent a tool that will unleash choices for someone else. If they don't, then their children will. Our mission as humans is not only to discover our fullest selves in the technium, and to find full contentment, but to expand the possibilities for others. Greater technology will selfishly unleash our talents, but it will also unselfishly unleash others: our children, and all children to come.

That means that as you embrace new technologies, you are indirectly working for future generations of Amish, and for the minimite homesteaders, even though they are not doing as much for you. Most of what you adopt they will ignore. But every once in a while your adoption of "something that doesn't quite work yet" (Danny Hillis's definition of technology) will evolve into an appropriate tool they can use. It might be a solar grain dyer; it might be a cure for cancer. Anyone who is inventing, discovering, and expanding possibilities will indirectly expand possibilities for others.

Nonetheless, the Amish and minimites have important lessons to teach us about selecting what we embrace. Like them, I don't want a lot of devices that add maintenance chores to my life without adding real benefits. I do want to be choosy about what I spend time mastering. I want to be able to back out of things that don't work out. I don't want stuff that closes off options for others (like lethal weapons). And I do want the minimum because I've learned that I have limited time and attention.

I owe the Amish hackers a large debt because through their lives I now see the technium's dilemma very clearly: To maximize our own contentment, we seek the minimum amount of technology in our lives. Yet to maximize the contentment of others, we must maximize the amount of technology in the world. Indeed, we can only find our own minimal tools if others have created a sufficient maximum pool of options we can choose from. The dilemma remains in how we can personally minimize stuff close to us while trying to expand it globally.

12

Seeking Conviviality

"So the whole question comes down to this: Can the human mind master what the human mind has made?" This, according to the French poet and philosopher Paul Valery, is the dilemma of the technium. Has the enormity and cleverness of our creation overwhelmed our ability to control or guide it? What choices do we have in navigating the technium when it charges ahead, pushed by the millennia of momentum behind it? Within the technium's imperative, do we have any freedom at all? And practically, where are the levers to pull?

We have lots of choices. But those choices are no longer simple, nor obvious. As technology increases its complexity, the technium demands more complex responses. For instance, the number of technologies to choose from so far exceeds our capacity to use them all that these days we define ourselves more by the technologies we *don't* use than by those we do. In the same way that a vegetarian has more of an identity than an omnivore, someone who chooses not to drive or use the internet stakes out a stronger technological stance than the ordinary consumer. Although we don't realize it, at the global scale, we opt out of more technology than we opt in to.

The pattern of our personal nonadoption is usually illogical and nonsensical. On first glance some Amish rejections of technology appear equally weird and nonsensical. They might use four horses to pull a noisy diesel-powered harvester because they reject motor vehicles. Outsiders point to that combo as hypocritical, but it really is no more hypocritical than a famous science-fiction writer I knew who surfed the web but did not do e-mail. It was a simple choice for him; he got what he wanted from the one technology but not the other. When I asked my friends about their own technological choices, I found one friend who e-mailed but did not fax; another who faxed but did not have a phone; a friend who had a phone but no TV; someone with TV who rejected microwave ovens; another with microwaves but no clothes dryer; one with a clothes dryer who had rejected air-conditioning; one who loves his AC but refuses to get a car; a car fanatic with no CD player (only vinyl records); a guy with CDs who refuses GPS navigation; someone who embraces GPS but not credit cards; and so on. To outsiders, these abstinences are idiosyncratic and, arguably, hypocritical, but they serve the same purpose as the choices made by the Amish, which is to sculpt the cornucopia of technology to suit our personal intentions.

The Amish, however, select or deny technology as a group. By contrast, in secular modern culture, particularly in the West, technology choices are made individually, as a personal decision. It is much easier to maintain a disciplined refusal of a popular technology when all your peers are doing likewise and much harder if they are not. Much of the success of the Amish is due to the unwavering community-wide support (bordering on social coercion) for their unorthodox technology lifestyle. In fact, this union of sympathy is so essential that Amish families won't move to an Amish-less region to pioneer new settlements until a sufficient number of other families join them for a critical mass.

Can collective choice work more broadly in a modern pluralistic society? Can we, together, as a nation—or even as a planet—successfully choose certain technologies and refuse others?

Over the centuries, societies have declared many technologies to be dangerous, economically upsetting, immoral, unwise, or simply too unknown for our good. The remedy to this perceived evil is usually a form of prohibition. The offending innovation may be taxed severely or legislated to narrow purposes or restricted to the outskirts or banned altogether. The list of offending inventions in history banned on a wide scale includes such major items as crossbows, guns, mines, nuclear bombs, electricity, automobiles, large sailing ships, bathtubs, blood transfusions, vaccines, copy machines, TVs, computers, and the internet.

Yet history shows that it is very hard for a society as a whole to say no to technology for very long. I recently examined all the cases of large-scale technology prohibitions that I could find in the last 1,000 years. I define "large-scale prohibition" as an official injunction against a specific technology made at the level of a culture, religious group, or nation, rather than as an individual or small locality. I am not counting technologies that are ignored, but only ones that are deliberately relinquished. I found about 40 cases that met these criteria. That is not very many cases for 1,000 years. In fact it's hard to come up with a list of anything else that has occurred only 40 times in 1,000 years!

Large-scale prohibitions against technologies are rare. They are hard to enforce. And my research shows most don't last much longer than the normal obsolescence cycle of accepted technology. A handful of prohibitions lasted several hundred years in an era when it took technology several hundred years to change. The gun was outlawed in Shogun Japan for three centuries, exploration ships in Ming China for three centuries,

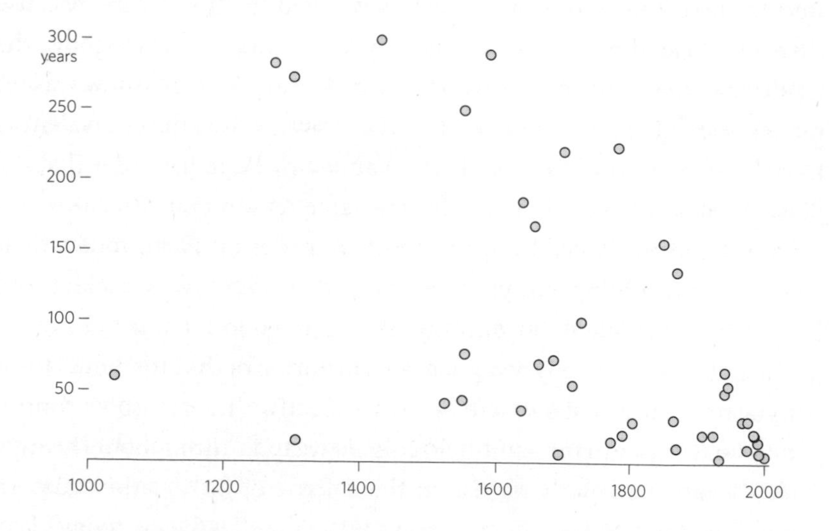

Duration of Prohibitions. The duration of a historical technological prohibition in years (vertical axis) plotted in the year of the initial prohibition. Durations are shortening over time.

and silk spinning in Italy for two centuries. Few others in history lasted as long. The guild of French scribes succeeded in delaying the introduction of printing into Paris, but only for 20 years. As the life cycle of technology sped up, the popularity of an invention could fade in a few years, and prohibitions against technology naturally shortened as well.

The chart on the previous page plots the duration of a prohibition against the year when the prohibition started. It includes only prohibitions that have concluded. As technology accelerates, so does the brevity of prohibition.

Bans may not last, but the question of whether they are effective during their duration is a much harder question to answer. Many earlier bans were based on economic considerations. The French banned the manufacture of machine-made cotton fabric for the same reason the English cottage weavers banned wide stocking-frame looms during the Luddite rebellion—it hurt their agrarian household businesses. Economic prohibitions can achieve their goals in the short term but often aggravate the inevitable transition to acceptance later.

Other prohibitions were made for security reasons. The ancient Greeks were the first to use crossbows, which they called "belly shooters" because they were loaded pried against the belly. Compared to the longbow, the traditional bow made of yew wood, the racket-assisted crossbow was far more powerful and far more deadly. The crossbow was the equivalent of today's AK-87 assault weapon. It was banned by Pope Innocent II at the Second Lateran Council in 1139 for the same reason that citizen-owned bazookas are prohibited by law in most countries on Earth today; their speedy, crowd-killing power is considered unnecessarily violent and broad for home defense or hunting. It's a tool good for war but not for peace. But according to David Bachrach, historian of the crossbow, "these bans against the crossbow were not at all effective. The crossbow continued to be the dominant hand-held missile weapon throughout the high middle ages particularly for use in the defense of fortifications and on ships." The 50-year ban on crossbows was as ineffective as today's ban on assault rifles has been in the underworld.

If we take a global view of technology, prohibition seems very ephemeral. While an item may be banned in one place, it will thrive in another. In 1299, officials in Florence prohibited their bankers from using Arabic numerals in their accounts. But the rest of Italy eagerly adopted them. In a global marketplace, nothing is eliminated. Where a technology is banned locally, it slips away to pool somewhere else on the globe.

Genetically modified foods have a reputation for being outlawed, and indeed some countries do ban them, but the acreage dedicated to growing genetically modified plant crops is increasing at 9 percent per year globally. Although prohibited by some nations, the amount of power delivered by nuclear power plants is increasing globally by 2 percent a year. The only worldwide relinquishment that seems to be working is the reduction of the nuclear weapon stockpile, which peaked at 65,000 units in 1986 and is now at 20,000. At the same time, the number of countries capable of making a nuclear weapon is increasing.

In a deeply connected world, the accelerated pace of technological succession—constant upgrades replacing former versions—renders even the most well-meaning ban unsustainable. Prohibitions are in effect postponements. Some, such as the Amish, find that delay useful enough. Others hope that a more desirable replacement technology might be found during the delay. That is possible. But wholesale prohibitions simply do not work to eliminate a technology that is considered subversive or morally wrong. Technologies can be postponed but not stopped.

Part of the reason that these widespread bans so rarely work is that we generally don't understand new inventions when they first appear. Every new idea is a bundle of uncertainty. No matter how sure the originator is that his or her newest idea will transform the world or end war or remove poverty or delight the masses, the truth is that no one knows what it will do. Even the short-term role of an idea is unclear. History is rife with cases of misguided technological expectations from the inventors themselves. Thomas Edison believed his phonograph would be used primarily to record the last-minute bequests of the dying. The radio was funded by early backers who believed it would be the ideal device for delivering sermons to rural farmers. Viagra was clinically tested as a drug for heart disease. The internet was invented as a disaster-proof communications backup. Very few great ideas start out headed toward the greatness they eventually achieve. That means that projecting what harm may come from a technology before it "is" is almost impossible.

With few exceptions technologies don't know what they want to be when they grow up. An invention requires many encounters with early adopters and collisions with other inventions to refine its role in the technium. Like people, young technologies often experience failure in their first careers before they find a better livelihood later. It's a rare technology that remains in its original role right from the start. More commonly a new invention is peddled by its inventors for one expected (and lucrative!) use, which is quickly proven wrong, and then advertised for a series of alternative (and less lucrative) uses, few of which work, until reality steers the technology toward a marginal unexpected use. Sometimes that marginal use blossoms into an exceptionally disruptive case that becomes the norm. When that kind of success happens, it obscures the earlier failures.

One year after Edison constructed the first phonograph, he was still trying to figure out what his invention might be used for. Edison knew more about this invention than anyone, but his speculations were all over the map. He thought his idea might birth dictation machines or audiobooks for the blind or talking clocks or music boxes or spelling lessons or recording devices for dying words or answering machines. In a list he drew up of possible uses for the phonograph, Edison added at the end, almost as an afterthought, the idea of playing recorded music.

Lasers were developed to industrial strength to shoot missiles down, but they are made in the billions primarily to read bar codes and movie DVDs. Transistors were created to replace vacuum tubes in room-sized computers, but most transistors manufactured today fill the tiny brains in cameras, phones, and communication equipment. Mobile phones began as . . . well, mobile phones. And for the first few decades that's what they were. But in its maturity, cell-phone technology is becoming a mobile computing platform for tablets, e-books, and video players. Switching occupations is the norm for technology.

The greater the number of ideas and technologies already in the world, the more possible combinations and secondary reactions there will be when we introduce a new one. Forecasting consequences in a technium where millions of new ideas are introduced each year becomes mathematically intractable.

We make prediction more difficult because our immediate tendency is to imagine the new thing doing an old job better. That's why the first cars were called "horseless carriages." The first movies were simply straightforward documentary films of theatrical plays. It took a while to realize the full dimensions of cinema photography as its own new medium that could achieve new things, reveal new perspectives, do new jobs. We are stuck in the same blindness. We imagine e-books today as being regular books that appear on electronic paper instead of as radically powerful threads of text woven into the one shared universal library. We think genetic testing is like blood testing, something you do once in your life to get an unchanging score, when sequencing our genes may instead become something we do hourly as our genes mutate, shift, and interact with our environment.

The predictivity of most new things is very low. The Chinese inventor of gunpowder most likely did not foresee the gun. William Sturgeon, the discoverer of electromagnetism, did not predict electric motors. Philo Farnsworth did not imagine the television culture that would burst forth from his cathode-ray tube. Advertisements at the beginning of the last century tried to sell hesitant consumers the newfangled telephone by stressing ways it could send messages, such as invitations, store orders, or confirmation of their safe arrival. The advertisers pitched the telephone as if it were a more convenient telegraph. None of them suggested having a conversation.

The automobile today, embedded in its matrix of superhighways, drive-through restaurants, seat belts, navigation tools, and digital hypermiling dashboards, is a different technology from the Ford Model T of 100 years ago. And most of those differences are due to secondary inventions rather than the enduring internal combustion engine. In the same way, aspirin today is not the aspirin of yesteryear. Put into the context of other drugs in the body, changes in our longevity and pill-popping habits (one per day!), cheapness, etc., it is a different technology from either the folk medicines derived from the essence of willow bark or the first synthesized version brought out by Bayer 100 years ago, even though they are all the same chemical, acetylsalicylic acid. Technologies shift as they thrive. They are remade as they are used. They unleash second- and third-order consequences as they disseminate. And almost always, they bring completely unpredicted effects as they near ubiquity.

On the other hand, most initial grand ideas for a technology fade into obscurity. An unfortunate few become an immense problem—a greatness wholly different from what their inventors intended. Thalidomide was a great idea for pregnant women but a horror for their unborn children. Internal combustion engines are great for mobility but awful for breathing. Freon kept things cold cheaply but took out the protective UV filter around the planet. In some cases this change in effect is a mere unintended side effect; in many cases it is a wholesale change of career.

If we examine technologies honestly, each one has its faults as well as its virtues. There are no technologies without vices and none that are neutral. The consequences of a technology expand with its disruptive nature. Powerful technologies will be powerful in both directions—for good and bad. There is no powerfully constructive technology that is not also powerfully destructive in another direction, just as there is no great idea that cannot be greatly perverted for great harm. After all, the most beautiful human mind is still capable of murderous ideas. Indeed, an invention or idea is not really tremendous unless it can be tremendously abused. This should be the first law of technological expectation: The greater the promise of a new technology, the greater its potential for harm as well. That's also true for new beloved technologies such the internet search engine, hypertext, and the web. These immensely powerful inventions have unleashed a level of creativity not seen since the Renaissance, but when (not if) they are abused, their ability to track and anticipate individual behavior will be awful. If a new technology is likely to birth a never-before-seen benefit, it will also likely birth a never-before-seen problem.

The obvious remedy for this dilemma is to expect the worst. That's the result of a commonly used approach to new technologies called the Precautionary Principle.

The Precautionary Principle was first crafted at the 1992 Earth Summit as part of the Rio Declaration. In its original form it advised that a "lack of full scientific certainty shall not be used as a reason for postponing cost-effective measures to prevent environmental degradation." In other words, even if you can't prove scientifically that harm is happening, this uncertainty should not prevent you from stopping the suspected harm. This principle of precaution has undergone many revisions and variations in the years since and has become more prohibitive over time. A recent version states: "Activities that present an uncertain potential for significant harm should be prohibited unless the proponent of the activity shows that it presents no appreciable risk of harm."

One version or another of the Precautionary Principle informs legislation in the European Union (it is included in the Maastricht Treaty) and appears in the United Nations Framework Convention on Climate Change. The U.S. Environmental Protection Agency (EPA) and Clean Air Act rely on the approach in establishing pollution control levels. The principle is also written into parts of the municipal codes of green cities such as Portland, Oregon, and San Francisco. It is a favorite standard for bioethicists and critics of rapid technological adoption.

All versions of the Precautionary Principle hold this axiom in common: A technology must be shown to do no harm before it is embraced. It must be proven to be safe before it is disseminated. If it cannot be proven safe, it should be prohibited, curtailed, modified, junked, or ignored. In other words, the first response to a new idea should be inaction until its safety is established. When an innovation appears, we should pause. Only after a new technology has been deemed okay by the certainty of science should we try to live with it.

On the surface, this approach seems reasonable and prudent. Harm must be anticipated and preempted. Better safe than sorry. Unfortunately, the Precautionary Principle works better in theory than in practice. "The precautionary principle is very, very good for one thing—stopping technological progress," says philosopher and consultant Max More. Cass R. Sunstein, who devoted a book to debunking the principle, says, "We must challenge the Precautionary Principle not because it leads in bad directions, but because read for all it is worth, it leads in no direction at all."

Every good produces harm somewhere, so by the strict logic of an absolute Precautionary Principle no technologies would be permitted. Even a more liberal version would not permit new technologies in a timely manner. Whatever the theory, as a practical matter we are unable to address all risks, independent of their low probability, while efforts to address all improbable risks hinders more likely potential benefits.

For example, malaria infects 300 million to 500 million people worldwide, causing 2 million deaths per year. It is debilitating to those who don't die and leads to cyclic poverty. But in the 1950s the level of malaria was reduced by 70 percent by spraying the insecticide DDT around the insides of homes. DDT was so successful as an insecticide that farmers eagerly sprayed it by the tons on cotton fields—and the molecule's by-products made their way into the water cycle and eventually into fat cells in animals. Biologists blamed it for a drop in reproduction rates for some predatory birds, as well as local die-offs in some fish and aquatic life species. Its use and manufacture were banned in the United States in 1972. Other countries followed suit. Without DDT spraying, however, malaria cases in Asia and Africa began to rise again to deadly pre-1950s levels. Plans to reintroduce programs for household spraying in malarial Africa were blocked by the World Bank and other aid agencies, who refused to fund them. A treaty signed in 1991 by 91 countries and the EU agreed to phase out DDT altogether. They were relying on the precautionary principle: DDT was probably bad; better safe than sorry. In fact DDT had never been shown to hurt humans, and the environmental harm from the miniscule amounts of DDT applied in homes had not been measured. But nobody could prove it did not cause harm, despite its proven ability to do good.

When it comes to risk aversion, we are not rational. We select which risks we want to contend. We may focus on the risks of flying but not driving. We may react to the small risks of dental X-rays but not to the large risk of undetected cavities. We might respond to the risks of vaccination but not the risks of an epidemic. We may obsess about the risks of pesticides but not the risks of organic foods.

Psychologists have learned a fair amount about risk. We now know that people will accept a thousand times as much risk for technologies or situations that are voluntary rather than mandatory. You don't have a choice where you get your tap water, so you are less tolerant in regard to its safety than you might be from using a cell phone of your choice. We also know that acceptance of a technology's risk is proportional to its corresponding perceived benefits. More gain is worth more risk. And, finally, we know that the acceptability of risk is directly influenced by how easy it is to imagine both the worst case and the best benefits, and that these are determined by education, advertising, rumor, and imagination. The risks that the public thinks are most significant are those in which it is easy to think of examples where the risk comes to fruition in a worst-case scenario. If it can plausibly lead to death, it's "significant."

In a letter Orville Wright wrote to his inventor friend Henry Ford, Wright recounts a story he heard from a missionary stationed in China. Wright told Ford the story for the same reason I tell it here: as a cautionary tale about speculative risks. The missionary wanted to improve the laborious way the Chinese peasants in his province harvested grain. The local farmers clipped the stalks with some kind of small hand shear. So the missionary had a scythe shipped in from America and demonstrated its superior productivity to an enthralled crowd. "The next morning, however, a delegation came to see the missionary. The scythe must be destroyed at once. What, they said, if it should fall into the hands of thieves; a whole field could be cut and carried away in a single night." And so the scythe was banished, progress stopped, because nonusers could imagine a possible—but wholly improbable—way it could significantly harm their society. (Much of the hugely disruptive theater around "national security" today is based on similarly improbable scenarios of worst-case dangers.)

In its efforts to be "safe rather than sorry," precaution becomes myopic. It tends to maximize only one value: safety. Safety trumps innovation. The safest thing to do is to perfect what works and never try anything that could fail, because failure is inherently unsafe. An innovative medical procedure will not be as safe as the proven standard. Innovation is not prudent. Yet because precaution privileges only safety, it not only diminishes other values but also actually reduces safety.

Big accidents in the technium usually don't start out as wings falling off or massive pipeline breaks. One of the largest shipping disasters in modern times began with a burning coffeepot in the crew kitchen. A regional electric grid can shut down not because a tower is toppled but because a gasket breaks in a minor pump. In cyberspace a rare, trivial bug on a web-page order form can take a whole site down. In each case the minor error triggers, or combines with, other unforeseen consequences in the system, also minor. But because of the tight interdependence of parts, minor glitches in the right improbable sequence cascade until the trouble becomes an unstoppable wave and reaches catastrophic proportions. Sociologist Charles Perrow calls these "normal accidents" because they "naturally" emerge from the dynamics of large systems. The system is to blame, not the operators. Perrow did an exhaustive minute-by-minute study of 50 large-scale technological accidents (such as Three Mile Island, the Bhopal disaster, Apollo 13, *Exxon Valdez*, Y2K, etc.) and concluded, "We have produced designs so complicated that we cannot anticipate all the possible interactions of the inevitable failures; we add safety devices that are deceived or avoided or defeated by hidden paths in the systems." In fact, Perrow concludes, safety devices and safety procedures themselves often create new accidents. Safety components can become one more opportunity for things to go wrong. For instance, adding security forces at an airport can increase the number of people with access to critical areas, which is a decrease in security. Redundant systems, normally a safety backup, can easily breed new types of errors.

These are called substitute risks. New hazards materialize directly as a result of attempts to reduce hazards. Fireproof asbestos is toxic, but most of its substitutes are equally if not more toxic. Furthermore, the removal of asbestos greatly increases its danger compared to the low risk of letting it remain in place in buildings. The Precautionary Principle is oblivious to the notion of substitute risks.

In general the Precautionary Principle is biased against anything new. Many established technologies and "natural" processes have unexamined faults as great as those of any new technology. But the Precautionary Principle establishes a drastically elevated threshold for things that are new. In effect it grandfathers in the risks of the old, or the "natural." A few examples: Crops raised without the shield of pesticides generate more of their own natural pesticides to combat insects, but these indigenous toxins are not subject to the Precautionary Principle because they aren't "new." The risks of new plastic water pipes are not compared with the risks of old metal pipes. The risks of DDT are not put in context with the old risks of dying of malaria.

The surest remedy for uncertainty is faster, better scientific studies. Science is a process of testing that will never eliminate uncertainty totally, and its consensus on particular questions will shift over time. But the consensus of evidence-based science is more reliable than anything else we have, including the hunches of precaution. More science, done openly by skeptics and enthusiasts, will enable us to sooner say: "This is okay to use" or "This is not okay to use." Once a consensus forms, we can regulate reasonably—as we have with lead in gasoline, tobacco, seat belts, and many other mandated improvements in society.

But in the meantime we should count on uncertainty. Even though we've learned to expect unintended consequences from every innovation, the particular unintended consequences are rarely foreseen. "Technology always does more than we intend; we know this so well that it has actually become part of our intentions," writes Langdon Winner. "Imagine a world in which technologies accomplish only the specific purposes one had in mind in advance and nothing more. It would be a radically constricted world and one totally unlike the world we now inhabit." We know technology will produce problems; we just don't know which new problems.

Because of the inherent uncertainties in any model, laboratory, simulation, or test, the only reliable way to assess a new technology is to let it run in place. An idea has to inhabit its new form sufficiently so that it can begin to express secondary effects. When a technology is tested soon after its birth, only its primary effects will be visible. But in most cases it is technology's unintended second-order effects that are the root of subsequent problems.

Second-order effects—the ones that usually overtake society—are rarely captured by forecasts, lab experiments, or white papers. Science-fiction guru Isaac Asimov made the astute observation that in the age of horses many ordinary people eagerly and easily imagined a horseless carriage. The automobile was an obvious anticipation since it was an extension of the first-order dynamics of a cart—a vehicle that goes forward by itself. An automobile would do everything a horse-pulled carriage did but without the horse. But Asimov went on to remark how difficult it was to imagine the second-order consequences of a horseless carriage, such as drive-in movie theaters, paralyzing traffic jams, and road rage.

Second-order effects often require a certain density, a semi-ubiquity, to reveal themselves. The main safety concern with the first automobiles centered on the safety of their occupants—the worry that the gas engines would blow up or that the brakes would fail. But the real challenge of autos emerged only in aggregate, when there were hundreds of thousands of cars—the accumulated exposure to their minute pollutants and their ability to kill others outside the car at high speeds, not to mention the disruptions of suburbs and long commutes—all second-order effects.

A common source of unforecastable effects of technologies stems from the way they interact with other technologies. In a 2005 debriefing that analyzed why the now-defunct U.S. Office for Technology Assessment, which existed from 1972 to 1995, did not have more of an impact in assessing upcoming technology, the researchers concluded:

> While plausible (although always uncertain) forecasts can be generated for very specific and fairly evolved technologies (e.g., the supersonic transport; a nuclear reactor; a particular pharmaceutical product), the radical transforming capacity of technology comes not from individual artifacts but from interacting subsets of technologies that permeate society.

In short, crucial second-order effects are absent from small, precise experiments and sincere simulations of new technologies, and so an emerging technology must be tested in action and evaluated in real time. In other words, the risks of a particular technology have to be determined by trial and error in real life.

The appropriate response to a new idea should be to immediately try it out. And to keep trying it out, and testing it, as long as it exists. In fact, contrary to the Precautionary Principle, a technology can never be declared "proven safe." It must be continuously tested with constant vigilance since it is constantly being reengineered by users and the coevolutionary technium it inhabits.

Technological systems "require continued attention, rebuilding and repair. Eternal vigilance is the price of artificial complexity," says Langdon Winner. Stewart Brand elevates constant assessment to the level of the vigilance principle in his book on ecopragmatism, *Whole Earth Discipline:* "The emphasis of the vigilance principle is on liberty, the freedom to try things. The correction for emergent problems is in ceaseless, fine-grained monitoring." He then suggests three categories that we might assign a probationary technology: "1) provisionally unsafe until proven unsafe; 2) provisionally safe until proven safe; 3) provisionally beneficial until proven beneficial." *Provisional* is the operative word. Another term for Brand's approach might be *eternally provisional.*

In his book about unintended consequences of technology, *Why Things Bite Back,* Edward Tenner spells out the nature of constant vigilance:

> Technological optimism means in practice the ability to recognize bad surprises early enough to do something about them. . . . It also requires a second level of vigilance at increasingly porous national borders against the world exchange of problems. But vigilance does not end there. It is everywhere. It is in the random alertness tests that have replaced the "dead man's pedal" for train operators. It is in the rituals of computer backup, the legally mandated testing of everything from elevators to home smoke alarms, routine X-ray screening, securing and loading new computer-virus definitions. It is in the inspection of arriving travelers for products that might harbor pests. Even our alertness in crossing the street, second nature to urbanites now, was generally unnecessary before the eighteenth century. Sometimes vigilance is more of a reassuring ritual than a practical precaution, but with any luck it works.

The Amish practice something very similar. Their approach to the technium is founded on their very fundamental religious faith; their theology drives their technology. Yet paradoxically, the Amish are far more scientific than most secular professionals about which technology they adopt. Typical nonreligious consumers tend to accept technology "on faith" based on what the media says, with no testing at all. In contrast, the Amish perform four levels of empirical testing on a potential technology. Instead of hypothetical worst-case-scenario precaution, the Amish employ evidence-based technological assessment.

First, they discuss among themselves (sometimes in councils of their elders) the expected community consequences of an upcoming innovation. What happens if farmer Miller starts using solar panels to pump water? Once he has the panels, will he be tempted to use the electricity to run his refrigerators? What then? And where do the panels come from? In short, the Amish develop a hypothesis of the technology's impact. Second, they closely monitor the actual effect of use among a small set of early adopters to see if their observations confirm their hypothesis. How do the Miller family and their interactions with neighbors change as they use the new stuff? And third, will the elders remove a technology if it appears to be undesirable based on observed effect and then assess the impact of its removal to further confirm their hypothesis? Was the community as a whole any better off without this technology? Last, they constantly reevaluate. Today, after 100 years of debate and observation, their communities are still discussing the merits of automobiles, electrification, and phones. None of this is quantitative; the results are compressed into anecdotes. Stories about what happened to so-and-so with such-and-such technology are retold in gossip or printed in the pages of their newsletters and become the currency of this empirical testing.

Technologies are nearly living things. Like all evolving entities, they must be tested in action, by action. The only way to wisely evaluate our technological creations is to try them out in prototypes, then refine them in pilot programs. In living with them we can adjust our expectations, shift, test, and rerelease. In action we monitor alterations, then redefine our aims. Eventually, by living with what we create, we can redirect technologies to new jobs when we are not happy with their outcomes. We move with them instead of against them.

The principle of constant engagement is called the Proactionary Principle. Because it emphasizes provisional assessment and constant correction, it is a deliberate counterapproach to the Precautionary Principle. This framework was first articulated by Max More, radical transhumanist, in 2004. More began with ten guidelines, but I have reduced his ten principles to five proactions. Each proaction is a heuristic to guide us in assessing new technologies.

The five proactions are:

1. Anticipation

Anticipation is good. All tools of anticipation are valid. The more techniques we use, the better, because different techniques fit different technologies. Scenarios, forecasts, and outright science fiction give partial pictures, which is the best we can expect. Objective scientific measurement of models, simulations, and controlled experiments should carry greater weight, but these, too, are only partial. Actual early data should trump speculation. The anticipation process should try to imagine as many horrors as glories, as many glories as horrors, and if possible to anticipate ubiquity; what happens if everyone has this for free? Anticipation should not be a judgment. The purpose of anticipation is not to accurately predict what will happen with a technology, because all precise predictions are wrong, but to prepare a base for the next four steps. It is a way to rehearse future actions.

2. Continual Assessment

Or eternal vigilance. We have increasing means to quantifiably test everything we use all the time, not just once. By means of embedded technology we can turn daily use of technologies into large-scale experiments. No matter how much a new technology is tested at first, it should be continuously retested in real time. Technology provides us with more precise means of niche testing. Using communication technology, cheap genetic testing, and self-tracking tools, we can focus on how innovations fare in specific neighborhoods, subcultures, gene pools, ethnic groups, or user modes. Testing can also be continual, 24/7, rather than the just on first release. Further, new technology such as social media (today's Facebook) allows citizens to organize their own assessments and do their own sociological surveys. Testing is active and not passive. Constant vigilance is baked into the system.

3. Prioritization of Risks, Including Natural Ones

Risks are real but endless. Not all risks are equal. They must be weighted and prioritized. Known and proven threats to human and environmental health are given precedence over hypothetical risks. Furthermore, the risks of inaction and the risks of natural systems must be treated symmetrically. In Max More's words: "Treat technological risks on the same basis as natural risks; avoid underweighting natural risks and overweighting human-technological risks."

4. Rapid Correction of Harm

When things go wrong—and they always will—harm should be remedied quickly and compensated in proportion to actual damages. The assumption that any given technology will create problems should be part of its process of creation. The software industry may offer a model for quick correction: Bugs are expected; they are not a reason to kill a product; instead they are employed to better the technology. Think of unintended consequences in other technologies, even fatal ones, as bugs that need to be corrected. The more sentient the technology, the easier it is to correct. Rapid restitution for harm done (which the software industry does not do) would also indirectly aid the adoption of future technologies. But restitution should be fair. Penalizing creators for hypothetical harm or even potential harm demeans justice and weakens the system, reducing honesty and penalizing those who act in good faith.

5. Not Prohibition but Redirection

Prohibition and relinquishment of dubious technologies do not work. Instead, find them new jobs. A technology can play different roles in society. It can have more than one expression. It can be set with different defaults. It can have more than one political cast. Since banning fails, redirect technologies into more convivial forms.

To return to the question at the beginning of this chapter: What choices do we have in steering the inevitable progress of the technium?

We have the choice of how we treat our creations, where we place them, and how we train them with our values. The most helpful metaphor for understanding technology may be to consider humans as the parents of our technological children. As we do with our biological children, we can, and should, constantly hunt for the right mix of beneficial technological "friends" to cultivate our technological offspring's best side. We can't really change the nature of our children, but we can steer them to tasks and duties that match their talents.

Take photography. If the processing of color photography is centralized (as it was for 50 years by Kodak), that applies a different tenor to photography than if the processing is done by chips in the camera itself. Centralization fosters a type of self-censorship of what pictures you take, and it also adds a time lag for displaying the results, which slows learning and discourages spontaneity. To be able to take a colorful picture of anything and then review it instantly and cheaply—that changed the character of the same glass lenses and shutter. Another example: It is easy to inspect the components in a motor, but not in a can of paint. But chemical products could be made to reveal their component ingredients with extra information, as if they were motor parts; the labeling could trace their manufacturing process back to their source as pigments in the Earth or in oil and thus make them more transparent to control and to interaction. This more open expression of paint technology would be different, and maybe more useful. Final example: Radio broadcasting—a very old and easily manufactured technology—is currently among the most heavily regulated technologies in most countries. This steep regulation by government has led to the current development of only a few bands of frequencies out of all those available, most of which remain underused. In an alternative system, radio spectrum could be allotted in a very different manner, potentially giving rise to cell phones that communicate directly with one another instead of through a local hub cell tower. The resulting alternative peer-to-peer broadcast system would yield a vastly different expression of radio.

Oftentimes the first job we assign to a technology is not at all ideal. For instance, DDT was an ecological disaster when assigned as an aerially sprayed insecticide on cotton crops. But restricted to the task of a household malaria remedy it shines as a public-health hero. Same technology, better job. It may take many tries, many jobs, many mistakes before we find a great role for a given technology.

The more autonomy our children (technological as well as biological) have, the more freedom they have to make mistakes. Our children's ability to create a disaster (or create a masterpiece) may even exceed our own, which is why parenting is both the most frustrating and the most rewarding thing we can do. By this measure our scariest offspring are forms of self-duplicating technology that already have significant potential autonomy. No creation of ours will test our patience and love as much as these will. And no technologies will test our ability to influence, steer, or guide the technium in the future as these will.

Self-duplication is old news in biology. It's the four-billion-year-old magic that allows nature to replenish herself, as one chicken hatches another chicken and so on. But self-duplication is a radical new force in the technium. The mechanical ability to make perfect copies of oneself and then occasionally create an improvement before copying, unleashes a type of independence that is not easily controlled by humans. Endless, ever-quickening cycles of reproduction, mutation, and bootstrapping can send a technological system into overdrive, leaving the rider far behind. As they zoom ahead, these technological creations will make new mistakes. Their unforeseeable achievements will amaze and terrify us.

The power of self-replication is now found in four fields of high technology: geno, robo, info, and nano. Geno stuff includes gene therapies, genetically modified organisms, synthetic life, and drastic genetic engineering of the human line. With genotechnology a new critter or new chromosome can be invented and released; it then reproduces forever, in theory.

Robo stuff is, of course, robots. Robots already work in factories making other robots, and at least one university lab has prototyped an autonomously self-assembling machine. Give this machine a pile of parts and it will assemble a copy of itself.

Info stuff is self-replicants such as computer viruses, artificial minds, and virtual personae built through data accumulation. Computer viruses have famously already mastered self-reproduction. Thousands infect hundreds of millions of computers. The holy grail of research into artificial learning and intelligence is, of course, to make an artificial mind smart enough to make another artificial mind smarter still.

Nano stuff is extremely tiny machines (as small as bacteria) that are designed for chores like eating oil or performing calculations or cleaning human arteries. Because they are so small, these tiny machines can work like mechanical computer circuits, and so in theory, they can be designed to self-assemble and reproduce like other computational programs. They would be a sort of like dry life, although this is many years away.

In these four areas the self-amplifying loops of self-duplication catapult the effects of these technologies into the future very quickly. Robots that make robots that make robots! Their accelerated cycles of creation can race so far ahead of our intentions that it is worrisome. Who's controlling the robo descendants?

In the geno world if we code changes into a gene line, for example, those changes can replicate down generations forever. And not just in family lines. Genes can easily migrate horizontally between species. So copies of new genes—bad or good—might disseminate through both time and space. As we know from the digital era, once copies are released they are hard to take back. If we can engineer an endless cascade of artificial minds inventing minds smarter than themselves (and us), what control do we have over the moral judgment of such creations? What if they start out with harmful prejudices?

Information shares this same avalanching property of replicating out of our control. Computer security experts claim that of the thousands of species of self-replicating worms and computer viruses invented by hackers to date, not one has gone extinct. They are here forever—or as long as two machines still run.

Finally, nanotechnology promises marvelous super-micro-thingies that are constructed at the precision of single atoms. The threat of these nano-organisms breeding without limit until they cover everything is known as the "gray goo" scenario. For a number of reasons, I think the gray goo is scientifically unlikely, though some kind of self-reproducing nanostuff is inevitable. But it is very likely that at least a few fragile species of nanotechnology (not goo) will breed in the wild, in narrow, protected niches. Once a nanobug goes feral, it could be indelible.

As the technium gains in complexity, it will gain in autonomy. What the current crop of self-duplicating GRIN (geno, robo, info, nano) technologies reveal is the way in which this rising autonomy demands our attention and respect. In addition to all the usual difficulties that new technologies present—shifting capabilities, unintended roles, hidden consequences—self-replicating technologies add two more: amplification and acceleration. Tiny effects rapidly escalate into major upheaval as one generation amplifies another, in the same way innocent feedback in a microphone whisper can burst into a deafening screech. And by the same cycles of self-generation, the speed at which a replicating technology impacts the technium keeps accelerating. The effects are pushed so far downstream that it complicates our ability to proactively engage and test and try the technology out in the present.

This is a replay of an old story. The amazing, uplifting power of life itself is rooted in its ability to leverage self-replication, and now that power is being born in technology. The most powerful force in the world will become much more powerful as it gains in ability to self-generate, but this liquid dynamite presents a grand challenge in managing it.

A common reaction to the out-of-control nature of geno-, robo-, info- and nanotechnology is to call for a moratorium on their development. Ban them. In 2000 Bill Joy, the pioneer computer scientist who invented several key programming languages that run the internet, called upon his fellow scientists in genetic, robotic, and computer sciences to relinquish GRIN technologies that could potentially be weaponized, to give them up the way we gave up biological weapons. Under the guidance of the Precautionary Principle, the Canadian watchdog group ETC called for a moratorium on all nanotechnological research. The German equivalent of the EPA demanded a ban on products containing silver nanoparticles (used in antimicrobial coatings). Others would like to ban autopiloted automobiles from public roads, outlaw genetically engineered vaccines in children, or halt human gene therapy until such time as each invention can be proven to cause no harm.

This is exactly the wrong thing to do. These technologies are inevitable. And they will cause some degree of harm. After all, to point to only one example above, human-piloted cars cause great harm, killing millions of people each year worldwide. If robot-controlled cars killed "only" half a million people per year, it would be an improvement!

Yet their most important consequences—both positive and negative—won't be visible for generations. We don't have a choice in whether there will be genetically engineered crops everywhere. There will be. We do have a choice in the character of the genetic food system—whether its innovations are publicly or privately held, whether it is regulated by government or industry, whether we engineer it for generational use or only the next business quarter. As inexpensive communication systems circle the globe, they knit a thin cloak of nervous material around the planet, making an electronic "world brain" of some kind inevitable. But the full downsides, or upsides, of this world brain won't be measurable until it is operating. The choice for humans is, What kind of world brain would we like to make out of this envelope? Is the participation default open or closed? Is it easy to modify procedures, and share, or is modification difficult and burdensome? Are the controls proprietary? Is it easy to hide from? The details of the web can go in a hundred different ways, although the technologies themselves will bias us in certain directions.

Yet how we express the inevitable global web is a significant choice we own. We can only shape technology's expression by engaging with it, by riding it with both arms around its neck.

To do that means to embrace those technologies now. To create them, turn them on, try them. This is the opposite of a moratorium. It's more like a try-atorium. The result would be a conversation, a deliberate engagement with the emerging technology. The faster these technologies spin into the future, the more essential it is that we ride them from the start.

Cloning, nanotech, network bots, and artificial intelligence (for just a few GRIN examples) need to be released within our embrace. Then we'll bend each this way and that. A better metaphor would be that we'll train the technology. As in the best animal and child training, positive aspects are reinforced with resources and negative aspects are starved until they diminish.

In one sense, self-amplifying GRIN-ologies are bullies, rogue technologies. They will need our utmost attention in order to be trained for consistent goodness. We need to invent appropriate long-term training technologies to guide them across the generations. The worst thing to do is banish and isolate them. Rather, we want to work with the bullying problem child. High-risk technologies need more chances for us to discover their true strengths. They need more of our investment and more opportunities to be tried. Prohibiting them only drives them underground, where their worst traits are emphasized.

There are already a few experiments to embed guiding heuristics in artificially intelligent systems as a means to make "moral" artificial intelligence, and other experiments to embed long-range control systems in genetic and nanosystems. We have an existing proof that such embedded principles work—in ourselves. If we can train our children—who are the ultimate power-hungry, autonomous, generational rogue beings—to be better than us, then we can train our GRINs.

As in raising our children, the real question—and disagreement—lies in what values we want to transmit over generations. This is worth discussing, and I suspect that, as in real life, we won't all agree on the answers.

The message of the technium is that any choice is way better than no choice. That's why technology tends to tip the scales slightly toward the good, even though it produces so many problems. Let's say we invent a hypothetical new technology that can give immortality to 100 people, but at the cost of killing 1 other person prematurely. We could argue about what the real numbers would have to be to "balance out" (maybe it is 1,000 who never die, or a million, for one who does) but this bookkeeping ignores a critical fact: Because this life-extension technology now exists, there is a new choice between 1 dead and 100 immortal that did not exist before. This additional possibility or freedom or choice—between immortality and death—*is good in itself.* So even if the result of this particular moral choice (100 immortal = 1 dead) is deemed a wash, the choice itself tips the balance a few percentage points to the good side. Multiply this tiny lean toward good by each of the million, 10 million, or 100 million inventions birthed in technology each year, and you can see why the technium tends to amplify the good slightly more than the evil. It compounds the good in the world because in addition to the direct good it brings, the arc of the technium keeps increasing choices, possibilities, freedom, and free will in the world, and that is an even greater good.

In the end, technology is a type of thinking; a technology is a thought expressed. Not all thoughts or technologies are equal. Clearly, there are silly theories, wrong answers, and dumb ideas. While a military laser and Gandhi's act of civil disobedience are both useful works of human imagination and thus both technological, there is a difference between the two. Some possibilities restrict future choices, and some possibilities are pregnant with other possibilities.

However, the proper response to a lousy idea is not to stop thinking. It is to come up with a better idea. Indeed, we should prefer a bad idea to no ideas at all, because a bad idea can at least be reformed, while not thinking offers no hope.

The same goes for the technium. The proper response to a lousy technology is not to stop technology or to produce no technology. It is to develop a better, more convivial technology.

Convivial is a great word whose roots mean "compatible with life."

In his book *Tools for Conviviality,* the educator and philosopher Ivan Illich defined convivial tools as those that "enlarge the contribution of autonomous individuals and primary groups. . . ." Illich believed that certain technologies were inherently convivial, while others, such as "multilane highways and compulsory education," were destructive no matter who ran them. In this way, tools were either good or bad for the living. But I am convinced by my study of the technium's imperative that conviviality resides not in the nature of a particular technology but in the job assignment, in the context, in the expression we construct for the technology. A tool's conviviality is mutable.

A convivial manifestation of a technology offers:

- Cooperation. It promotes collaboration between people and institutions.
- Transparency. Its origins and ownership are clear. Its workings are intelligible to nonexperts. There is no asymmetrical advantage of knowledge to some of its users.
- Decentralization. Its ownership, production, and control are distributed. It is not monopolized by a professional elite.
- Flexibility. It is easy for users to modify, adapt, improve, or inspect its core. Individuals may freely choose to use it or give it up.
- Redundancy. It is not the only solution, not a monopoly, but one of several options.
- Efficiency. It minimizes impact on ecosystems. It has a high efficiency for energy and materials and is easy to reuse.

Living organisms and ecosystems are characterized by a high degree of indirect collaboration, transparency of function, decentralization, flexibility and adaptability, redundancy of roles, and natural efficiency; these are all traits that make biology useful to us and the reasons why life can sustain its own evolution indefinitely. So the more lifelike we train our technology to be, the more convivial it becomes for us and the more sustainable the technium becomes in the long run. The more convivial a technology is, the more it aligns with its nature as the seventh kingdom of life.

It is true that some technologies are more inclined toward certain traits than others. Certain technologies will easily be decentralized, while others will tend to centralize. Some will take to transparency naturally, others lean to obscurity, perhaps requiring great expertise to use. But every technology—no matter its origin—can be channeled toward more transparency, greater collaboration, increased flexibility, and greater openness.

And this is where our choice comes in. The evolution of new technologies is inevitable; we can't stop it. But the character of each technology is up to us.

PART FOUR

DIRECTIONS

13

Technology's Trajectories

So what does technology want? Technology wants what we want—the same long list of merits we crave. When a technology has found its ideal role in the world, it becomes an active agent in increasing the options, choices, and possibilities of others. Our task is to encourage the development of each new invention toward this inherent good, to align it in the same direction that all life is headed. Our choice in the technium—and it is a real and significant choice—is to steer our creations toward those versions, those manifestations, that maximize that technology's benefits, and to keep it from thwarting itself.

Our role as humans, at least for the time being, is to coax technology along the paths it naturally wants to go.

But how do we know just where it wants to go? If certain aspects of the technium are preordained and certain aspects contingent upon our choices, how do we know which are which? Systems theorist John Smart has suggested that we need a technological version of the Serenity Prayer. Popular among participants in 12-step addiction recovery programs, the prayer, probably written in the 1930s by the theologian Reinhold Niebuhr, goes:

> God grant me the serenity
> To accept the things I cannot change;
> Courage to change the things I can;
> And wisdom to know the difference.

So how do we acquire the wisdom to discern the difference between the inevitable stages of technological development and the volitional forms that are up to us? What technique makes the inevitable obvious?

I think that tool is our awareness of the technium's long-term cosmic trajectories. The technium wants what evolution began. In every direction, technology extends evolution's four-billion-year path. By placing technology in the context of that evolution, we can see how those macroimperatives play out in our present time. In other words, technology's inevitable forms coalesce around the dozen or so dynamics common to all exotropic systems, including life itself.

I propose that the greater the number of exotropic traits we observe in a particular expression of technology, the greater its inevitability and its conviviality. If we want to compare, say, a vegetable-oil steam-powered automobile versus a rare-Earth-metal solar electric car, we could inspect the extent to which each of these mechanical manifestations supports these trends—not just follows the trends, but extends them. A technology's alignment with the trajectory of exotropic forces becomes the Serenity Prayer filter.

Extrapolated, technology wants what life wants:

Increasing efficiency
Increasing opportunity
Increasing emergence
Increasing complexity
Increasing diversity
Increasing specialization
Increasing ubiquity
Increasing freedom
Increasing mutualism
Increasing beauty
Increasing sentience
Increasing structure
Increasing evolvability

This list of exotropic trends can serve as a sort of checklist to help us evaluate new technologies and predict their development. It can guide us in guiding them. For instance, at this particular phase in the technium, at the turn of the 21st century, we are building many intricate, complex systems of communications. This wiring up of the planet can happen in a number of ways, but my modest prediction is that the most sustainable technological arrangements will be those manifestations that tend toward the greatest increases in diversity, sentience, opportunity, mutuality, ubiquity, etc. We can compare two competing technologies to see which one favors more of these exotropic qualities. Does it open up diversity or close it down? Does it bank on increasing opportunities or assume they wither? Is it moving toward embedded sentience or ignoring it? Does it blossom in ubiquity or collapse under it?

Using this perspective we might ask, Is large-scale petrol-fed agriculture inevitable? This highly mechanical system of tractors, fertilizers, breeders, seed producers, and food processors provides the abundant cheap food that is the foundation of our leisure to invent other things. It feeds our longevity to keep inventing, and ultimately this food system fuels the increase in population that generates increasing numbers of new ideas. Does this system support the trajectories of the technium more than the food-production schemes that preceded it—both subsistence farming and animal-powered mixed farming at its peak? How does it compare to hypothetical alternative food systems we might invent? I would say as a rough first pass that mechanized farming was inevitable in that it increased the merits of energy efficiency, complexity, opportunities, structure, sentience, and specialization. It does not, however, support increasing diversity or beauty.

According to many food experts, the problem with the current food-production system is that it is heavily dependent on monocultures (not diverse) of too few staple crops (five worldwide), which in turn require pathological degrees of intervention with drugs, pesticides, and herbicides, soil disturbance (reduced opportunities), and overreliance on cheap petro fuels for both energy and nutrients (reduced freedoms).

Alternative scenarios that can scale up to the global level are hard to imagine, but there are hints that a decentralized agriculture, with less reliance on politically motivated government subsidies or petroleum or monocultures, might work. This evolved system of hyperlocal, specialized farms might be manned either by a truly globally mobile migrant labor force or by smart, nimble worker robots. In other words, instead of highly technological mass-production farms, the technium would run on highly technological personal or local farms. Compared to the industrial factory farm, as found in the corn belt of Iowa, this type of advanced gardening would lean toward more diversity, more opportunities, more complexity, more structure, more specialization, more choices, and more sentience.

This new, more convivial agriculture would sit "on top" of industrial agriculture in the same way industrial farming sits on top of subsistence farming, which is still the norm for most of the farmers alive today (most of them living in the developing world). Petrol-based farming will inevitably remain the largest global producer of food for many decades. The trajectories of the technium point toward a more sentient, diverse agriculture intelligently layered over it, much as the tiny region of our language skills sits on top of the bulk of our animal brain. In this way a more heterogeneous, decentralized agriculture is inevitable.

But if the trajectories of the technium are long trains of inevitability, why should we bother encouraging them? Won't they just roll along on their own? In fact, if these trends are inevitable, we couldn't stop them even if we wanted to, right?

Our choices can slow them down. Postpone them. We can work against them. As the dark skies of North Korea show, it is very possible to opt out of the inevitable for a while. On the other hand, there are several good reasons for hastening the inevitable. Imagine what a different world it would be if 1,000 years ago people had accepted the inevitability of political self-governance, or massive urbanization, or educated women, or automation. It is possible an early embrace of these trajectories could have accelerated the arrival of the Enlightenment and science, lifting millions of people out of poverty and increasing longevity centuries earlier. Instead, each of these movements was resisted, delayed, or actively suppressed in different parts of the world at different periods. Those efforts succeeded in crafting societies without these "inevitabilities." From inside these systems these trends did not seem inevitable at all. Only in retrospect do we agree they are clearly long-term trends.

Of course, long-term trends are not equivalent to inevitabilities. Some argue that these particular trends still are not "inevitable" in the future; at any moment a dark age could descend and reverse their course. That is a possible scenario.

They are really only inevitable in the long term. These tendencies are not ordained to appear at a given time. Rather, these trajectories are like the pull of gravity on water. Water "wants" to leak out of the bottom of a dam. Its molecules are constantly seeking a way down and out, as if overcome with an obsessive urge. In a certain sense it is inevitable that someday the water *will* leak out—even though it may be retained by the dam for centuries.

Technology's imperative is not a tyrant ordering our lives in lock-step. Its inevitabilities are not scheduled prophesies. They are more like water behind a wall, an incredibly strong urge pent up and waiting to be released.

It may seem like I am painting a picture of a supernatural force, akin to a pantheistic spirit roaming the universe. But what I am outlining is almost the opposite. Like gravity, this force is embedded in the fabric of matter and energy. It follows the path of physics and obeys the ultimate law of entropy. The force that is waiting to erupt into the technologies of the technium was first pushed by exotropy, built up by self-organization, and gradually thrown from the inert world into life, and from life into minds, and from minds into the creations of our minds. It is an observable force found in the intersection of information, matter, and energy, and it can be repeated and measured, though it has only recently been surveyed.

The trends cataloged here are 13 facets of this urge. This list is not meant to be comprehensive. Other people may draw a different profile.

I would also expect that as the technium expands in coming centuries and our understanding of the universe deepens, we will add more facets to this exotropic push.

In preceding chapters, I've sketched out three of these tendencies and shown how they display themselves in biological evolution and are now extending themselves into the growing technium. In chapter four, I traced the long-term increase in energy density from celestial bodies to the current champ of energy efficiency, the PC chip. In chapter six, I described the way the technium expands possibilities and opportunities. In chapter seven, I retold the story of life's rise as the story of increasing emergence, showing how "higher" levels of organization crystallize out of "lower" parts. In the sections that follow, I will briefly describe the other 10 universal tendencies carrying us forward.

COMPLEXITY

Evolution manifests a number of tendencies, but the most visible of these trends is the long-term move toward complexity. If asked to describe the history of the universe in plain language, most people today would outline this great story: Creation moves from the ultimate simplicity after the big bang to a slow buildup of molecules in a few hot spots till the first tiny spark of life appears, and then an ever-increasing parade of more complex beings, from single cells to monkeys, and then the rush from simple brains to complex technology.

For most observers, the increasing complexity of life, mind, and technology feels intuitive. In fact, modern citizens need no argument to convince them that things have been getting more complex for 14 billion years. That trend seems to parallel the apparent increase in complexity they have seen in their own life spans, so it is easy to believe it has been going on a while.

But our notions of complexity are still ill defined, elusive, and mostly unscientific. What's more complex, a Boeing 747 or a cucumber? The answer right now is we don't know. We intuitively sense that the organization of a parrot is much more complicated than that of a bacterium, but is it 10 times more complicated or a million times? We have no testable way to measure the difference in organization between the two creatures, and we don't even have a good working definition of complexity to help us frame the question.

A favorite mathematical theory of the moment relates complexity to the ease of "compressing" the subject's information content. The more it can be abbreviated without losing its essence, the less complex it is; the less it can be compressed, the more complex. This definition has its own difficulty: Both an acorn and an immense 100-year-old oak tree contain the same DNA, which means both can be compressed, or abbreviated, to the same minimal string of informational symbols. Therefore, both the nut and the tree have the same depth of complexity. But we sense the spreading tree—all those unique crenulated leaves and crooked branches—to be more complex than the acorn. We'd like a better definition. Physicist Seth Lloyd counts 42 other theoretical definitions of complexity, all of them equally inadequate in real life.

While we await a practical mathematical definition of complexity, there is plenty of factual evidence that intuitive "complexity"—loosely defined—exists and is increasing. Some of the most prominent evolutionary biologists don't believe there is an innate long-term trend toward complexity in evolution—or in fact any direction to evolution whatsoever. But a relatively new group of renegade biologists and evolutionists has amassed a convincing case for the broad rise of complexity across all epochs of evolutionary time.

Seth Lloyd, among others, suggests that effective complexity did not begin with biology but began at the big bang. I made the same argument in previous chapters. In Lloyd's informational perspective, fluctuations of quantum energy within the first femtoseconds of the cosmos caused matter and energy to clump. Amplified over time by gravity, these clumps are responsible for the large-scale structure of galaxies—which in their organization display effective complexity.

In other words, complexity preceded biology. Complexity theorist James Gardner calls this "the cosmological origins of biology." The slow ratchet of biological complexity was imported from antecedent structures such as galaxies and stars. Like life, these exotropic self-organized systems teeter on the edge of persistent disequilibrium. They don't burn out like a chaotic flame or explosion (they are persistent) but rather sustain their flux (disequilibrium) over long periods of time without settling into predictable patterns or equilibrium. Their order is neither chaotic nor periodic but semiregular, like a DNA molecule. This type of long-lasting, nonrandom, nonrepeating complexity found in, say, the stable atmosphere of a planet served as the platform for the long-lasting, nonrandom, nonrepeating order in life. In exotropic forms of organization, whether in a star or in genes, effective complexity accrues over time. The complexity of a system rises in a series of steps, where each higher level congeals into a new wholeness. Think of a mass of stars swirling as one galaxy or a mass of cells becoming a multicellular organism. Like with a ratchet, exotropic systems rarely reverse, devolve, or become simpler.

The irreversible ladder of ratcheting complexity and autonomy can be seen in Smith and Szathmary's eight major transitions in organic evolution (discussed in chapter three). Evolution began with "self-replicating molecules" transitioning to the more complex self-sustaining structure of "chromosomes." Then evolution passed through the further complexifying change of cells "from prokaryotes to eukaryotes." After a few more phase changes, the last ratcheting self-organization moved life from languageless societies to those with language.

Each transition shifted the unit that replicated (and upon which natural selection worked). At first, molecules of nucleic acid duplicated themselves, but once they self-organized into a set of linked molecules, they replicated together as a chromosome. Then evolution worked on the whole of both nucleic acid and chromosomes. Later, these chromosomes, housed in primitive prokaryote cells like bacteria, joined together to form a larger autonomous cell (the component cells became organelles of the new one), and now their information was structured and replicated via the complex eukaryote host cell (like an amoeba). Evolution began to work on three levels of organization: genes, chromosomes, cells. These first eukaryote cells reproduced by division on their own, but eventually some (like the protozoan *Giardia*) began to replicate sexually, so now life required a diverse sexual population of similar cells to evolve.

A new level of effective complexity was added: Natural selection began to operate on populations as well. Populations of early single-cell organisms could survive on their own, but many lines self-assembled into multicellular organisms and so replicated as a whole, like a mushroom or seaweed. Now natural selection operated on multicelled creatures, in addition to all the lower levels. Some of these multicellular organisms (such as ants, bees, and termites) gathered into superorganisms and could only reproduce within a colony or society; here evolution emerged at the society level as well. Later, language in human societies gathered individual ideas and culture into a global technium, so humans and their technology could only prosper and replicate together, presenting another autonomous level—society—for evolution and effective complexity.

At each escalating step, the logical, informational, and thermodynamic depth of the resulting organization increased. It became more difficult to compress the structure, and at the same time it contained less randomness and less predictable order. Each step was also irreversible. In general, multicellular lineages do not re-evolve into single-cell organisms; sexual reproducers rarely evolve into parthenogens; social insects rarely unsocialize; and to the best of our knowledge, no replicator with DNA has ever given up genes. Nature will sometimes simplify, but it rarely devolves down a level.

Just to clarify: Within one level of organization trends are uneven. A movement over time toward larger body size or longer life span or higher metabolism may be found only in a minority of species within a family. And directions of change can be inconsistent across taxa. For instance, in mammals, horses may tend to get larger over time, while rodents may get smaller. The trend toward greater effective complexity is primarily visible only in the accumulation of new levels of organization over macrotime. So complexification may not be visible within ferns, say, but it appears between ferns and flowering plants (going from spores to sexual fertilization).

Not every evolutionary species line will proceed up the escalator of complexity (and why should they?), but those that do advance will unintentionally gain new powers of influence that can alter the environment far beyond them. And as with a ratchet, once a branch of life moves up a level, it does not move back. In this way there is an irreversible drift toward greater effective complexity.

This arc of complexity flows from the dawn of the cosmos into life. But the arc continues through biology and now extends itself forward through technology. The very same dynamics that shape complexity in the natural world shape complexity in the technium.

Just as in nature, the number of simple manufactured objects continues to increase. Brick, stone, and concrete are some of the earliest and simplest technologies, yet by mass they are the most common technologies on Earth. And they compose some of the largest artifacts we make: cities and skyscrapers. Simple technologies fill the technium in the way bacteria fill the biosphere. There are more hammers made today than at any time in the past. Most of the visible technium is, at its core, noncomplex technology.

But as in natural evolution, a long tail of ever-complexifying arrangements of information and materials fills our attention, even if

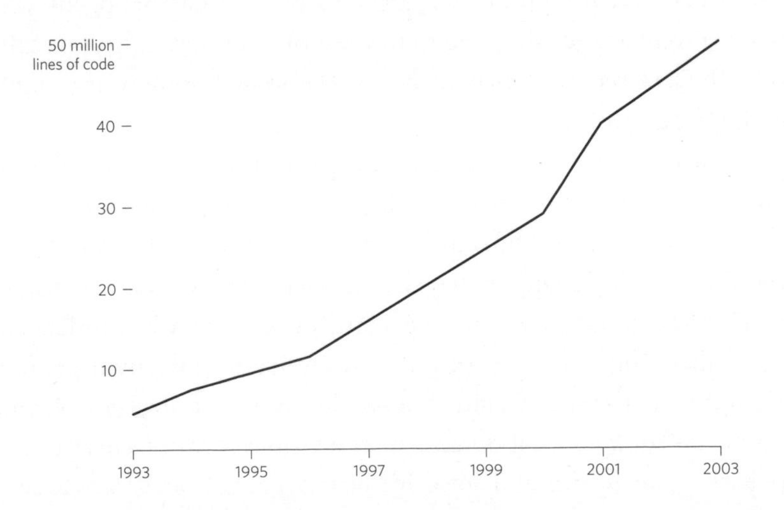

Complexity of Software. The number of lines of codes used by each release of Microsoft Windows between 1993 and 2003.

those complex inventions are small in mass. (Indeed, demassification is one avenue of complexification.) Complex inventions stack up information rather than atoms. The most complex technologies we make are also the lightest and least material. For instance, software, in principle, is weightless and disembodied. It has been complexifying at a rapid rate. The number of lines of code in a basic tool such as Microsoft's Windows operating system has increased tenfold in thirteen years. In 1993, Windows entailed 4 to 5 million lines of code. In 2003, Windows Vista contained 50 million lines of code. Each of those lines of code is the equivalent of a gear in a clock. The Windows OS is a machine with 50 million moving pieces.

Throughout the technium, lineages of technology are restructured with additional layers of information to yield more complex artifacts. For the past 200 years (at least), the number of parts in the most complex machines has been increasing. The diagram below is a logarithmic chart of the trends in complexity of mechanical apparatuses. The first prototype turbo jet had several hundred parts, while a modern turbo jet has over 22,000 parts. The space shuttle has tens of millions of physical parts, yet it contains most of its complexity in its software, which is not included in this assessment.

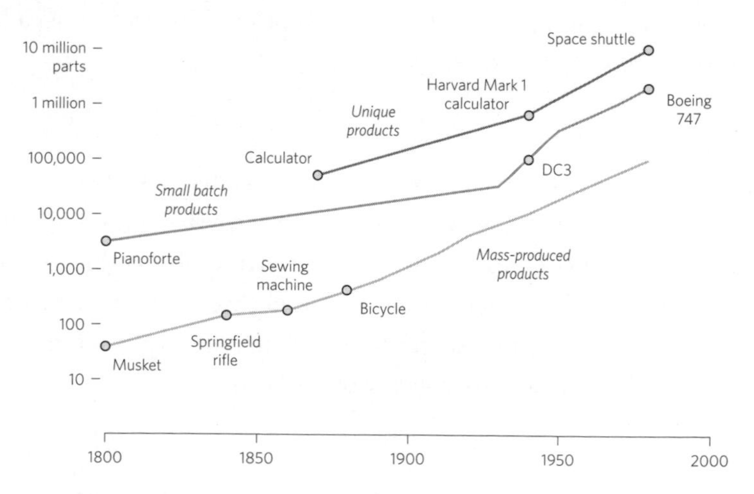

Complexity of Manufactured Machines. The number of parts (shown as powers of 10) used in the most complicated machines of each era over two centuries.

Our refrigerators, cars, and even doors and windows are more complex than two decades ago. The strong trend for complexification in the technium provokes the question, how complex can it get? Where does the long arc of complexity take us? The thrust of 14 billion years of increasing complexity cannot stop today. But when we try to imagine a technium with another million years of complexity accruing at the current rate, we shudder.

There are several different ways technology's complexity can go.

Scenario #1

As in nature, the bulk of technology remains simple, basic, and primeval because it works. And the primitive works well as a foundation for the thin layer of complex technology built upon it. Because the technium is an ecosystem of technologies, most of it will remain in its equivalent microbial stage: brick, wood, hammers, copper wires, electric motors, and so on. We could design nanoscale keyboards that reproduced themselves, but they wouldn't fit our fingers. For the most part, humans will deal with simple things (as we do now) and only interact with the dizzily more complex occasionally, just as we now do. (For most of our day our hands touch relatively coarse artifacts.) Cities and houses remain similar, populated with a veneer of fast-evolving gadgets and screens on every surface.

Scenario #2

Complexity, like all other factors in growing systems, plateaus at some point, and some other quality we had not noticed earlier (perhaps quantum entanglement) takes its place as the prime observable trend. In other words, complexity may simply be the lens we see the world through at this moment, the metaphor of the era, when in reality it is a reflection of us rather than an actual property of evolution.

Scenario #3

There is no limit to how complex all things can get. Everything is complexifying over time, headed toward that omega point of ultimate complexity. The bricks in our building will become smart; the spoon in our hand will adapt to our grip; cars will be as complicated as jets are today. The most complex things we use in a day will be beyond any single person's comprehension.

If I had to, I would bet on scenario #1 and dismiss #2 as unlikely. The bulk of technology will remain simple or semisimple, while a smaller portion will continue to complexify greatly. I expect our cities and homes a thousand years hence to be recognizable, rather than unrecognizable. As long as we inhabit bodies approximately our size—a few meters and 50 kilograms—the bulk of the technology that will surround us need not be crazily more complex. And there is good reason to expect we'll remain the same size, despite intense genetic engineering. Our body size is, weirdly, almost exactly in the middle of the size of the universe. The smallest things we know about are approximately 30 orders of magnitude smaller than we are, and the largest structures in the universe are about 30 orders of magnitude bigger. We inhabit a middle scale that is sympathetic to sustainable flexibility in the universe's current physics. Bigger bodies encourage rigidity; smaller ones encourage empheralization. As long as we own bodies—and what happy being does not want to be embodied?—the infrastructure technology we already have will continue (in general) to work: roads of stone, buildings of modified plant material and Earth, elements not that different from our cities and homes 2,000 years ago. Some visionaries might imagine complex living buildings in the future, for instance, and some of these may happen, but most average structures are unlikely to be composed of materials more complex than the formerly living plants we already use. They don't need to be. I think there is a "complex enough" restraint. Technologies need not complexify to be useful in the future. Danny Hillis, computer inventor, once confided that he believed that there's a good chance that 1,000 years from now computers might still be running programming code from today, say a UNIX kernel. They almost certainly will be binary digital. Like bacteria or cockroaches, these simpler technologies remain simple, and remain viable, because they work. They don't have to get more complex.

On the other hand, the acceleration of the technium could speed up complexity so that even the technological equivalent of bacteria evolve. That's scenario #3, where the entire technosphere zooms away in complexity. Stranger things have happened.

In any of the scenarios, there is no limit to the most complex things we will make. We'll dazzle ourselves with new complexity in many directions. This will complexify our lives further, but we'll adapt to it. There is no going back. We'll hide this complexity with beautiful "simple" interfaces, as elegant as the round ball of an orange. But behind this membrane our stuff will be more complex than the cells and biochemistry of an orange. To keep up with this complexification, our language, tax codes, government bureaucracies, news media, and daily lives will all become more complex as well.

It's a trend we can count on. The long arc of complexity began before evolution, worked through the four billion years of life, and now continues through the technium.

DIVERSITY

The diversity of the universe has been increasing since the beginning of time. In its very first seconds the universe contained only quarks, which began to assemble into a variety of subatomic particles within minutes. By the end of the first hour, the universe contained dozens of types of particles but only two elements, hydrogen and helium. Over the next 300 million years, drifting hydrogen and helium bound themselves together into masses of growing nebulae that eventually collapsed into fiery stars. Star fusion built up dozens of new heavier elements, so the diversity of the chemical universe increased. Eventually, some "metallic" stars exploded into supernovae, spewing their heavy elements into space to be swept up again over millions of years into new stars. In a kind of pumping action, these second- and third-round star furnaces added yet more neutrons to metallic elements to create more varieties of heavy metals until all 100 or so varieties of stable elements were created. The increasing diversity of elements and particles also created an increasing variety of star species, galaxy types, and kinds of orbiting planets. On planets with active tectonic crusts, new kinds of minerals increased in time, as geologic forces reworked and rearranged the elements into new crystals and rocks. The diversity of crystallized minerals on Earth, for instance, increased threefold with the advent of bacterial life. Some geologists believe biochemical processes, and not geological alone, are responsible for the bulk of the 4,300 mineral species we find today.

The invention of life greatly accelerated the diversity in the universe. From a very few species 4 billion years ago, the number and variety of living species on Earth has increased dramatically over geological time to the 30 million now present. This rise has been uneven in several ways. At certain times in Earth's history large-scale cosmological disruptions (such as asteroid hits) have wiped out gains in diversity. And in specific branches of life, diversity sometimes did not advance very much, or even retreated temporarily. But overall, in life as a whole over geologic time, diversity has widened. In fact, life's diversity of taxonomic forms has doubled since the dinosaurian era, only 200 million years ago. The growth of biological differences, as a whole, is expanding exponentially, and this rocketing increase can be seen in vertebrates, plants, and insects.

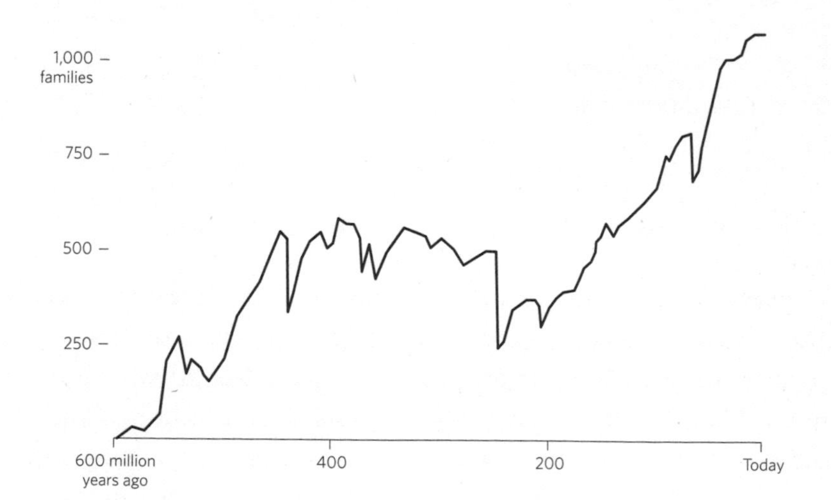

Total Diversity of Life. The increasing diversity of species on Earth, as measured by the number of taxonomic families over the last 600 million years.

The trend toward diversity is further accelerated by the technium. The number of species of technology invented every year is increasing at an increasing rate. It's difficult to precisely count the varieties of technological invention because innovations don't have the defined borders of breeding that most living organisms do. We might count ideas, which underlie each invention. Each scientific article represents at least one new idea. The number of journal articles has exploded in the last 50 years. Each patent is also a species of idea. At last count there were 7 million patents issued in the United States alone, and their total has been increasing exponentially as well.

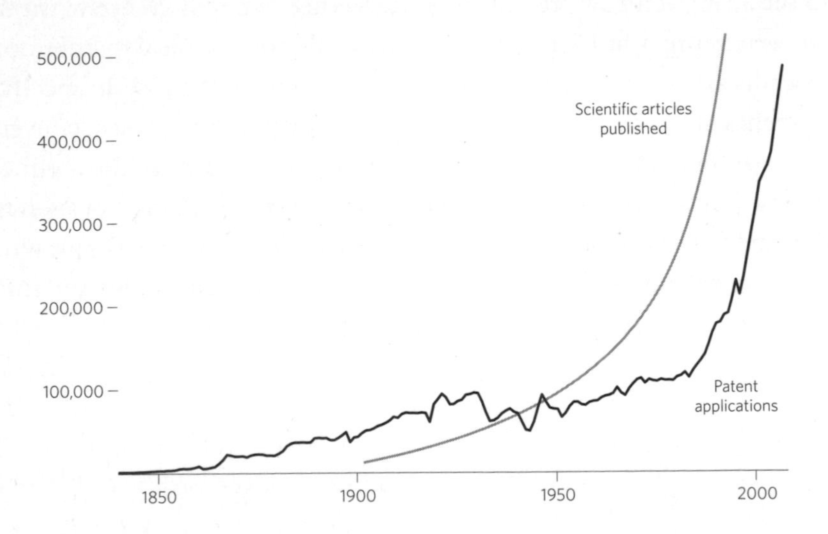

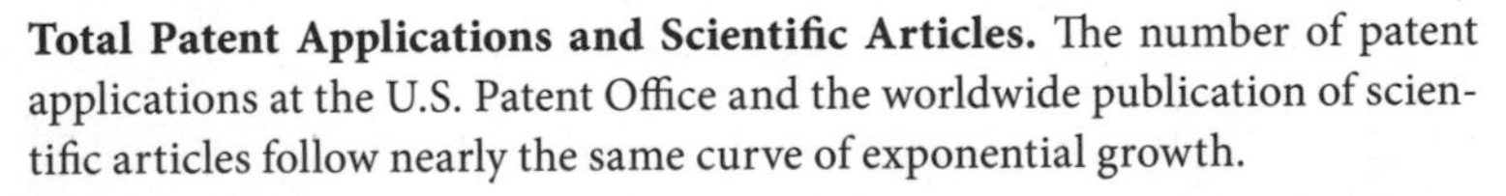
Total Patent Applications and Scientific Articles. The number of patent applications at the U.S. Patent Office and the worldwide publication of scientific articles follow nearly the same curve of exponential growth.

Everywhere we look in the technium we see increased diversity. Manufactured species of underwater organisms, such as 70-foot-long submarines, parallel living organisms, such as a blue whale. Airplanes mimic birds. Our houses are but better nests. But the technium explores niches that the born never ventured into. We know of no living organism using radio waves, yet the technium has produced hundreds of varieties of radio-communicating species. While moles have been digging up Earth for millions of years, two-story tunnel-digging contraptions are so much larger, faster, and less daunted by solid rock than anything born, that we can truly say these synthetic moles occupy a new niche on Earth. X-ray machines have a type of sight unknown among the living. And there is simply no biological analog to an Etch A Sketch, a glow-in-the-dark digital watch, or a space shuttle, to name a few examples. Increasingly, the diversity of the technium has no counterpart in biological evolution, so the technium has truly increased diversity.

The diversity of the technium has already surpassed our skills of recognition. There are so many varieties of things that one individual can't name them. Cognitive researchers have discovered there are about 3,000 easily recognizable noun categories in modern life. This total includes manufactured objects and living organisms, for example, elephant, airplane, palm tree, telephone, chair. These are things that are readily discernable in a flash without thinking. Researchers came up with the estimate of 3,000 categories based on several clues: the number of nouns listed in dictionaries; how many objects are found in the vocabulary of an average six-year-old child; and the number of objects that a primitive artificial learning machine can recognize. They estimated there are, on average, 10 named varieties for each noun category. Ten kinds of chairs, 10 kinds of fish, 10 kinds of phones, 10 kinds of beds that ordinary people might describe. That gives a rough estimate of 30,000 objects in most people's lives, or at least 30,000 that they would recognize. Even when we name a form, most of the variety of life and the technium goes by us without a specific name. We may recognize a bird, but not which species of bird. We know a grass, but not which grass. We know it is a cell phone, but not what model. When pressed we can discern a chef's knife from a Swiss Army knife from a spear point, but we may or may not be able to distinguish a fuel pump from a water pump.

There are branches of the technium where the diversity of technological species is dwindling; today there are fewer innovations in spark catchers, buggy whips, handlooms, and oxcarts. I doubt anyone has invented a new manual butter churn in the last 50 years (although many people are still inventing "better" mousetraps). Handlooms will always be around for art. Oxcarts are not extinct and will probably never go extinct globally as long as oxen are born. But because oxcarts encounter no new demands, they are remarkably stable inventions, continuing over time unchanged, like horseshoe crabs. Most artifacts hovering near obsolescence show a similar constancy. But technological backwaters like these are overwhelmed by the mind-numbing avalanche of innovation, ideas, and artifacts throughout the rest of the expanding technium.

The online retailer Zappos carries 90,000 different varieties of shoes. One hardware wholesaler in the United States, McMaster-Carr, lists over 480,000 products in its catalog. There you can find 2,432 varieties of wood screws alone (yes, I added them up). Amazon carries 85,000 different cell phones and cell-phone products. So far humans have created 500,000 different movies and about one million TV episodes. At least 11 million different songs have been recorded. Chemists have cataloged 50 million different chemicals. Historian David Nye reports, "In 2004, the Ford F-150 pickup truck was available in 78 different configurations that included variations in the cab, the bed, the engine, the drive train, and the trim as well as in the colors of the upholstery and the exterior paint. And once a vehicle was purchased, the owner could customize it further to the point that it literally was one of a kind." If the current rates of inventiveness continue, in 2060 there will be 1.1 billion unique songs and 12 billion different kinds of products for sale.

A few iconoclasts believe this ultradiversity is toxic to humans. In *The Paradox of Choice*, psychologist Barry Schwartz argues that the 285 varieties of cookies, 175 kinds of salad dressing, and 85 brands of crackers for sale in the typical supermarket today are paralyzing consumers. Shoppers enter the store looking for crackers, see a bewildering wall of cracker choices, become overwhelmed with trying to make an informed decision, and finally walk out not purchasing any crackers at all. "Whether people are choosing jam in a grocery store or essay topics in a college class, the more options people have, the less likely they are to make a choice," says Schwartz. Similarly, in trying to choose a medical-benefits plan with hundreds of options, many consumers give up because the complexity of choice is mind numbing and instead withdraw from the program, whereas programs that included a default choice (no decision necessary) had much higher enrollments. Schwartz concludes: "As the number of choices grows further, the negatives escalate until we become overloaded. At this point, choice no longer liberates, but debilitates. It might even be said to tyrannize."

It is true that too many choices may induce regret, but "no choice" is a far worse option. Civilization is a steady migration away from "no choice." As always, the solution to the problems that technology brings, such as an overwhelming diversity of choices, is better technologies. The solution to ultradiversity will be choice-assist technologies. These better tools will aid humans in making choices among bewildering options. That is what search engines, recommendation systems, tagging, and a lot of social media are all about. Diversity, in fact, will produce tools to handle diversity. (Diversity-taming tools will be among the wildly diversity-making 821 million patents that current rates predict will have been filed in the U.S. Patent Office by 2060!) We are already discovering how to use computers to augment our choices with information and web pages (Google is one such tool), but it will take additional learning and technologies to do this with tangible stuff and idiosyncratic media. At the dawn of the web, some very smart computer scientists declared that it would be impossible to select from a billion web pages using a keyword search, but we routinely do just that on 100 billion web pages today. No one is asking for fewer web pages.

Not too long ago the stereotypical image of a technological future was one of standard products, worldwide sameness, and unwavering uniformity. Yet paradoxically, diversity can be unleashed by a type of uniformity. The uniformity of a standard writing system (like an alphabet or script) unleashes the unexpected diversity of literature. Without uniform rules, every word has to be made up, so communication is localized, inefficient, and thwarted. But with a uniform language, sufficient communication transpires in large circles so that a novel word, phrase, or idea can be appreciated, caught, and disseminated. The rigidity of an alphabet has done more to enable creativity than any unhinged brainstorming exercise ever invented.

The standard 26 letters in English have produced 16 million different books in English. Words and language will keep evolving, of course, but their evolution rides on basic fundamentals that are conserved and shared; unvarying (over the short term) letters, spelling, and grammar rules enable creativity in ideas. Increasingly, the technium will converge upon a few universal standards—perhaps basic English, modern musical notation, the metric system (except in the United States!), and mathematical symbols, but also widely adopted technical protocols, from the metric system to ASCII and Unicode. The infrastructure of the world today is built upon a shared system woven from these kinds of standards. That is why you can order machine parts in China to be used in factories in South Africa or have research done in India for drugs released in Brazil. This convergence of fundamental protocols is also why the youth of today can speak to one another directly in a way not possible even a decade ago. They use cell phones and netbooks running common operating systems, but they also employ standard abbreviations and increasingly share common cultural touchstones by watching the same movies, listening to the same music, studying the same subjects and textbooks in school, and pocketing the same technology. In a curious way, the homogenization of shared universals allows them to transmit the diversity of cultures.

In a world of converging global standards, a recurring fear among minority cultures is that their niche differences will be lost. They need not be. In fact, the increasingly common carrier of global communication can heighten the value of their differences. The distinctive foods, medicinal knowledge, and child-rearing practices of, say, the Yanomamo tribe in the Amazon or the San Bushmen in Africa were only esoteric, local knowledge before. Their diversity constituted a difference that did not make a difference outside the tribe, because their knowledge was not connected to other human cultures. But once connected to standard roads, electricity, and communications, their differences can potentially make a difference to others. Even if their knowledge can be applied only in their local environment, wider knowledge of their knowledge makes a difference. Where do wealthy people travel to? Places that retain differences. What eateries attract customers? The ones with distinctive characteristics. What products sell in a global market? Ones that think different.

If such local diversity can remain distinctively different while it is connected (and this is a very big *if*), then that difference becomes steadily more valuable in a global matrix. Maintaining that balance of connected-but-different is a challenge, of course, because much of this cultural difference and diversity originated via isolation, and in the new mix it no longer will be isolated. Cultural differences that thrive without isolation (even if they were born out of it) will compound in value as the world becomes standardized. One example is Bali, Indonesia. The rich, distinguished Balinese culture seems to deepen even as it becomes interconnected with the contemporary world. Like other inhabitants of old and new, the Balinese may wield English as their universal second language while speaking their own tongue at home. They make their ritual flower offerings in the morning and study science at school in the afternoon. They do gamelan and Google.

But how does widening diversity square with the equally pervasive trend I addressed earlier: the inevitable sequence of technologies and the convergence of the technium upon certain forms? At first glance, it would seem as if the channeling of the technium's direction would work against its outward spreading in new directions. If technology converges into a single global sequence of innovations, in what way does this encourage technological diversity?

The sequence of the technium is akin to the development of an organism as it grows through a scripted series of stages. All brains, for instance, progress through a growth pattern from infancy to maturity. But anywhere along that line the brain can generate a remarkable diversity of thoughts.

For the most part, technology will converge to uniform usage around the globe, but occasionally some group or subgroup will devise and refine a type of technology or technique that has limited appeal to a fringe group or marginal use. Very occasionally, this fringe diversity will triumph in the mainstream and overwhelm the existing paradigm, thus rewarding the processes the technium has of encouraging diversity.

Anthropologist Pierre Petrequin once noted that the Meervlakte Dubele and Iau tribes in Papau New Guinea had been using steel axes and beads for many decades but their use had not been adopted by the

Wano tribe a "mere day's walk away." This is still true today. Cell-phone use is significantly broader, deeper, and changing faster in Japan than in the United States. Yet the same factories make the gear for both countries. Similarly, automobile use is broader, deeper, and changing faster in the United States than in Japan.

This pattern is not new. Since the birth of tools, humans have preferred some forms of technology over others for irrational reasons. They may avoid one version or one invention—even when it appears to be more efficient or productive—simply as an act of identity: "Our clan does not do it that way" or "Our tradition does it this way." People may skip an obvious technical improvement because the new way does not feel right or comfortable, even though it is more utilitarian. Anthropologist of technology Pierre Lemonnier has reviewed such patchy interruptions in history and says, "Time and again, people exhibit technical behaviors that do not correspond with any logic of material efficiency or progress."

The Anga tribesmen of Papua New Guinea have hunted wild pigs for thousands of years. To kill a wild pig, which may weigh as much as a man, the Anga construct a trap using little more than sticks, vines, rocks, and gravity. Over time the Anga have refined and modified trap technology to fit their terrain. They have devised three general styles. One is a trench lined with sharp stakes camouflaged under leaves and branches; one is a row of sharpened stakes hidden behind a low barrier protecting bait; and one is a deadfall—a heavy weight suspended above a path that is tripped and released by a passing pig.

Technical know-how of this sort passes easily from village to village in the West Papua highlands. What one community knows, all know (at least over decades, if not centuries). You have to travel many days before variation in knowledge is felt. Most groups of Anga can set any of the three varieties of traps as needed. However, one particular group, the Langimar, ignore the common knowledge of the deadfall trap. According to Lemonnier, "Members of this group can name without difficulty the ten pieces that make up the dead-fall trap, they can describe its functioning, and they can even make a rough sketch; but they do not use the device." Right across the river, the houses of the neighboring Menye tribe can be seen; they use this type of trap—which is a very good technology. Two hours' walk away, the Kapau tribe uses the deadfall, yet the Langimar choose not to. As Lemonnier notes, sometimes "a perfectly understood technology is voluntarily ignored."

It's not as if the Langimar are backward. Further north of the Langimar, some Anga tribes make their wooden arrow tips barbless, selectively ignoring the critical technology of injurious barbs that the Langimar use, despite the fact that the Anga "have had many occasions to note the superiority of the barbed arrows shot at them by their enemies." Neither the available wood type nor the available type of game hunted explains this ethnic dismissal.

Technologies have a social dimension beyond their mere mechanical performance. We adopt new technologies largely because of what they do for us, but also in part because of what they mean to us. Often we refuse to adopt technology for the same reason: because of how the avoidance reinforces or shapes our identity.

Whenever researchers look closely at the dispersal patterns of technology, both modern and ancient, they see patterns of ethnic adoption. Sociologists have noticed that one group of Sami rejected one of the two known types of reindeer lassos, while other Laplanders used both forms. A peculiarly inefficient type of horizontal waterwheel spread all over Morocco, but nowhere else in the world, even though the physics of waterwheels are constant.

We should expect people to continue to exhibit ethnic and social preferences. Groups or individuals will reject all kinds of technologically advanced innovations simply because they can. Or because everyone else accepts them. Or because they clash with their self-conception. Or because they don't mind doing things with more effort. People will choose to abstain from or forsake particular global standards of technology as a form of idiosyncratic distinction. In this way while the planetary culture slides toward convergence of technologies, billions of technology users will diverge in their personal choices as they edge toward using smaller and more eccentric selections of available stuff.

Diversity powers the world. In an ecosystem, increasing diversity is a sign of health. The technium, too, runs on diversity. From the dawn of creation the tide of diversity has risen, and as far as we can look into the future, it will continue to diverge without end.

SPECIALIZATION

Evolution moves from the general to the specific. The first version of the cell was a general-purpose survival-machine blob. Over time, evolution honed that one generality into multiple specialties. In the beginning, the domain of life was restricted to warm ponds. But most of the planet was far more extreme: volcanoes and glaciers. Evolution devised cells that specialized in living in boiling hot water or within freezing ice and special cells that could eat oil or trap heavy metals. Specialization enabled life to colonize these major, but varied, extreme habitats and also to fill millions of niche environments—such as the insides of other organisms or the dimples of dust particles in the air. Very soon, every possible environment on the planet sprouted a specialized variety of life making a living there. Presently there are no sterilized places anywhere on the planet, except in a very few temporary spots within a hospital setting. The cells of life keep specializing.

The trend toward specialization holds for multicellular organisms as well. Cells within an organism specialize. The human body has 210 different types of cells, including the specialized cells in the liver and kidneys. It has distinctive heart muscle cells, different from ordinary skeletal muscle cells. The original omnipotent egg cell that initiates every animal divides into cells with greater specificity, until after less than 50 mitotic cell divisions you and I wind up with a unified assemblage of 10^{15} bone cells, skin cells, and brain cells.

Over evolutionary time, there is a significant rise in the number of cell types in the most complex organisms. In fact, these organisms are more complex in part because they contain more specialized parts. So specialization follows the arc of complexity.

The organism itself also tends toward great specialization. Over the course of time, for one example, barnacles (comprised of 50 specialty cell types) evolve into specialty barnacles: The six-plated barnacle spe-

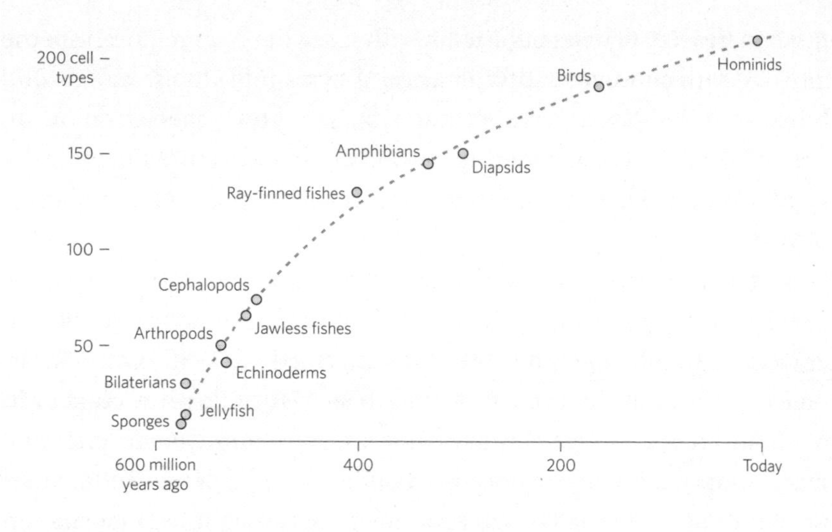

Specialized Cell Types. The maximum number of different cell types found in an organism has increased over evolutionary time.

cializes in extreme high-tide locations that are flooded (with food to eat) only several times a month. The *Sacculina* barnacle grows only inside the egg sac of a living crab. Birds focus into specialized types of seed eaters with specialized beaks: fine ones for small seeds, big fat beaks for hard seeds. A few plants (we call them weeds) are opportunistic and will occupy any disturbed soil, but most plants dedicate their survival skills to a particular niche: dark tropical swamps or dry, windy alpine peaks. Koala bears are famously specialized on eucalyptus trees, and pandas on bamboo.

The trend toward specialization in life is propelled by an arms race. More specialized organisms (such as a clam thriving on sulfuric emissions in lightless deep-sea vents) present more specialized environments for competitors and prey (such as crabs that feed on the sulfuric clams), which breed more specialized strategies (such as parasites on the crabs) and solutions and in the end yet more specialized organisms.

This urge to specialize extends into the technium. The original tool of the hominins, a roundish rock with a broken edge, was a general purpose object used for scraping, cutting, and hammering. Once taken up by Sapiens, it morphed into specialty tools: a separate scraper, cutter, and hammer. The variety of tool species increased over time as specialty tasks increased. Sewing required needles; sewing hide required special needles, sewing woven fabric another. When simple tools were recombined into composite tools (string + stick = bow), specialization increased further. The astounding diversity of manufactured items today is primarily driven by the need for specialized parts of complicated devices.

At the same time, just as in organic life, tools tend to start out being useful for many things and then evolve toward specific tasks. The first camera with photographic film was invented in 1885. Once incarnated, the idea of the camera started to specialize. Within years of its birth, inventors devised tiny spy cameras, extra-large panoramic cameras, compound-lens cameras, high-speed flash cameras. Today there are hundreds of specialty cameras, including those for use deep underwater, those designed for the vacuum of space, and those able to capture the infrared or the ultraviolet. While one can still purchase (or make) the original general-purpose camera, those count for an increasingly small fraction of cameradom.

This sequence from general to specific holds true for most technologies. Automobiles start off appealing broadly, and over time they evolve to specific models, while the general-purpose variety fades. You can choose among compacts, vans, sporty models, sedans, pickups, hybrids, and so on. Scissors are specified for hair, paper, carpet, mesh, or flowers.

As we look into the future, specialization will continue to increase. The first gene sequencer sequenced any gene. The next step is a specialized human DNA sequencer that does only the DNA of humans or another specific species, say, the mouse for researchers. Then we'll see sequencers that specialize in racial genomes (say, for African Americans or Chinese) or extremely portable ones or ones that are extremely fast and sequence in real time, letting a person know whether pollutants are damaging their genes right now. The first commercial virtual-reality consoles will serve up virtual realities for all purposes, but over time, VR consoles will evolve special versions with special gear for games or military practice or movie rehearsals or shopping.

At the moment, computers seem to be headed in the opposite direction, toward becoming ever more general-purpose machines, as they swallow more and more functions. Entire occupations and their workers' tools have been subsumed by the contraptions of computation and networks. Computers have already absorbed calculators, spreadsheets, typewriters, film, telegrams, telephones, walkie-talkies, compasses and sextants, television, radio, turntables, draft tables, mixing boards, war games, music studios, type foundries, flight simulators, and many other vocational instruments. You can no longer tell what a person does by looking at their workplace, because they all look the same: a personal computer; 90 percent of employees are using the same tool. Is that the desk of the CEO, the accountant, the designer, or the receptionist? This convergence is amplified by cloud computing, where the actual work is done on the net as a whole and the tool at hand merely becomes a portal to the work. All portals have become the simplest possible window: a flat screen of some size.

This convergence is temporary. We are still in the early stages of computerization—or rather, intelligenation. Everywhere we currently apply our own personal intelligence (in other words, everywhere we work and play) we are rapidly applying artificial and collective intelligence as well, and rapidly overhauling our tools and expectations. We've intelligenized bookkeeping, photography, financial trading, metal machining, and airplane piloting, among thousands of other tasks. We are about to computerize automobile driving, medical diagnosis, and speech understanding. In our rush toward large-scale intelligenation, we first installed the general-purpose PC, with its mass-produced small brain, midsize screen, and conduit to the net. So all chores get the same tool. To complete the dispersion of intelligenation into all occupations will probably require another decade. Silly as it now sounds, we will put artificial intelligence into hammers, dental picks, forklifts, stethoscopes, and frying pans. All these tools will gain new powers by sharing the universal intelligence of the network. But as their newly augmented roles become clear, the tools will specialize. We can see the first glimmers in the iPhone, Kindle, Wii, tablets, and netbooks. As display and battery technology catches up to chips, the interface to ubiquitous intelligenation will diverge and specialize. Soldiers and other athletes who use their full body want large-scale, enveloping screens, while mobile road warriors will want small ones. Gamers want minimal latency; readers want maximum legibility; hikers want waterproofing; kids want indestructibility. The portals into the grid of computation, or the net, will specialize to a remarkable degree. The keyboard, for one, will lose its monopoly. Speech and gesture input will gain a major role. Spectacle and eyeball screens will supplement walls and flexible surfaces.

With the advent of rapid fabrication (machines that can fabricate things on demand in quantities of one) specialization will leap ahead so that any tool can be customized to an individual's personal needs or desires. Very niche-y functions may summon devices that are assembled for only one task and then unassembled. Ultraspecialized artifacts may live for only a day, like a mayfly. The "long tail" of niches and personal customization is a characteristic not merely of media but of technological evolution itself.

We can forecast the future of almost any invention working today by imagining it evolving into dozens of narrow uses. Technology is born in generality and grows to specificity.

UBIQUITY

The consequence of self-reproduction in life, as well as in the technium, is an inherent drive toward ever-presence. Given a chance, dandelions or raccoons or fire ants will replicate till they cover the Earth. Evolution equips a replicant with tricks to maximize its spread no matter the constraints. But because physical resources are limited and competition relentless, no species can ever reach full ubiquity. Yet all life is yearning in that direction. Technology, too, wants to be ubiquitous.

Humans are the reproductive organs of technology. We multiply manufactured artifacts and spread ideas and memes. Because humans are limited (only six billion alive at the moment) and there are tens of millions of species of technology or memes to spread, few gadgets can reach full 100 percent ubiquity, although several come close.

Nor do we really want all technology to be ubiquitous. Preferably, we would engineer away the need for replacement artificial hearts through genetics or pharmaceuticals or diet. In the same way, the remedial technology of carbon sequestration (removing carbon from the atmosphere) would ideally never become ubiquitous. Much better would be the ubiquity of low-carbon energy sources in the first place, using the technologies of photons (solar), fusion (nuclear), wind, or hydrogen. The problem with remedial technologies is that once their niches are filled, they lead nowhere else. A vaccine has no future if it is universally successful. In the long run, convivial technologies that open up other technologies tend to ascend to ubiquity fastest.

From the perspective of the planetary biosphere, the most ubiquitous technology on Earth is agriculture. The steady surplus of high-quality food from agriculture is vigorously open-ended in that this abundance enabled civilization and birthed its millions of technologies. The spread of agriculture is the largest-scale engineering project on the planet. One third of Earth's land surface has been altered by the mind and hand of humans. Native plants have been displaced, soil moved, and domesticated crops planted in their stead. Great stretches of Earth's surface have been semidomesticated into pastureland. The most drastic of these changes—such as uninterrupted tracts of giant farms—are visible from space. Measured in number of square kilometers, the most ubiquitous technology on the planet are the five major domesticated crops: maize, wheat, rice, cane sugar, and cows.

The second most plentiful planetary technology is roads and buildings. Simple clearings for the most part, dirt roads extend their rootlike tentacles into most watersheds, crisscrossing valleys and winding their way up many mountains. The web of constructed roads forms a reticulated cloak around the continents of this planet. A string of buildings follow along the dendritic branches of roads. These artifacts are made of cut tree fiber (wood, thatch, bamboo) or molded Earth (adobe, brick, stone, concrete). At the hubs of roads stand magnificent stone and silica megalopolises, which have rerouted the flow of materials so that much of the technium circulates through them. Rivers of food and raw materials flow in, and debris flows out. Every person living in a developed urban area moves 20 tons of material annually.

Not as visible, but perhaps more pervasive at the planetary level, are the technologies of fire. Controlled burning of carbon fuels, particularly mined coal and oil, has led to changes in the Earth's atmosphere. Reckoned in total mass and converge, these furnaces (which often travel along the roads as engines in automobiles) are dwarfed by roads. Though smaller in scale than the roads they ride on or the homes and factories they burn in, these tiny, deliberate fires are able to shift the composition of the globe's voluminous atmosphere. It is possible that this collective burning, tiny in footprint, may be the largest-scale technological impact on the planet.

Then there are the things we surround ourselves with. In daily human life, the list of near-ubiquitous technologies includes cotton cloth, iron blades, plastic bottles, paper, and radio signals. These five technological species are within reach of nearly every human alive today, both in the cities and in the most remote rural villages. Each of these

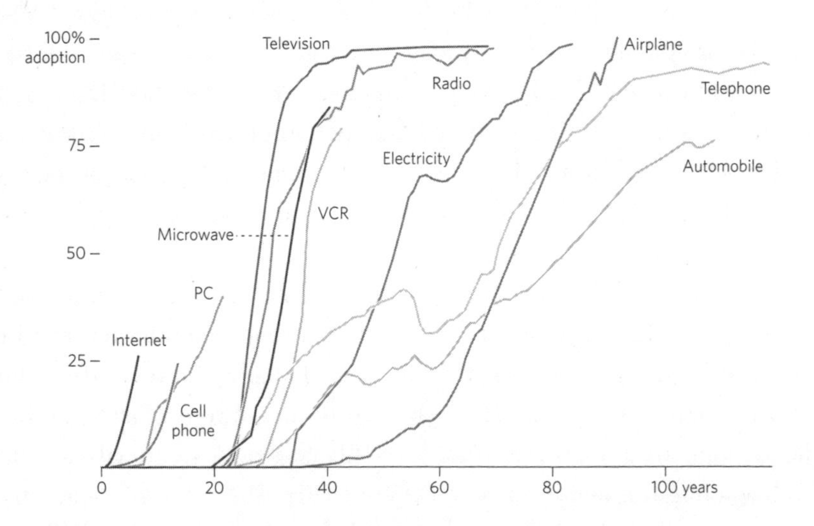

Accelerating Pace of Technology Adoption. The percentage of U.S. consumers owning or using a particular technology plotted over the number of years since its invention.

technologies opens up vast new territories of possibilities: paper—cheap writing, printing, and money; metal blades—art, craft, gardening, and butchering; plastic—cooking, water, and medicines; radio—connection, news, and community. Fast on their heels follow the nearly ubiquitous species of metal pots, matches, and cell phones.

Total ubiquity is the end point that all technologies tend toward but never reach. But there is a practical ubiquity of near saturation that is sufficient to flip the dynamic of a technology onto another level. In urban places everywhere, the speed at which new technologies disperse to the point of saturation has been increasing.

Whereas it took electrification 75 years to reach 90 percent of U.S. residents, it's taken only 20 years for cell phones to reach the same penetration. The rate of diffusion is accelerating.

And more is different. Something strange happens with ubiquity. A few automobiles roaming along a few roads is fundamentally different from a few automobiles for every person. And not just because of the increased noise and pollution. A billion operating cars spawn an emergent system that creates its own dynamics. Ditto for most inventions. The first few cameras were a novelty. Their impact was primarily to fire painters from the job of recording the times. But as photography became easier to use, common cameras led to intense photojournalism, and eventually they hatched movies and Hollywood alternative realities. The diffusion of cameras cheap enough that every family had one in turn fed tourism, globalism, and international travel. The further diffusion of cameras into cell phones and digital devices birthed a universal sharing of images, the conviction that something is not real until it is captured on camera, and a sense that there is no significance outside of the camera view. The still further diffusion of cameras embedded into the built environment, peeking from every city corner and peering down from every room's ceiling, forces a transparency upon society. Eventually, every surface of the built world will be covered with a screen and every screen will double as an eye. When the camera is fully ubiquitous, everything is recorded for all time. We have a communal awareness and memory. These effects powered by ubiquity are a long way from simply displacing painting.

Again and again ubiquity changes everything.

One thousand automobiles open up mobility, create privacy, supply adventure. One *billion* automobiles create suburbia, eliminate adventure, erase parochial minds, trigger parking problems, birth traffic jams, and remove the human scale of architecture.

One thousand live, always-on cameras make downtowns safe from pickpockets, nab stoplight runners, and record police misbehavior. One *billion* live, always-on cameras serve as a community monitor and memory, they give the job of eyewitness to amateurs, they restructure the notion of the self, and they reduce the authority of authorities.

One thousand teleportation stations rejuvenate vacation travel. One *billion* teleportation stations overturn commutes, reimagine globalism, introduce tele-lag sickness, reintroduce the grand spectacle, kill the nation-state, and end privacy.

One thousand human genetic sequences jump-start personalized medicine. One *billion* genetic sequences every hour enable real-time genetic damage monitoring, upend the chemical industry, redefine illness, make genealogies hip, and launch "ultraclean" lifestyles that make organic look filthy.

One thousand screens the size of buildings keep Hollywood going. One *billion* screens everywhere become the new art, create a new advertising medium, revitalize cities at night, accelerate locative computing, and rejuvenate the commons.

One thousand humanoid robots revamp the Olympics and give a boost to entertainment companies. One *billion* humanoid robots cause massive shifts in employment, reintroduce slavery and its opponents, and demolish the status of established religions.

In the course of evolution every technology is put to the question, What happens when it becomes ubiquitous? What happens when everyone has one?

Usually what happens to a ubiquitous technology is that it disappears. Shortly after their invention in 1873, modern electric motors propagated throughout the manufacturing industry. Each factory stationed one very large, expensive motor in the place where a steam engine had formerly stood. That single engine turned a complex maze of axles and belts, which in turn spun hundreds of smaller machines scattered throughout the factory. The rotational energy twirled through the buildings from that single source.

Ubiquitous Motors. Machinery for grinding crankshafts at the Ford Motor Company, 1915.

By the 1910s, electric motors had started their inevitable spread into homes. They had been domesticated. Unlike a steam engine, they did not smoke or belch or drool. Just a tidy, steady whirr from a five-pound

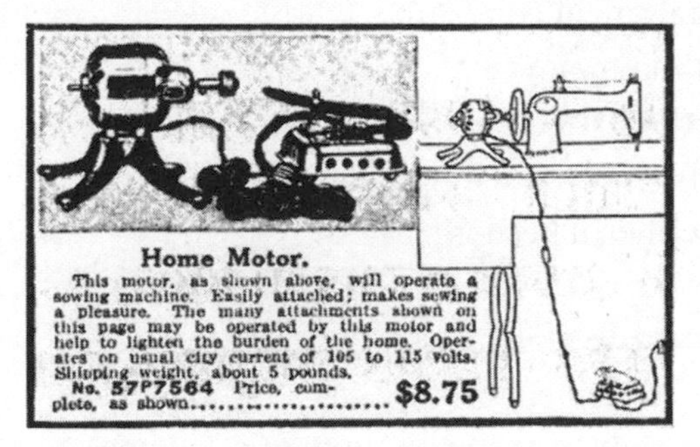

Ad for the Home Motor. A 1918 magazine advertisement for the Sears Home Motor.

hunk. As in factories, these single "home motors" were designed to drive all the machines in one home. The 1916 Hamilton Beach "Home Motor" had a six-speed rheostat and ran on 110 volts. Designer Donald Norman points out a page from the 1918 Sears, Roebuck and Co. catalog advertising the Home Motor for $8.75 (which is equivalent to about $100 these days). This handy motor would spin your sewing machine. You could also plug it into the Churn and Mixer Attachment ("for which you will find many uses") and the Buffer and Grinder Attachments ("will be found very useful in many ways around the home"). The Fan Attachment "can be quickly attached to Home Motor," as well as the Beater Attachment to whip cream and beat eggs.

One hundred years later, the electric motor has seeped into ubiquity and invisibility. There is no longer one home motor in a household; there are dozens of them, and each is nearly invisible. No longer stand-alone devices, motors are now integral parts of many appliances. They actuate our gadgets, acting as the muscles for our artificial selves. They are everywhere. I made an informal census of all the embedded motors I could find in the room I am sitting in while I write:

5 spinning hard disks
3 analog tape recorders
3 cameras (move zoom lenses)
1 video camera
1 watch
1 clock
1 printer
1 scanner (moves scan head)
1 copier
1 fax machine (moves paper)
1 CD player
1 pump in radiant floor heat

That's 20 home motors in one room of my home. A modern factory or office building has thousands. We don't think about motors. We are unconscious of them, even though we depend on their work. They rarely fail, but they have changed our lives. We aren't aware of roads and electricity either because they are ubiquitous and usually work. We don't think of paper and cotton clothing as technology because their reliable presences are everywhere.

In addition to a deep embeddedness, ubiquity also breeds certainty. The advantages of new technology are always disruptive. The first version of an innovation is cumbersome and finicky. It is, to repeat Danny Hillis's definition of technology, "stuff that does not work yet." A newfangled type of plow, waterwheel, saddle, lamp, phone, or automobile can offer only uncertain advantages in exchange for certain trouble. Even after an invention has been perfected elsewhere, when it is first introduced into a new zone or culture it requires the retraining of old habits. The new type of waterwheel may require less water to run but also require a different type of milling stone that is hard to find, or it may produce a different quality of flour. A new plow may speed tilling but demand planting seed later, thus disrupting ancient traditions. A new kind of automobile may have a longer range but less reliability or greater efficiency but less range, altering driving and fueling patterns. The first version is almost always only marginally better than what it hopes to displace. That is why only a few eager pioneers are inclined to adopt an innovation at first, because the new primarily promises headaches and the unknown. As an innovation is perfected, its benefits and education are sorted out and illuminated, it becomes less uncertain, and the technology spreads. That diffusion is neither instantaneous nor even.

In every technology's life span, then, there will be a period of haves and have-nots. Clear advantages may flow to the individuals or societies who first take a risk with unproven guns or the alphabet or electrification or laser eye surgery over those who do not. The distribution of these advantages may depend on wealth, privilege, or lucky geography as much as desire. This divide between the haves and the have-nots was most recently and most visibly played out at the turn of the last century when the internet blossomed.

The internet was invented in the 1970s and offered very few benefits at first. It was primarily used by its inventors, a very small clique of professionals fluent in programming languages, as a tool to improve itself. From birth the internet was constructed in order to make talking about the idea of an internet more efficient. Likewise, the first ham radio operators primarily broadcast discussions about ham radio; the early world of CB radio was filled with talk about CB; the first blogs were about blogging; the first several years of twitterings concerned Twitter. By the early 1980s, early adopters who mastered the arcane commands of network protocols in order to find kindred spirits interested in discussing this tool moved onto the embryonic internet and told their nerdy friends. But the internet was ignored by everyone else as a marginal, teenage male hobby. It was expensive to connect to; it demanded patience, the ability to type, and a willingness to deal with obscure technical languages; and very few other nonobsessive people were online. Its attraction was lost on most people.

But once the early adaptors modified and perfected the tool to give it pictures and a point-and-click interface (the web), its advantages became clearer and more desirable. As the great benefits of digital technology became apparent, the question of what to do about the have-nots became a contested issue. The technology was still expensive, requiring a personal computer, a telephone line, and a monthly subscription fee—but those who adopted it acquired power through knowledge. Professionals and small businesses grasped its potential. The initial users of this empowering technology were—on the global scale—the same set of people who had so many other things: cars, peace, education, jobs, opportunities.

The more evident the power of the internet as an uplifting force became, the more evident the divide between the digital haves and have-nots. One sociological study concluded that there were "two Americas" emerging. The citizens of one America were poor people who could not afford a computer, and of the other, wealthy individuals equipped with PCs who reaped all the benefits. During the 1990s, when technology boosters like me were promoting the advent of the internet, we were often asked: What are we going to do about the digital divide? My answer was simple: nothing. We didn't have to do anything, because the natural history of a technology such as the internet was self-fulfilling. The have-nots were a temporary imbalance that would be cured (and more) by technological forces. There was so much profit to be made connecting up the rest of the world, and the unconnected were so eager to join, that they were already paying higher telecom rates (when they could get such service) than the haves. Furthermore, the costs of both computers and connectivity were dropping by the month. At that time most poor in America owned televisions and had monthly cable bills. Owning a computer and having internet access was no more expensive and would soon be cheaper than TV. In a decade, the necessary outlay would become just a $100 laptop. Within the lifetimes of all born in the last decade, computers of some sort (connectors, really) will cost $5.

This was simply a case, as computer scientist Marvin Minsky once put it, of the "haves and have-laters." The haves (the early adopters) overpay for crummy early editions of technology that barely work. They purchase flaky first versions of new goods that finance cheaper and better versions for the have-laters, who will get things that work for dirt cheap not long afterward. In essence, the haves fund the evolution of technology for the have-laters. Isn't that how it should be, that the rich fund the development of cheap technology for the poor?

We saw this have-later cycle play out all the more clearly with cell phones. The very first cell phones were larger than bricks, extremely costly, and not very good. I remember an early-adopter techie friend who bought one of the first cell phones for $2,000; he carried it around in its own dedicated briefcase. I was incredulous that anyone would pay that much for something that seemed more toy than tool. It seemed equally ludicrous at that time to expect that within two decades, the $2,000 devices would be so cheap as to be disposable, so tiny as to fit in a shirt pocket, and so ubiquitous that even the street sweepers of India would have one. While internet connection for sidewalk sleepers in Calcutta seemed impossible, the long-term trends inherent in technology aim it toward ubiquity. In fact, in many respects the cell coverage of these "later" countries overtook the quality of the older U.S. system, so that the cell phone became a case of the haves and have-sooners, in that the later adopters got the ideal benefits of mobile phones sooner.

The fiercest critics of technology still focus on the ephemeral have-and-have-not divide, but that flimsy border is a distraction. The significant threshold of technological development lies at the boundary between commonplace and ubiquity, between the have-laters and the "all have." When critics asked us champions of the internet what we were going to do about the digital divide and I said "nothing," I added a challenge: "If you want to worry about something, don't worry about the folks who are currently offline. They'll stampede on faster than you think. Instead you should worry about what we are going to do when *everyone* is online. When the internet has six billion people, and they are all e-mailing at once, when no one is disconnected and always on day and night, when everything is digital and nothing offline, when the internet is ubiquitous. That will produce unintended consequences worth worrying about."

I would say the same today about DNA sequencing, GPS location tracking, dirt-cheap solar panels, electric cars, or even nutrition. Don't worry about those who don't own a personal fiber-optic cable to their school; worry what happens when everyone does. We were so focused on those who don't have plenty to eat that we missed what happens when everyone does have plenty. A few isolated manifestations of a technology can reveal its first-order effects. But not until technology saturates a culture do the second- and third-order consequences erupt. Most of the unintended consequences that so scare us in technology usually arrive in ubiquity.

And most of the good things as well. The trend toward embedded ubiquity is most pronounced in technologies that are convivially open-ended: communications, computation, socialization, and digitization. There appears to be no end to their possibilities. The amount of computation and communication that can be crowded into matter and materials seems infinite. There is nothing we have invented to date about which we've said, "It's smart enough." In this way the ubiquity of this type of technology is insatiable. It constantly stretches toward a pervasive presence. It follows the trajectory that pushes all technology into ubiquity.

FREEDOM

As with other things, our free wills are not unique. Unconscious free-will choice exists in primeval modes in animals. Every animal has primitive wants and will make choices to satisfy them. But free will precedes even life. Some theoretical physicists, including Freeman Dyson, argue that free will occurs in atomic particles, and therefore free choice was born in the great fire of the big bang and has been expanding ever since.

As an example Dyson notes that the exact moment when a subatomic particle decays or changes the direction of its spin must be described as an act of free will. How can this be? Well, all the other microscopic motions of that cosmic particle are absolutely predetermined from the particle's previous position/state. If you know where a particle is and its energy and direction, you can predict exactly, without fail, where it will be in the next moment. This utter allegiance to a path predetermined by its previous state is the foundation of the "laws of physics." Yet a particle's spontaneous dissolution into subparticles and energy rays is not predictable, nor predetermined by laws of physics. We tend to call this decay into cosmic rays a "random" event. Mathematician John Conway proposed a proof arguing that neither the mathematics of randomness nor the logic of determinism can properly explain the sudden (why right now?) decay or shift of spin direction in cosmic particles. The only mathematical or logical option left is free will. The particle simply chooses in a way that is indistinguishable from the tiniest quantum bit of free will.

Theoretical biologist Stuart Kauffman argues that this "free will" is a result of the mysterious quantum nature of the universe, by which quantum particles can be two places at once, or be both wave and particle at once. Kauffman points out that when physicists shoot photons of light (which are wave/particles) through two tiny parallel slits (a famous experiment), the photon can pass through only as either a wave or a particle, but not both. The photon particle must "choose" which form it manifests. But the weird and telling thing about this experiment, which has been done many times, is that the wave/particle only chooses its form (either a wave or a particle) *after* it has already passed through the slit and is measured on the other side. According to Kauffman, the particle's shift from undecided state (called quantum decoherence) to the decided state (quantum coherence) is a type of volition and thus the source of free will in our own brains, since these quantum effects happen in all matter.

As John Conway writes,

> Some readers may object to our use of the term "free will" to describe the indeterminism of particle responses. Our provocative ascription of free will to elementary particles is deliberate, since our theorem asserts that if experimenters have a certain freedom, then particles have exactly the same kind of freedom. Indeed, it is natural to suppose that this latter freedom is the ultimate explanation of our own.

The tiny specks of quantum choice inherent in particles were leveraged by the vast increases in organization whipped up by life. A spontaneous "volitional" decay of a cosmic particle might pass through a cell and on its way trigger a mutation in the highly ordered structure of its DNA molecule. Let's say it knocks a hydrogen atom off a cytosine base; then that indirect volition (what biologists used to call a random mutation) could give birth to an innovative protein sequence. Of course, most particle choices only bring death to the cell sooner, but with luck a mutation will confer a survival advantage to the whole organism. Since beneficial traits are retained and built upon by the DNA system, the positive effects of free will can accumulate. Volitional cosmic rays also trigger synapse firings in neurons, which introduce novelty signals (ideas) into nerves and brain cells, some of which indirectly nudge an organism to do this or that. By the complex machinery of evolution, these remotely induced "choices" are captured, retained, and amplified as well. Mutations triggered by the free will of particles, in the aggregate and over billions of years, evolve organisms with more senses, more limbs, more degrees of freedom. As usual, this is a virtuous self-amplifying circle.

As evolution rises, "choicefulness" increases. A bacterium has a few choices—perhaps to slide toward food or to divide. A plankton, with more complexity, more cellular machinery, has more options. A starfish can wiggle its arms, flee (fast or slow?) or fight a rival, choose a meal or a mate. A mouse has a million choices to make in its life. It has a longer list of things it can move (whiskers, eyeballs, eyelids, tail, toes) and a wider range of environments to exert its will upon, as well as a longer duration of life to decide in. More complexity expands the number of possible choices.

A mind, of course, is a choice factory, constantly inventing new ways to choose. "With more choices, we have more opportunities," declared Emmanuel Mesthene, a technology philosopher at Harvard. "With more opportunities, we can have more freedom, and with more freedom we can be more human."

A major consequence of creating cheap and ubiquitous artificial minds is to infuse higher levels of free will into our built environment. Of course we'll put minds into robots, but we'll also implant cars, chairs, doors, shoes, and books with slivers of choice-making intelligence, and all these expand the realm of those making free choices, even if those choices are only particle sized.

Where there are free wills there are mistakes. When we unleash inanimate objects from their shackles of hereditary inertness and give them particles of choice, we give them freedom to make mistakes. We can think of each new crumb of artificial sentience as a new way to make mistakes. To do stupid things. To make errors. In other words, technology teaches us how to make innovative kinds of mistakes we could not make before. In fact, asking ourselves how humanity might make entirely new kinds of mistakes is probably the best metric we have for discovering new possibilities of choice and freedom. Engineering our genome is primed to create a new kind of mistake and therefore indicates a new level of free will. Geoengineering the planet's climate might also indicate a new arena of mistakes and therefore choice. Connecting every person to every other person alive in real time via cell phone or wires also unleashes new powers of choice and incredible potential for mistakes.

All inventions widen the space of what is possible and thereby stretch the parameters in which choices can be made. But just as important, the technium creates new mechanisms that can exercise unconscious free will. Whenever you send an e-mail, invisible fancy algorithms on data servers decide the path your message will hop along in the global network in order to arrive with minimal congestion and maximum speed. Quantum choice probably does not play a role in these choices. Rather, a billion interacting deterministic factors influence it. Because unraveling these factors is an intractable problem, these choices are in practice free-will decisions of the network, and the internet is making billions of them every day.

Fuzzy-logic appliances make real choices. Their tiny chip brains weigh competing factors, and in a nondeterministic way the fuzzy-logic circuits make a decision about when to turn off the dryer or to what temperature to heat the rice. Many kinds of complex, adaptive contraptions—for example, the sophisticated computerized autopilot that flew the 747 jet you rode on the other day—expand the range of free will by generating new kinds of behaviors out of reach of either humans or other living creatures. An experimental robot at MIT can catch a tennis ball using a brain and arm that is thousands of times faster than a human brain/arm combo. This robot shifts so fast while deciding where to put its hand that our eyes can't even see it move. Here free will has expanded into a new realm of speed.

When you type a keyword into Google, it considers approximately a trillion documents before it chooses (and "choose" is the correct word) the page it guesses you want. No human can possibly encompass that planetary volume of material. In this way, a search engine gives free choice a scale way beyond the human. Once our machines unleashed possibilities as fast as we could think them up; now they unleash possibilities without waiting for us.

In the world of tomorrow, high-tech automobiles that park themselves will make as many free-will choices as we do when we park. To varying degrees, technology will practice free will at greater levels than it does today.

First the technium expands the range of possible choices, and then it expands the range of agents that can make choices. The more powerful a new technology is, the greater the new freedoms it opens up. Multiplying options goes hand in hand with multiplying liberty. Nations of the world with plenty of economic choices, abundant communication options, and high education possibilities tend to rank highest in available liberty. But this expansion includes possible abuse as well. Present in every new technology is the potential to make new mistakes. The freedom to choose increases in many ways as the technium grows.

MUTUALISM

More than half of the living species on this planet are parasitic. That is, they depend upon another species for their survival in at least one phase of their life. At the same time, biologists believe that every organism alive (including parasites themselves) hosts at least one parasite. This makes the natural world a hotbed of shared existence.

Parasitism is just a single degree along a wide continuum of mutualism. At one end there is the fact that any living creature depends on others (its parents directly and others indirectly) for its life; at the other end is the symbiotic embrace of two distinct species, algae and fungi, which together present as one species of lichen. In between are multiple varieties of parasitism, some of which do the host no harm at all and others (such as ants on an acacia bush) where the parasite aids its host.

Three strands of increasing mutualism weave through evolution, or what is properly called coevolution.

1. As life evolves, it becomes increasingly dependent on other life. The oldest bacteria eke out their livelihood from lifeless rock, water, and volcanic fumes. They touch only inert matter. Later, more complex microbes, such as *E. coli,* will spend their entire life inside our guts, surrounded by our living cells, eating our food. They touch only other living things. Over time, the home environment for a creature is more likely to be living rather than inert. The

entire animal kingdom is a fine example of this trend. Why bother to produce food from the elements yourself when you can just steal it from other living organisms? Animals are more mutualistic than plants in this way.

2. As life evolves, nature creates more opportunities for dependencies *between species.* Every organism that creates a successful niche for itself also creates potential niches for other species (all those potential parasites!). Let's say an alpine meadow enriches its mix over time with an additional new species of bee to pollinate the crocus. That addition increases the numbers of possible relationships between all the meadow creatures.

3. As life evolves, possibilities for cooperation between members of the *same species* increase. The superorganism of an ant colony or beehive is an extreme case of intraspecies cooperation and mutualism. Greater sociality among organisms is a stabilizing ratchet in evolution. Once socialization is acquired, it is rarely let go.

Human life is immersed in all three mutualisms. First, we are remarkably dependent on other life for survival. We eat plants and other animals. Second, there is no other species on this planet that uses the variety and number of other living species that we do to stay healthy and prosperous. And third, we are famously a social animal, requiring others of our species to raise us, teach us how to survive, and keep us sane. In this way our life is deeply symbiotic; we live inside of other life. The technium pushes these three varieties of mutualism even further.

Most machines today never touch the Earth, or water, or even the air. The tiny heart of a microcircuit beating in the core of the PC scripting these words I am writing is sealed from the elements and is completely surrounded by other manufactured artifacts. This microscopic artifact feeds off energy generated by a huge turbine (or on a sunnier day by the solar panels on my roof), sends its output to another machine (my cinema display monitor), and if it is lucky will be digested for precious elements by other machines when it is dead.

There are plenty of machine parts that never touch human hands. They are made by robots and inserted inside devices (such as the bearings in an automobile water pump) that are then placed within larger technological contraptions. A little while ago my son and I disassembled the innards of an old CD player. I am certain that when we opened the laser housing, we were the first nonmechanical beings to see that intricate inner piece. Until then it had only been touched by machines.

The technium is moving toward increased symbiosis between humans and machines. This is the subject of thrilling Hollywood sci-fi blockbusters, but it also plays out in a million small ways in real life. It is very clear that we are creating a symbiotic memory with the web and Google-like technologies. When Google (or one of its descendants) is able to understand ordinary spoken questions and is living in a layer of our clothing, we will quickly absorb this tool into our minds. We will depend on it, and it will depend on us—both to continue to exist and to continue getting smarter, because the more people use it the smarter it gets.

Some people find this technological symbiosis scary, or even horrifying, but it is not much different from our use of paper and pencil in long division. For most ordinary humans, dividing long numbers without technology is impossible. Our brains are simply not wired to accomplish this naturally. We use the technologies of writing and tricks of arithmetic to divide, multiply, or manipulate large or multiple numbers. We can do it in our heads in a fashion, but only by watching ourselves virtually write the problem out on virtual paper in our mind. My wife grew up using an abacus to do arithmetic. An abacus is a 4,000-year-old analog calculator, a technological aid for doing calculations faster than with a pencil. When there is no abacus around, she does the same thing, virtually moving the virtual beads with her fingers in order to arrive at an answer. Somehow, being totally dependent on technology to add and subtract doesn't spook us, but being dependent on the web to remember facts sometimes does.

The technium is also pushing the increased mutualism among machines. The majority of telecommunications traffic in the world is not messages flowing between humans but messages between machines. Nearly 75 percent of the world's total nonsolar energy—in other words, the energy created through technological means and flowing through the pipes and wires of the technium—is used for the benefit of moving, housing, and maintaining our machines. Most trucks, trains, and planes are not moving people but freight. Most heating and cooling is not conditioning humans but other stuff. The technium spends only one quarter of its energy on human comfort, food, and travel needs; the rest of the energy is made by technology for technology.

We are just starting our journey of increasing mutualism between the technium and ourselves. Mastering this commensalism, like adding with pen and paper, will take some education. The most visible aspect of the exotropic trend toward mutualism is the way in which the technium increases the sociability between humans. I'd like to sketch out this trajectory because it is most immediate. For the next 10 to 20 years, the socializing aspects of the technium will be one of its major traits and a major event for our culture.

There is a natural progression of increased connectivity among humans. Groups of people start off simply sharing ideas, tools, creations, and then progress to cooperation, collaboration, and finally collectivism. At each step the amount of coordination increases.

Today, online masses have an incredible willingness to share. The number of personal photos posted on Facebook and MySpace is astronomical. It's a safe bet that the overwhelming majority of photos taken with a digital camera are shared in some fashion. Wikipedia is another remarkable example of symbiotic technology in operation—and not just Wikipedia, but wikiness at large. There are 145 other wiki engines today, each one powering myriad sites that allow users to collaboratively write and edit material. Then there are status updates, map locations, half thoughts posted online. Add to this the six billion videos delivered by YouTube each month in the United States alone and the millions of fan-created stories deposited on fan-fiction sites. The list of sharing organizations is almost endless: Yelp for reviews, Loopt for locations, Delicious for bookmarks.

Sharing serves as the foundation for the next higher level of communal engagement: cooperation. When individuals work together toward a large-scale goal, this effort produces results that emerge at the group level. Not only have amateurs shared more than three billion photos on Flickr, but they have cooperatively tagged them with categories, labels, and keywords. Others in the community cull the pictures into sets. The popularity of Creative Commons licensing means that communally, if not outright communistically, your picture is my picture. Anyone can use a photo, just as a communard might use the community wheelbarrow. I don't have to shoot yet another photo of the Eiffel Tower, because the community can provide a better one than I can take myself.

Evolution engineers mutualism into biology because its benefits are win-win. Individuals gain and the group gains. The same is happening in digital technology today on several levels. First, the tools of social media in aggregator sites such as Facebook and Flickr benefit users directly, letting them tag, bookmark, rank, and archive their own material for their own improved access. They spend time categorizing their photos because it makes it easier for themselves to find old ones. That is individual gain. Second, other users benefit from an individual's tags, bookmarks, and so on. That individual's work makes it easier for them to use the images. In this way, the whole group benefits at the same time that the individual benefits. With more highly evolved technology, additional value can emerge from the group's efforts as a whole. For instance, tagged photo snapshots of the same tourist scene from different angles by different tourists can be assembled into a stunning three-dimensional rendering of the original location. No individual would bother to make that.

Serious amateur writers contributing to a community-built news site add far more value than they could ever get in return individually, but they keep contributing, in part because of the cultural power these cooperative instruments wield. A contributor's influence extends way beyond a lone vote, and the community's collective influence can be far out of proportion to the number of contributors. That is the whole point of social organization—the sum outperforms the parts. This is the emergent power that technology nurtures.

Additional technical innovation can boost ad hoc cooperation to a type of deliberate collaboration. Just look at any of hundreds of open-source software projects, such as Wikipedia. In these endeavors, finely tuned communal tools generate high-quality products from the coordinated work of thousands or tens of thousands of members. One study estimates that 60,000 man-years of work were poured into the release of the Fedora Linux 9 software. Altogether, roughly 460,000 people around the world are currently working on an amazing 430,000 different open-source projects. That's almost twice the size of General Motors' workforce, but without any bosses. Collaborative technology works so well that many of these collaborators have never met and may live in distant countries.

The drift toward mutualism in the technium is moving us toward an old dream: to maximize both individual human autonomy and the power of people working together. Who would have believed that poor farmers could secure $100 loans from perfect strangers on the other side of the planet—and pay them back? That is what Kiva does, with peer-to-peer mutual lending employing the mutualistic technology of an internet social website. Every public health-care expert declared confidently that sharing was fine for photos, but no one would share their medical records. But PatientsLikeMe, where patients pool results of treatments to better their own care, proved that collective action can trump both doctors and privacy fears. The increasingly common habit of sharing what you're thinking (Twitter), what you're reading (StumbleUpon), your finances (Wesabe), your everything (the web) is becoming a foundation of our technium.

Collaboration, which is not new, was once hard to do en masse. Cooperation, not new, was hard to scale into the millions. Sharing, as old as humans, is difficult to maintain among strangers. The extension of increasing mutualism from biology into the technium points to yet more sociality and mutualism to come. Right now we are using technology to collaboratively build encyclopedias, news agencies, video archives, and software in groups that span continents. Can we build bridges, universities, and charter cities the same way?

Every day over the past century someone asked, What can't free markets do? We took a long list of problems that seemed to require rational planning or paternal government and instead applied the astoundingly powerful invention of marketplace logic. In most cases, the market solution worked significantly better. Much of the prosperity in recent decades was gained by unleashing market forces into the technium.

Now we're trying the same trick with the emerging technologies of collaboration, applying these techniques to a growing list of wishes—and occasionally to problems that the free market couldn't solve—to see if they work. We are asking ourselves, What can't technological mutualism do? So far, the results have been startling. At nearly every turn, the powers of socialization—sharing, cooperation, collaboration, openness, and transparency—have proven to be more practical than anyone thought possible. Each time we try it, we find that the power of mutuality is greater than we imagined. Each time we reinvent something, we'll make it yet more mutualistic.

BEAUTY

Most evolved things are beautiful, and the most beautiful are the most highly evolved. Every living organism today has benefited from four billion years of evolution, so every creature alive—from a spherical diatom to a jellyfish to a jaguar—displays a depth that we see as beauty. This is why we are attracted to natural organisms and materials and why it is so hard to create synthetic objects with a similar glow. (Facial beauty in humans is a different phenomenon entirely. The closer a face hews to an ideal average human face, the more attractive we find it.) The complex history of a living creature gives it a patina that holds up to inspection no matter how close we get.

My friends in the Hollywood special effects business who create the lifelike virtual creatures for movies like *Avatar* and the *Star Wars* series say the same thing. They first engineer their made-up creature to follow the logic of physics, and then they make it beautiful by layering on history. The monster on the ice planet in the 2009 film *Star Trek* was once white (in its virtual evolution), but after it became the top predator in its snowy white world, camouflage was no longer necessary, so parts of its body shifted to bright red to display its dominance. The same creature once had thousands of eyes not visible in the movie, but these organs shaped its form and behavior. Watching it on the screen, we "read" the results of this fantasy evolution as authentic and beautiful. Sometimes directors will even transfer the development of a creature from one designer to another, so that it does not acquire a homogenous style but feels deeper, more layered, move evolved.

The world-making wizards create beautiful artifacts in the same way. They give a prop the convincing heft of reality by layering on "greeblies," or intricate surface details that reflect a fictitious past history. To produce a stunning cinematic city in one recent movie, they took photographic bits of decaying Detroit buildings and added modern structures around the ruins according to a backstory of past disasters and rebirth. The resolution of the detail was not as important as historically meaningful layers.

Real cities display this same principle of evolutionary beauty. Throughout history, humans have found new cities ugly. For years people recoiled from young Las Vegas. Many centuries ago the first few versions of London were considered heinous eyesores. But over generations, every urban block in London was tested by daily use. The parks and streets that worked were retained; those that failed were demolished. The height of buildings, the size of a plaza, the rake of an overhang were all adjusted by variations to suit current needs. But not all imperfections were removed, nor can they be, since many aspects of a city—say, the width of streets—cannot be changed easily. So urban work-arounds and architectural compensations are added over generations, upping the city's complexity. In most real cities, such as London or Rome or Shanghai, the tiniest alleyway is hijacked and then utilized for public space, the smallest nook becomes a store, the dampest arch under a bridge is filled in with a home. Over centuries, this constant infilling, ceaseless replacement, renewal, and complexification—in other words, evolution—creates a deeply satisfying aesthetic. The places most renowned for their beauty (Venice, Kyoto, Esfahan) are those that reveal intersecting deep layers of time. Every corner carries the long history of the city embedded in it like a hologram, glimpses of which unfold as we stroll by.

Evolution is not just about complications. One pair of scissors can be highly evolved and beautiful, while another is not. Both scissors entail two swinging pieces of metal joined at their center. But in the highly evolved scissors, the accumulated knowledge won over thousands of years of cutting is captured by the forged and polished shape of the scissor halves. Tiny twists in the metal hold that knowledge. While our lay minds can't decode why, we interpret that fossilized learning as beauty. It has less to do with smooth lines and more to do with smooth continuity of experience. The attractive scissors and the beautiful hammer and the gorgeous car all carry in their form the wisdom of their ancestors.

The beauty of evolution has put a spell on us. According to psychologist Erich Fromm and famed biologist E. O. Wilson, humans are endowed with biophilia, an innate attraction to living things. This hardwired genetic affinity for life and life processes ensured our survival in the past by nurturing our familiarity with nature. In joy we

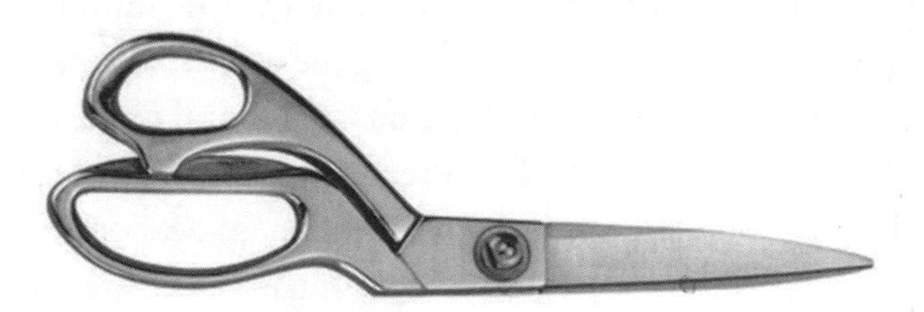

Ergonomic Scissors. A highly evolved tailor's scissors for cutting cloth on a table.

learned the secrets of the wild. The aeons that our ancestors spent walking in the woods finding coveted herbs or stalking a rare green frog were bliss; ask any hunter-gatherer about their time in the wilds. In love we discovered the bounty each creature could provide and the great lessons organic forms had to teach us. This love still simmers in our cells. It is why we keep pets and potted plants in the city, why we garden when supermarket food is cheaper, and why we are drawn to sit in silence under towering trees.

But we are likewise embedded with technophilia, an attraction for technology. Our transformation from smart hominin into Sapiens was midwifed by our tools, and at our human core we harbor an innate affinity for made things, in part because we are made. Also in part because every technology is our child, and so we love our children—all of them. We are embarrassed to admit it, but we love technology. At least sometimes.

Craftsmen have always loved their tools, birthing them in ritual and guarding them from the uninitiated. They were very personal things. As the scale of technology outgrew the hand, machines became a communal experience. By the age of industry, lay folk had many occasions to encounter complexifying technology larger than any natural organism they had ever seen, and they began to fall under its sway. In 1900, the historian Henry Adams visited and revisited the Great Exposition in Paris, where he haunted the hall showcasing the amazing new electric dynamos, or motors. Writing about himself in the third person, he recounts his initiation:

> [To Adams] the dynamo became a symbol of infinity. As he grew accustomed to the great gallery of machines, he began to feel the forty-foot dynamos as a moral force, much as the early Christians felt the Cross. The planet itself seemed less impressive, in its old-fashioned, deliberate, annual or daily revolution, than this huge wheel, revolving within an arm's-length at some vertiginous speed, and barely murmuring—scarcely humming an audible warning to stand a hair's-breadth further for respect of power—while it would not wake the baby lying close against its frame. Before the end, one began to pray to it.

Almost 70 years later California writer Joan Didion made a pilgrimage to the Hoover Dam, a trip she recounts in her anthology, *The White Album*. She, too, felt the heart of a dynamo.

> Since the afternoon in 1967 when I first saw Hoover Dam, its image has never been entirely absent from my inner eye. I will be talking to someone in Los Angeles, say, or New York, and suddenly the dam will materialize, its pristine concave face gleaming white against the harsh rusts and taupes and mauves of that rock canyon hundreds or thousands of miles from where I am.
>
> . . . Once when I revisited the dam I walked through it with a man from the Bureau of Reclamation. We saw almost no one. Cranes moved above us as if under their own volition. Generators roared. Transformers hummed. The gratings on which we stood vibrated. We watched a hundred-ton steel shaft plunging down to that place where the water was. And finally we got down to that place where the water was, where the water sucked out of Lake Mead roared through thirty-foot penstocks and then into thirteen-foot penstocks and finally into the turbines themselves. "Touch it," the Reclamation man said, and I did, and for a long time I just stood there with my hands on the turbine. It was a peculiar moment, but so explicit as to suggest nothing beyond itself.
>
> . . . I walked across the marble star map that traces a sidereel revolution of the equinox and fixes forever, the Reclamation man had told me, for all time and for all people who can read the stars, the date the dam was dedicated. The star map was, he had said, for when we were all gone and the dam was left. I had not thought much of it when he said it, but I thought of it then, with the wind whining and the sun dropping behind a mesa with the finality of a sunset in space. Of course that was the image I had seen always, seen it without quite realizing what I saw, a dynamo finally free of man, splendid at last in its absolute isolation, transmitting power and releasing water to a world where no one is.

Of course, dams inspire dread and disgust as well as awe and admiration. Soaring, breathtaking dams frustrate the return of single-minded salmon and other spawning fish, and they indiscriminately flood homelands. In the technium, revulsion and reverence often go hand in hand. Our biggest technological creations are like people in that way; they elicit our deepest loves and hates. On the other hand, no one has ever been revolted by a cathedral of redwoods. In reality no dam, even Hoover Dam, is eternal under the stars because rivers have a mind of their own; they pile up silt behind the dam's wedge so that eventually their waters can crawl over it. But while it stands, the artificial wins our admiration. We can identify with the dynamo revolving forever, as we feel our living hearts must do.

Passions for the made run wide. Almost anything manufactured will have adoring fans. Cars, guns, cookie jars, fishing reels, tableware, you name it. The "wild elaboration, passion and utility" of clocks snag some. For others the beauty of suspension bridges or of high-speed aircraft such as the SR71 or V2 is the apex of the made.

MIT sociologist Sherry Turkle calls a particular specimen of technology that is revered by an individual an "evocative object." These bits of the technium are totems that serve as a springboard for identity or for reflection or for thinking. A doctor may love her stethoscope, as both badge and tool; a writer might cherish a special pen and feel its smooth weight pushing the words on its own; a dispatcher can love his ham radio, relishing its hard-won nuances as a magical door to other realms that opens to him alone; and a programmer can easily love the root operating code of a computer for its essential logical beauty. Turkle says, "We think with the objects we love, and we love the objects we think with." She suspects that most of us have some kind of technology that acts as our touchstone.

I am one of them. I am no longer embarrassed to admit that I love the internet. Or maybe it's the web. Whatever you want to call the place we go to while we are online, I think it is beautiful. People love places and will die to defend a place they love, as our sad history of wars proves. Our first encounters with the internet/web portrayed it as a very widely distributed electronic dynamo—a thing one plugs into—and that it is. But the internet as it has matured is closer to the technological equivalent of a place. An uncharted, almost feral territory where you can genuinely get lost. At times I've entered the web just to get lost. In that lovely surrender, the web swallows my certitude and delivers the unknown. Despite the purposeful design of its human creators, the web is a wilderness. Its boundaries are unknown, unknowable, its mysteries uncountable. The bramble of intertwined ideas, links, documents, and images creates an otherness as thick as a jungle. The web smells like life. It knows so much. It has insinuated its tendrils of connection into everything, everywhere. The net is now vastly wider than I am, wider than I can imagine; in this way, while I am in it, it makes me bigger, too. I feel amputated when I am away from it.

I find myself indebted to the net for its provisions. It is a steadfast benefactor, always there. I caress it with my fidgety fingers; it yields to my desires, like a lover. Secret knowledge? Here. Predictions of what is to come? Here. Maps to hidden places? Here. Rarely does it fail to please, and more marvelous, it seems to be getting better every day. I want to remain submerged in its bottomless abundance. To stay. To be wrapped in its dreamy embrace. Surrendering to the web is like going on an aboriginal walkabout. The comforting illogic of dreams reigns. In dream time you jump from one page, one thought, to another. First on the screen you are in a cemetery, looking at an automobile carved out of solid rock; the next moment, there's a man in front of a blackboard writing the news in chalk, then you are in jail with a crying baby, then a woman in a veil gives a long speech about the virtues of confession, then tall buildings in a city blow their tops off in a thousand pieces in slow motion. I encountered all those dreamy moments this morning within the first few minutes of my web surfing. The net's daydreams have touched my own and stirred my heart. If you can honestly love a cat, which can't give you directions to a stranger's house, why can't you love the web?

Our technophilia is driven by the inherent beauty of the technium. Admittedly, this beauty has been previously hidden by a primitive phase of development that was not very pretty. Industrialization was dirty, ugly, and dumb in comparison to the biological matrix it grew from. A lot of that stage of the technium is still with us, spewing its ugliness. I don't know whether this ugliness is a necessary stage of the technium's growth or whether a smarter civilization than we could have tamed it earlier, but the arc of technology's origins from life's evolution, now accelerated, means that the technium contains all of life's inherent evolutionary beauty—waiting to be uncovered.

Technology does not want to remain utilitarian. It wants to become art, to be beautiful and "useless." Since technology is born out of usefulness, this is a long haul. As utilitarian technologies age, they tend to become recreational. Witness sailboats, open convertible cars, fountain pens, and fireplaces. Who would have guessed anyone would burn candles when lightbulbs are so cheap? But burning candles is now a mark of luxuriant uselessness. Some of our hardest-working technology today will achieve beautiful uselessness in the future. Perhaps a hundred years from now people will carry around "phones" simply because they like to carry things, even though they may be connected to the net by something they wear.

In the future, we'll find it easier to love technology. Machines win our hearts with every step they take in evolution. Like it or not, animal-like robots (at the level of pets, at first) will gain our affections, as even minimally lifelike ones do already. The internet provides a hint of the passion possible. Like many loves, it begins with infatuation and obsession. The global internet's nearly organic interdependence and emerging sentience make it wild, and its wildness draws our affections. We are deeply attracted to its beauty, and its beauty resides in its evolution.

Humans are the most complex, highly evolved organisms we have encountered, so we fixate on imitations of this form (quite naturally), but our technophilia is fundamentally not for anthropy, but for anything highly evolved.

Humanity's most advanced technology will soon leave imitation behind and create obviously nonhuman intelligences and obviously nonhuman robots and obviously non-Earthlike life, and all these will radiate an evolved attractiveness that will dazzle us.

As it does, we'll find it easier to admit that we have an affinity for it. In addition, the accelerated arrival of tens of millions more artifacts will deposit more layers onto the technium, polishing existing technology with more history and deepening the strata of embedded knowledge.

Year by year, as it advances, technology, on average, will increase in beauty. I am willing to bet that in the not-too-distant future the magnificence of certain patches of the technium will rival the splendor of the natural world. We will rhapsodize about this or that technology's charms and marvel at its subtlety. We will travel to it with children in tow to sit in silence beneath its towers.

SENTIENCE

The rock ant is tiny, even for an ant. Individually, each ant is the size of a comma on this page. Their colonies are small, too. Numbering about 100 workers, plus one queen, they normally nest between slivers of crumbling rock, hence their common name. Their entire society can fit into the glass case of a watch or between the one-inch covers of a microscope slide, which is where they are usually bred in laboratories. The brain of a rock ant contains fewer than 100,000 neurons and is so small as to be invisible. Yet a rock ant mind can perform an amazing feat of calculation. To assess the potential of a new nesting site, rock ants will measure the dimensions of the room in total darkness and then calculate—and that is the proper word—the volume and desirability of it. For many millions of years, rock ants have used a mathematical trick that was only discovered by humans in 1733. Rock ants can estimate the volume of a space, even an irregularly shaped one, by laying a scent trail across the floor of the space, "recording" the length of that line, and then counting the number of times they encounter that scented line during additional diagonal runs across the floor. The calculated area is inversely proportional to the frequency of intersections times length. In other words, the ants discovered an approximate value for pi derived by intersecting diagonals, a technique now known in mathematics as Buffon's Needle. Headroom in the potential ant house is measured by the ants with their bodies and then "multiplied" with the calculated area to give an approximate volume of their hole.

But these incredible tiny ant minds do more. They measure the width and numbers of entrances, the amount of light, the proximity of neighbors, and the degree of hygiene of the room. Then they tally these variables and calculate a desirability score for the potential nest by a process that resembles a "weighted additive" fuzzy-logic formula in computer science. All in 100,000 neurons.

The minds of animals are legion, and even fairly dumb ones can evoke amazement. Asian elephants will strip away branches to construct a fly switch to keep pesky flies away from their hind parts. Beavers, mere rodents, have been known to stockpile construction materials before starting to build their dams, thus displaying the ability to anticipate a future intent. They can even outwit humans trying to prevent their dams from flooding fields. Squirrels, another thinking rodent, continually outwit very smart college-degree suburbanites over control of their backyard bird feeders. (I've been battling my own black squirrel Einstein.) The honeyguide bird in Kenya lures humans to wild bee nests so that the birds can feast on the remaining bee brood after the humans remove the honey; sometimes, according to ornithologists, the honeyguide will "deceive" the hunters about the actual distance to a deep forest nest if it is more than two kilometers away, so as not to discourage them.

Plants, too, possess a decentralized type of intelligence. As biologist Anthony Trewavas argues in his remarkable paper, "Aspects of Plant Intelligence," plants demonstrate a slow version of problem solving that fits most of our definitions of animal intelligence. They perceive their environment in great detail, they assess threats and competition, then take action to either adapt or remedy the problems, and they anticipate future states. Time-lapse motion pictures that speed up the action of vine tendrils probing their neighborhood make it clear that plants are closer to animals in their behavior than our fast lives permit us to see. Charles Darwin may have been the first to observe this. He wrote in 1822, "It is hardly an exaggeration to say that the tip of the root acts like the brain of one of the lower animals." Like sensitive fingers, roots will caress the soil, seeking out moisture and nutrients much as a nose or trunk of a herbivore might dig in the earth. The ability of a leaf to follow the sun (heliotropism) to gain optimal light exposure can be replicated in a machine, but only by using a fairly sophisticated computer chip as a brain. A plant thinks without a brain. It uses a vast network of transducing molecular signals instead of electronic nerves to carry and process information.

Plants exhibit all the characteristics of intelligence, except they do it without a centralized brain and in slow motion. Decentralized minds and slow minds are actually quite common in nature and occur at many levels throughout the six kingdoms of life. A slime-mold colony can solve the shortest distance to food in a maze, much like a rat. The animal immune system, whose primary purpose is to distinguish between self and nonself, retains a memory of outside antigens it has encountered in the past. It learns in a Darwinian process and in a sense also anticipates future variations of antigens. And throughout the animal kingdom collective intelligence is expressed in hundreds of ways, including the famous hive minds of social insects.

The manipulation, storage, and processing of information is a central theme of life. Learning erupts over and over again in the history of evolution, as if it were a force waiting to be released. A charismatic version of intelligence—the kind of anthropomorphic smartness we associate with apes—evolved not just in primates but in at least two other unrelated taxa: whales and birds.

Stories of dolphin intelligence are famous. Dolphins and whales not only demonstrate intelligence, but they also occasionally give hints that they share a style of intelligence with us, the hairless apes. For instance, captive dolphins have been known to train other dolphins new to the pool. Yet the most recent common ancestor for apes, whales, and dolphins was 250 million years ago. In between apes and dolphins are many families of animals without this variety of thought. We can only surmise that this style of intelligence evolved independently.

The same can be said for birds. Measured by their intelligence, crows, ravens, and parrots are the "primates" of birds. Their forebrains are as relatively large as those of nonhuman apes, and the ratio of their brain weight to body weight is in the same line as apes. Like primates, crows live long and in complex social groups. New Caledonian crows, like chimpanzees, craft tiny spears to fish for grubs in crevices. Sometimes they save the manufactured spears and carry them around. In experiments with scrub jays, researchers discovered that that jays would re-hide their food later if another bird was watching them when they first hid it, but only if the jays had been robbed before. Naturalist David Quammen suggests that crow and raven behavior is so clever and peculiar that they should be evaluated "not by an ornithologist but by a psychiatrist."

Thus, charismatic intelligence evolved independently three times: in birds on wing, in mammals that returned to the sea, and in primates.

Still, charismatic intelligence is relatively rare. But smartness is a competitive advantage everywhere. We see the widespread recurrence and reinvention of intelligence because the living universe is a place where learning makes a difference. Up and down the six kingdoms of life, minds have evolved many times. So many times, in fact, that minds seem inevitable. Yet as inordinately fond as nature is of minds, the technium is even more so. The technium is rigged to birth minds. All the inventions we have constructed to assist our own minds—our many storage devices, signal processing, flows of information, and distributed communication networks—all these are also essential ingredients for producing new minds. And so new minds spawn in the technium in inordinate degrees. Technology wants mindfulness.

This yearning for increasing sentience reveals itself in three different ways in the technium:

1. Mind infiltrates matter as ubiquitously as possible.
2. Exotropy continues to organize more complex types of intelligences.
3. Sentience diversifies into as many types of minds as possible.

The technium is primed to hijack matter and rearrange its atoms to infiltrate it with sentience. There seems to be no place a mind can't be born or inserted. These mind children will be small, dim, and dumb at first, but tiny minds keep getting better and more abundant. In 2009 there were 1 billion electronic "brains" etched into silicon. Many of these tiny minds contain a billion transistors each, which the global semiconductor industry is manufacturing at the speed of 30 billion per second! The smallest silicon brain has a minimum of 100,000 transistors, about as many neurons as the brain of the rock ant. They, too, can do surprising feats. Tiny synthetic minds no bigger than an ant's know where on Earth they are and how to get back to your home (GPS); they remember the names of your friends and translate foreign languages. These dim minds are finding their way into everything: shoes, doorbells, books, lamps, pets, beds, clothes, cars, light switches, kitchen appliances, and toys. If the technium continues to prevail, some level of sentience will find its way into everything it creates. The smallest bolt or plastic knob will contain as many decision-making circuits as a worm, elevating it from the inert to the animate. Unlike the billions of minds in the wild, the best of these technological minds (in aggregate) are getting smarter by the year.

We are blind to this massive eruption of minds into the technium because humans have a chauvinistic bias against any kind of intelligence that does not precisely mirror our own. Unless an artificial mind behaves exactly like a human one, we don't count it as intelligent. Sometimes we dismiss it by calling it "machine learning." So while we weren't watching, billions of tiny, insectlike artificial minds spawned deep into the technium, doing invisible, low-profile chores like reliably detecting credit-card fraud or filtering e-mail spam or reading text from documents. These proliferating microminds run speech recognition on the phone, assist in crucial medical diagnosis, aid stock-market analysis, power fuzzy-logic appliances, and guide automatic gearshifts and brakes in cars. A few experimental minds can even drive a car autonomously for a hundred miles.

The future of the technium at first seems to point to bigger brains. But a bigger computer is not necessarily smarter, more sentient. And even when intelligence is demonstrably greater in biological minds, it is only weakly correlated to how many brain cells are present. In nature, animal computers come in all sizes. An ant brain is a 100th-of-a-gram speck; the 8-kilogram brain of a sperm whale is 100,000 times bigger. But it is not clear that a whale is 100,000 times smarter than an ant or that humans are only three times as smart as chimpanzees, as the specifications of pure numbers of cells might suggest. Our large human brain, with its endless ideas, is only one-sixth the size of a sperm whale brain. It is even slightly smaller than the average Neanderthal brain. On the other hand, recently discovered minihumans on Flores Island had brains one-third the size of ours and may have been no dumber. The correlation between the absolute scale of the brain and smartness is not significant.

The architecture of our own brain suggests the future of artificial sentience may reside in a different kind of big. Until recently, conventional wisdom held that specialized big-brain supercomputers would first host artificial intelligences, and then perhaps we'd get mini ones at home or add them to the heads of our personal robots. They would be bounded entities. We would know where our thoughts ended and theirs began.

However, the snowballing success of search engines such as Google this past decade suggests the coming AI will most likely not be confined in a stand-alone supercomputer but will be birthed in the superorganism of a billion CPUs known as the web. It will run on the global megacomputer that encompasses the internet, all its services, all peripheral chips and affiliated devices from scanners to satellites, and the billions of human minds entangled in this global network. Any device that touches this web AI will share—and contribute to—its intelligence.

This gargantuan machine already exists in a primitive form today. Consider the virtual supercomputer of all the world's computers online. There are one billion online PCs, which is about as many transistors as are in an Intel chip in one computer. All the transistors in all the computers connected together add up to about 100 quadrillion (10^{17}) transistors. In many ways, this global virtual network acts like a very large computer that operates at approximately the clock speed of an early PC.

This supercomputer processes three million e-mails each second, which essentially means network e-mail runs at 3 megahertz. Instant messaging runs at 162 kilohertz, SMS at 30 kilohertz. In any one second, 10 terabits of information can be coursing through its backbone, and each year it generates nearly 20 exabytes of data.

This planetary computer embraces more than just laptops. Today it contains approximately 2.7 billion cell phones, 1.3 billion land phones, 27 million data servers, and 80 million wireless PDAs. Each device is a differently shaped screen that peers into the global computer. It takes a billion windows to glimpse what it is thinking.

The web holds about a trillion pages. The human brain holds about a hundred billion neurons. Each biological neuron sprouts synaptic links to thousands of other neurons, while each web page on average links to 60 other pages. That adds up to a trillion "synapses" between the static pages on the web. The human brain has about 100 times that number of links—but brains are not doubling in size every few years. The global machine is.

And who is writing the software that makes this contraption useful and productive? We are, each of us, every day. When we post and then tag pictures on the community photo album Flickr, we are teaching the machine to give names to images. The thickening links between caption and picture form a neural net that can learn. Think of the 100 billion times *per day* humans click on one web page or another as a way of teaching the web what we think is important. Each time we forge a link between words, we teach it an idea. We think we are merely wasting time when we surf mindlessly or blog an item, but each time we click a link we strengthen a node somewhere in the supercomputer's mind, thereby programming the machine by using it.

Whatever the nature of this large-scale sentience, it won't even be recognized as intelligence at first. Its very ubiquity will hide it. We'll use its growing smartness for all kinds of humdrum chores—data mining, memory archive, simulations, forecasting, pattern matching—but because the smartness lives on thin bits of code spread across the globe in windowless boring warehouses, and it lacks a unified body, it will be faceless. You can reach this distributed intelligence in a million ways, through any digital screen anywhere on Earth, so it will be hard to say where it is. And because this synthetic intelligence is a combination of human intelligence (all past human learning, all current humans online) and digital memory, it will be difficult to pinpoint just what it is. Is it our memory or a consensual agreement? Are we searching it, or is it searching us?

Someday we might meet other intelligences in the galaxies. But long before then we will manufacture millions of new kinds of minds on our own world. This is the third vector of evolution's long-term trajectory toward increased sentience. First, insinuate intelligence into all matter. Second, bring all those embedded minds together. Third, increase the diversity of minds. There may be as many species of intelligence possible as there are species of beetles, which is saying a lot.

There are a million and one reasons to build a million and one different types of artificial intelligences. Specialized intelligences will perform specialized tasks; other AIs will be general-purpose intelligences that accomplish familiar tasks differently from how we do. Why? Because difference makes progress. The one kind of mind I doubt we'll make many of is an artificial mind just like a human. The only way to reconstruct a viable human species of mind is to use tissue and cells—and why bother when making human babies is so easy?

Some problems will require multiple *kinds* of minds to crack, and our job will be to discover new methods of thinking and to set this diversity of intelligences loose in the universe. Planetary-scale problems will require some kind of planetary-scale mind; complex networks made of trillions of active nodes will require network intelligences; routine mechanical operations will need nonhuman precision in calculations. Since our own brains are such poor thinkers in terms of probability, we'd really benefit by discovering an intelligence at ease with statistics.

We'll need all varieties of thinking tools. An off-the-grid stand-alone AI will be handicapped compared with a hive-mind supercomputer. It can't learn as fast, as broadly, or as smartly as one plugged into six billion human minds, several quintillion online transistors, hundreds of exabytes of real-life data, and the self-correcting feedback loops of the entire civilization. But a consumer may still choose to pay the penalty of lesser smarts in order to have the mobility of an isolated AI in distant places, or for privacy reasons.

Currently we are prejudiced against machines, because all the machines we have met so far have been uninteresting. As they gain in sentience, that won't be true. But we won't find all types of artificial minds equally attractive. Just as we find some natural creatures more charismatic than others, some minds will be charismatic (attractive to our way of thinking) and some won't. In fact, we might be repulsed by the alien nature of many of the most powerful types of intelligences. For instance, the ability to remember *everything* can be scary.

What technology wants is increasing sentience. This does not mean evolution will move us only toward one universal supermind. Rather, over the course of time the technium tends to self-organize into as many varieties of mind as is possible.

The primary thrust of exotropy is to uncover the full diversity of intelligences. Each type of thinking, no matter how large it is scaled up, can only understand in a limited way. The universe is so huge, so vast in its available mysteries, that it will require every possible type of mind to comprehend it. The technium's job is to invent a million, or a billion, varieties of comprehension.

This is not as mystical as it sounds. Minds are highly evolved ways of structuring the bits of information that form reality. That is what we mean when we say a mind understands; it generates order. As exotropy pushes through history, self-organizing matter and energy into greater complexity and possibilities, minds are the fastest, most efficient, most exploratory technology so far for creating order. By now our planet owns the dim minds of plants, the multiple manifestations of a common animal mind, and the restless self-consciousness of human minds. Just a second ago, cosmically speaking, human minds began to invent a second generation of sentience. They installed their inventiveness in the most powerful force in the world—technology—and are trying to clone their own tricks. Most of these newly invented minds are no more intelligent than plants, a few are as smart as insects, and a couple hint at greater thoughts to come. All the while, the technium assembles brainlike networks at scales way beyond individual humans.

The trajectory of the technium is pointed toward a million more minds inhabiting the least bits of matter, in a million new varieties of thinking, subsumed with our own multiple minds into a planetary thought—on the way to comprehending itself.

STRUCTURE

It took Sapiens several million years to evolve from an apelike ancestor. During that transition to humanity, our DNA changed by a few million bits. So the natural rate of biological evolution in humans, in terms of information accumulation, is about one bit per year. Now, after almost four billion years of bit-by-bit biological evolution, we have unleashed a new type of evolution, one that creates rivers of mutations using language, writing, printing, and tools—what we call technology. Compared to the one bit per year we made as apes, we are adding 400 exabytes of new information to the technium each year, so the rate of our technological evolution is a billion billion times as fast as the evolution of DNA. As humans it takes us less than a second to process the same amount of information that our DNA took a billion years to process.

We are accumulating information so rapidly that it is the fastest increasing quantity on this planet. The amount of mail sent through the U.S. postal system has been doubling every 20 years for 80 years. The number of photographic images (a very dense information platform) has risen exponentially since the medium was invented in the 1850s. The total number of telephone-call minutes each day likewise has followed an exponential curve for over 100 years. There's no stream of information that is lessening.

According to a calculation Hal Varian, an economist at Google, and I made, total worldwide information has been increasing at the rate of 66 percent per year for many decades. Compare that explosion to the rate of increase in even the most prevalent manufactured stuff—such as concrete or paper—which averages only 7 percent annually over decades. At 10 times faster than the growth of any other manufactured product on this planet, the rate of growth of information may even be faster than any biological growth at the same scale.

The quantity of scientific knowledge, as measured by the number of scientific papers published, has been doubling approximately every 15 years since 1900. If we measure simply the number of journals published, we find that they have been multiplying exponentially since the 1700s, when science began. Everything we manufacture produces an item and information about that item. Even when we create something that is information based to start with, it will generate yet more information about its own information. The long-term trend is simple: The information about and from a process will grow faster than the process itself. Thus, information will continue to grow faster than anything else we make.

The technium is fundamentally a system that feeds off the accumulation of this explosion of information and knowledge. Similarly, living organisms are also systems that organize the biological information flowing through them. We can read the technium's evolution as the deepening of the structure of information begun by natural evolution.

Nowhere is this increasing structure as visible as in science. Despite its own rhetoric, science is not built to increase either the "truthfulness" or the total volume of information. It is designed to increase the order and organization of knowledge we generate about the world. Science creates "tools"—techniques and methods—that manipulate information such that it can be tested, compared, recorded, recalled in an orderly fashion, and related to other knowledge. "Truth" is really only a measure of how well specific facts can be built upon, extended, and interconnected.

We casually talk about the "discovery of America" in 1492 or the "discovery of gorillas" in 1856 or the "discovery of vaccines" in 1796. Yet vaccines, gorillas, and America were not unknown before their "discovery." Native peoples had been living in the Americas for 10,000 years before Columbus arrived, and they had explored the continent far better than any European ever could. Certain West African tribes were intimately familiar with the gorilla and many more primate species yet to be "discovered." Dairy farmers in Europe and cow herders in Africa had long been aware of the protective inoculative effect that related diseases offered, although they did not have a name for it. The same argument can be made about whole libraries' worth of knowledge—herbal wisdom, traditional practices, spiritual insights—that are "discovered" by the educated, but only after having been long known by native and folk peoples. These supposed "discoveries" seem imperialistic and condescending—and often are. Yet there is one legitimate way in which we can claim that Columbus discovered America, and the French-American explorer Paul du Chaillu discovered gorillas, and Edward Jenner discovered vaccines. They "discovered" previously locally known knowledge by adding it to the growing pool of structured global knowledge. Nowadays we would call that accumulating of structured knowledge *science.* Until Du Chaillu's adventures in Gabon any knowledge about gorillas was extremely parochial; the local tribes' vast natural knowledge about these primates was not integrated into all that science knew about all other animals. Information about "gorillas" remained outside the structured known. In fact, until zoologists got their hands on Paul du Chaillu's specimens, gorillas were scientifically considered to be a mythical creature similar to Bigfoot, seen only by uneducated, gullible natives. Du Chaillu's "discovery" was actually science's discovery. The meager anatomical information contained in the killed animals was fitted into the vetted system of zoology. Once their existence was "known," essential information about gorillas' behavior and natural history could be annexed. In the same way, local farmers' knowledge about how cowpox could inoculate against smallpox remained local knowledge and was not connected to the rest of what was known about medicine. The remedy therefore remained isolated. When Jenner "discovered" the effect, he took what was known locally and linked its effect to medical theory and all the little science knew of infection and germs. He did not so much "discover" vaccines as "link in" vaccines. Likewise America. Columbus's encounter put America on the map of the globe, linking it to the rest of the known world, integrating its own inherent body of knowledge into the slowly accumulating, unified body of verified knowledge. Columbus joined two large continents of knowledge into a growing consilient structure.

The reason science absorbs local knowledge and not the other way around is because science is a machine we have invented to connect information. It is built to integrate new knowledge with the web of the old. If a new insight is presented with too many "facts" that don't fit into what is already known, then the new knowledge is rejected until those facts can be explained. (This is an oversimplification of Thomas Kuhn's theory of the overthrow of scientific paradigms.) A new theory does not need to have every unexpected detail explained (and rarely does) but it must be woven to some satisfaction into the established order. Every strand of conjecture, assumption, observation is subject to scrutiny, testing, skepticism, and verification.

Unified knowledge is constructed by the technical mechanics of duplication, printing, postal networks, libraries, indexing, catalogs, citations, tagging, cross-referencing, bibliographies, keyword search, annotation, peer review, and hyperlinking. Each epistemic invention expands the web of verifiable facts and links one bit of knowledge to another. Knowledge is thus a network phenomenon, with each fact a node. We say knowledge increases not only when the number of facts increases, but also, and more so, when the number and strength of relationships between facts increases. It is that relatedness that gives knowledge its power. Our understanding of gorillas deepens and becomes more useful as their behavior is compared to, indexed with, aligned with, and related to the behavior of other primates. The structure of knowledge is expanded as gorillas' anatomy is related to other animals', as their evolution is integrated into the tree of life, as their ecology is connected to the other animals coevolving with them, as their existence is noted by many kinds of observers, until the facts of gorillahood are woven into the encyclopedia of knowledge in thousands of crisscrossing and self-checking directions. Each strand of enlightenment enhances not only the facts of gorillas, but also the strength of the whole cloth of human knowledge. The strength of those connections is what we call truth.

Today there remain many unconnected pools of knowledge. The unique wealth of traditional wisdom won by indigenous tribes in their long, intimate embrace of their natural environment is very difficult (if not impossible) to move out of their native context. Within their system, their sharp knowledge is tightly woven, but it is disconnected from the rest of what we collectively know. A lot of shamanic knowledge is similar. Currently science has no way to accept these strands of spiritual information and weave them into the current consilience, and so their truth remains "undiscovered." Certain fringe sciences, such as ESP, are kept on the fringe because their findings, coherent in their own framework, don't fit into the larger pattern of the known. But in time, more facts are brought into this structure of information. More important, the methods whereby knowledge is structured are themselves evolving and being restructured.

The evolution of knowledge began with relatively simple arrangements of information. The most simple organization was the invention of the fact. Facts, in fact, were invented. Not by science but by the European legal system, in the 1500s. In court lawyers had to establish agreed-upon observations as evidence that could not shift later. Science adopted this useful innovation. Over time, the novel ways in which knowledge could be ordered increased. This complex apparatus for relating new information to old knowledge is what we call science.

The scientific method is not one uniform "method." It is a collection of scores of techniques and processes that has evolved over centuries (and continues to evolve). Each method is one small step that incrementally increases the unity of knowledge in society. A few of the more seminal inventions in the scientific method include:

280 B.C.E. *Cataloged library with index* (at Alexandria), a way to search recorded information
1403 *Collaborative encyclopedia,* a pooling of knowledge from more than one person
1590 *Controlled experiment,* used by Francis Bacon, wherein one changes a single variable in a test
1665 *Necessary repeatability,* Robert Boyle's idea that results of an experiment must be repeatable to be valid
1752 *Peer-review-refereed journal,* adding a layer of confirmation and validation over shared knowledge
1885 *Blinded, randomized design,* a way to reduce human bias; randomness as a new kind of information
1934 *Falsifiable testability,* Karl Popper's notion that any valid experiment must have some testable way it can fail
1937 *Controlled placebo,* a refinement in experiments to

remove the effect of biased knowledge of the participant

1946 *Computer simulations,* a new way of making a theory and generating data

1952 *Double-blind experiment,* a further refinement to remove the effect of knowledge of the experimenter

1974 *Meta-analysis,* a second-level analysis of all previous analysis in a given field

Together these landmark innovations create the modern practice of science. (I am ignoring various alternative claims of priority because for my purposes the exact dates don't matter.) A typical scientific discovery today will rely on facts and a falsifiable hypothesis; be tested in repeatable, controlled experiments, perhaps with placebos and double-blind controls; and be reported in a peer-reviewed journal and indexed in a library of related reports.

The scientific method, like science itself, is accumulated structure. New scientific instruments and tools add new ways to organize information. Recent methods build upon earlier techniques. The technium keeps adding connections among facts and more complex relations among ideas. As this short timeline makes clear, many of the key innovations of what we now think of as "the" scientific method are relatively recent. The classic double-blind experiment, for instance, in which neither the subject nor the tester is aware of what treatment is being given, was not invented until the 1950s. The placebo was not used in practice until the 1930s. It is hard to imagine science today without these methods.

This recency makes one wonder what other "essential" method in science will be invented next year. The nature of science is still in flux; the technium is rapidly discovering new ways to know. Given the acceleration of knowledge, the explosion of information, and the rate of progress, the nature of the scientific process is on a course to change more in the next 50 years than it has in the last 400 years. (A few probable additions: inclusion of negative results, computer proofs, triple-blind experiments, wiki journals.)

At the core of science's self-modification is technology. New tools enable new ways of discovery, different ways of structuring information. We call that organization knowledge. With technological innovations the structure of our knowledge evolves. The achievement of science is to discover new things; the evolution of science is to organize the discoveries in new ways. Even the organization of our tools themselves is a type of knowledge. Right now, with the advance of communication technology and computers, we have entered a new way of knowing. The thrust of the technium's trajectory is to further organize the avalanche of information and tools we are generating and to increase the structure of the made world.

EVOLVABILITY

Natural evolution is a way for an adaptive system—in this case, life—to search for new ways to survive. Life tries this or that size cell, round or long torso, slow or fast metabolism, without legs or with wings. Most forms it encounters live only a short time. But over aeons, the system of life settles on very stable forms—say, a spherical cell or DNA chromosome—that become stable platforms to experiment upon for more innovations. Evolution searches for designs that will keep the game of searching going. In this way, evolution wants to evolve.

The evolution of evolution? That sounds like a bad case of doubletalk. At first glance, this idea may seem oxymoronic (self-contradictory) or tautological (needlessly repetitive). But on close inspection, the "evolution of evolution" is no more tautological than, say, a "network of networks," which is what the internet is.

Life kept evolving for four billion years because it discovered ways to increase its own evolvability. At the start, the space of possible life was very small. Room to change was limited. For instance, early bacteria could mutate their genes, change the length of their genome, and swap genes with one another. Several billion years of evolution later, cells could still mutate and swap genes, but they could also repeat entire modules (like repeating segments in an insect), and they could manage their own genome, turning select genes off or on. When evolution discovered sexual reproduction, entire genetic "words" in a cell's genome could be recombined in a mix-and-match method that achieved far faster improvement than merely altering genetic "letters" one at a time.

At the start of life, natural selection operated on molecules, later on population of molecules, and eventually on cells and colonies of cells. Eventually, evolution selected organisms out of a population, favoring the most fit. So over biological aeons, the focus of evolution shifted upward to more complex structures. In other words, over time, the process of evolution became a conglomeration of many different forces working at many levels. By slow accumulation of tricks, the system of evolution acquired a diversity of ways to adapt and create. Imagine a shape-shifter that can change the areas in which it changes! Who could keep up with it? In this way evolution has gathered itself up and ceaselessly remakes itself over and over again.

But this description doesn't quite capture the full power of this trend. Yes, life has gained more ways to adapt, but what is really changing is its evolvability—its propensity and agility to create change. Think of this as changeability. Not only is the aggregate process of evolution evolving, but it is evolving more ability to evolve, or greater evolvability. Gaining evolvability is much like a video game where you find a door that opens up another whole level that is much more complex, faster, and full of unexpected powers.

A natural organism, such as a chicken, is the mechanism for its genes to propagate more genes. From the selfish genes' point of view, the greater the number of organisms (chickens) they can produce and keep alive, the more those genes can spread themselves. We can also view an ecosystem as the vehicle for evolution to propagate itself and grow. Without a cornucopia of diverse organisms, evolution cannot evolve more evolvability. So evolution generates complexity and diversity and millions of beings to give itself material and room to evolve into a more powerful evolver.

If we think of each living species as an answer to the question "How does something survive in this environment?," then evolution is a formula that provides concrete answers that are embodied in matter and energy. We might say that evolution is a search method for living solutions; it searches by endlessly trying out possibilities until it finds a design that works.

Of all the tricks that evolution came up with to find solutions in its first four billion years, none compared to minds. Sentience—and not just human sentience—bestows on life a greatly accelerated way to learn and adapt. This should not be surprising, because minds are built to find answers, and one of the key things to answer might be how to learn better and more quickly in order to survive. If what minds are good for is learning and adaptation, then learning how to learn will accelerate your learning. So the presence of sentience in life vastly increased its evolvability.

The most recent extension of this expansion of evolvability is technology. Technology is how human minds explore the space of possibilities and change the methods of searching for solutions. It is almost a cliche to point out that technology has brought as much change on this planet in the last 100 years as life has in the last billion years.

When we look at technology, we tend to see pipes and blinking lights. But in the long-term view, technology is simply the further evolution of evolution. The technium is a continuation of a four-billion-year-old force that pursues more ability to evolve. The technium has discovered entirely new forms in the universe, such as ball bearings, radios, and lasers, that organic evolution could never invent. Likewise, the technium has discovered wholly new ways to evolve, methods that were unreachable by biology. And just as evolution did with life, technological evolution uses its fecundity to evolve more widely and faster. The "selfish" technium generates millions of species of gadgets, techniques, products, and contraptions in order to give it sufficient material and room to keep evolving its power to evolve.

The evolution of evolution is change squared. There's a visceral sense now that changes are happening so fast in technology that we cannot possibly imagine what will happen in 30 years, let alone 100. The technium can feel like a black hole of uncertainty sometimes. But humanity has passed through several similar evolutionary transitions already.

The first, as I mentioned earlier, was the invention of language. Language shifted the burden of evolution in humans away from genetic inheritance (the only line of evolutionary learning for most other creatures) and allowed our language and culture to carry our species' aggregate learning as well. The second invention, writing, changed the speed of learning in humans by easing the transmission of ideas across territories and across time. Solutions could be archived and transmitted on durable paper. This vastly accelerated humanity's evolution.

The third transition is science, or rather, the structure of the scientific method. This is the invention that enables greater invention. Instead of depending on random hit or miss, or trial and error, the scientific method methodically explores the cosmos and systematically delivers novel ideas. It has accelerated discovery a thousandfold, if not a millionfold. The evolution of the scientific method is responsible for the exponential rise in progress we now enjoy. Without a doubt science has uncovered possibilities—and new ways of finding them—that neither biological nor cultural evolution could have invented alone.

But at the same time, the technium has also accelerated the speed of human biological evolution. Swelling populations of humans in denser cities upped the contagion of disease and hastened the rate of our biological adaptation. Humans are smart and very mobile and so select mates from a much larger pool of candidates. New foods also sped the evolution of our bodies. For instance, the adult ability to drink milk evolved and spread quickly once humans succeeded in domesticating herbivores. Today, according to research on the mutations in our DNA, our genes are evolving 100 times faster than in preagricultural times.

Now, in just the last few decades, science has evolved yet another manner of evolution. We are reaching deep within ourselves to adjust the master knob. We are messing with our source code, including the code that grows our brains and makes our minds. Gene splicing, genetic engineering, and gene therapy have given our minds direct control of our genes, ending a four-billion-year hegemony of Darwinian evolution. Now the inheritance of acquired and desirable traits in human lines is possible. The technium will be completely liberated from the tyranny of slow-moving DNA. The consequences of this new symbiotic evolution are so immense that they silence us.

All the while each technological innovation creates new opportunities for the technium to change in new ways. And every kind of new problem caused by technology also creates a chance for new kinds of solutions and new paths to find those solutions—which is a type of cultural evolution. As the technium expands, it accelerates the rate of evolution first begun with life, so that it now evolves the idea of change itself. This is more than simply the most powerful force in the world; the evolution of evolution is the most powerful force in the universe.

These broad sweeps—of increasing opportunities, emergence, complexity, diversity, and so on—are one answer to where technology is going. On the much smaller day-to-day scale, predicting the future of technology is impossible. It's too hard to filter out the random noise of commerce. We will have better luck extrapolating historical trends that in some cases go back billions of years to see how they arc through technology today. These trends are subtle, nudging technologies in a slow drift in one direction that may not even be visible in the blink of a year.

They move slowly because they are not driven by human events. Instead these tendencies are biases generated by the tangle of the technium's system. Their momentum is like the gravity of the Moon, a weak, persistent, insensible pull that can eventually move oceans. Over the span of generations these trends overcome the churning noise of human infatuation, fads, and financial trends to push and pull technologies in certain ingrained directions.

Rather than a series of meandering lines that travel into a set future, picture these arrows of technological trends exploding outward from the present. Just as space is expanding away from us in every direction, opening up the universe, these rising forces are like ballooning spheres that create the territory they are expanding into. The technium is an explosion of information, organization, complexity, diversity, sentience, beauty, and structure that is changing itself as it expands.

This exhilarating self-acceleration resembles the mythical snake Uroboros grabbing its own tail and turning itself inside out. It is rife with paradox—and promise. Indeed, the expanding technium—its cosmic trajectories, its ceaseless reinvention, its inevitabilities, its self-generation—is an open-ended beginning, an infinite game calling us to play.

14

Playing the Infinite Game

Technology wants us, but what does it want *for* us? What do we get out of its long journey?

When Henry David Thoreau spied engineers constructing a long-distance telegraph along the railroad tracks that ran past his hermitage on Walden Pond, he wondered if humans had anything important enough to say to warrant the engineers' considerable effort.

From his family farm in Kentucky, Wendell Berry watches how technology such as steam engines have taken over the manual work of farmers and wonders if machines have anything to teach humans: "The nineteenth century thought that machinery was a moral force and would make men better. How could the steam-engine make men better?"

It's a good question. The technium is reinventing us, but does any of this complicated technology make us any better as humans? Are there any manifestations of human thought anywhere than can make men better?

An answer that Wendell Berry might agree with is that the technology of law makes men better. A system of laws keeps men and women responsible, urges them toward fairness, restrains undesirable impulses, breeds trust, and so on. The elaborate system of law that undergirds Western societies is not very different from software. It's a complex set of code that runs on paper instead of in a computer, and it slowly calculates fairness and order (ideally). Here, then, is a technology that has bettered us—although, really, nothing can *make* us better. We can't be forced to do good, but we can be given opportunities.

I think Berry can't appreciate the gifts of the technium because his idea of technology is too small. He gets stuck on the cold, hard, yucky stuff, such as steam engines, chemicals, and hardware, which may be the mere juvenile stage of more mature things. Viewed from a wider perspective, where steam engines are merely a tiny part of the whole, convivial forms of technology really do allow us to be better.

How can technology make a person better? Only in this way: by providing each person with chances. A chance to excel at the unique mixture of talents he or she was born with, a chance to encounter new ideas and new minds, a chance to be different from his or her parents, a chance to create something his or her own.

I will be the first to add that by themselves—without any context around them—these possibilities are insufficient for human happiness, let alone betterment. Choice works best when it has values to guide it. But if one has spiritual values, Wendell Berry seems to say, you don't even need technology to be happy. In other words, he asks, is technology really necessary at all for human betterment?

Because I believe both the technium and civilization are rooted in the same bootstrapping cosmic trends, I think another way to ask the question is this: Is civilization necessary for human betterment?

When I trace the full course of the technium, I would say, definitely, yes. The technium is necessary for human betterment. How else are we going to change? A special subset of humans will find the constrained choices available in, say, a monastery cell or the tiny opportunities in a hermit's hut on the edge of a pond or in the deliberately restricted horizon of a wandering guru to be the ideal path to betterment. But most humans, at most moments in history, see the accumulating pile of possibilities in a rich civilization as something that makes them better people. That's why we make civilization/technology. That's why we have tools. They produce choices, including the choice for good.

Choices without values yield little, this is true; but values without choices are equally dry. We need the full spectrum of choices won by the technium to unleash our own maximum potential.

What technology brings to us individually is the possibility of finding out who we are, and more important, who we might be. During his or her lifetime, each person acquires a unique combination of latent abilities, handy skills, nascent insights, and potential experiences that no one else shares. Even twins—who share common DNA—don't share the same life. When people maximize their set of talents, they shine because no one can do what they do. People fully inhabiting their unique mixture of skills are inimitable, and that is what we prize about them. Talent unleashed doesn't mean that everyone will sing on Broadway or play in the Olympics or win a Nobel Prize. Those high-profile roles are merely three well-worn ways of being a star, and by deliberate design those particular opportunities are limited. Popular culture wrongly fixates on proven star roles as the destiny of anyone successful. In fact, those positions of prominence and stardom can be prisons, straitjackets defined by how someone else excelled.

Ideally, we would find a position of excellence tailored specifically for everyone born. We don't normally think of opportunities this way, but these possibilities for achievement are called "technology." The technology of vibrating strings opened up (created) the potential for a virtuoso violin player. The technology of oil paint and canvas unleashed the talents of painters through the centuries. The technology of film created cinematic talents. The soft technologies of writing, lawmaking, and mathematics all expanded our potential to create and do good. Thus in the course of our lives as we invent things and create new works that others may build on, we—as friends, family, clan, nation, and society—have a direct role in enabling each person to optimize their talents—not in the sense of being famous but in the sense of being unequaled in his or her unique contribution.

However, if we fail to enlarge the possibilities for other people, we diminish them, and that is unforgivable. Enlarging the scope of creativity for others, then, is an obligation. We enlarge others by enlarging the possibilities of the technium—by developing more technology and more convivial expressions of it.

If the best cathedral builder who ever lived was born now, instead of 1,000 years ago, he would still find a few cathedrals being built to spotlight his glory. Sonnets are still being written and manuscripts still being illuminated. But can you imagine how poor our world would be if Bach had been born 1,000 years before the Flemish invented the technology of the harpsichord? Or if Mozart had preceded the technologies of piano and symphony? How vacant our collective imaginations would be if Vincent van Gogh had arrived 5,000 years before we invented cheap oil paint? What kind of modern world would we have if Edison, Greene, and Dickson had not developed cinematic technology before Hitchcock or Charlie Chaplin grew up?

Missing Technologies. The boy Mozart before the piano was invented, Alfred Hitchcock before movie cameras, and my son Tywen before the next big thing.

How many geniuses at the level of Bach and Van Gogh died before the needed technologies were available for their talents to take root? How many people will die without ever having encountered the technological possibilities that they would have excelled in? I have three children, and though we shower them with opportunities, their ultimate potential may be thwarted because the ideal technology for their talents has yet to be invented. There is a genius alive today, some Shakespeare of our time, whose masterworks society will never own because she was born before the technology (holodeck, wormhole, telepathy, magic pen) of her greatness was invented. Without these manufactured possibilities, she is diminished, and by extension all of us are diminished.

For most of history, the unique mix of talents, skills, insights, and experiences of each person had no outlet. If your dad was a baker, you were a baker. As technology expands the possibility of space, it expands the chance that someone can find an outlet for their personal traits. We thus have a moral obligation to increase the best of technology. When we enlarge the variety and reach of technology, we increase options not just for ourselves and not just for others living but for all those to come as the technium ratchets up complexity and beauty over generations.

A world with more opportunities produces more people capable of producing yet more opportunities. That's the strange loop of bootstrapping creation, which constantly makes offspring superior to itself. Every tool in the hand presents civilization (all those alive) with another way of thinking about something, another view of life, another choice. Every idea that is made real (technology) enlarges the space we have to construct our lives. The simple invention of a wheel unleashed a hundred new ideas of what to do with it. From it issued carts, pottery wheels, prayer wheels, and gears. These in turn inspired and enabled millions of creative people to unleash yet more ideas. And many people along the way found their story through these tools.

This is what the technium is. The technium is the accumulation of stuff, of lore, of practices, of traditions, and of choices that allow an individual human to generate and participate in a greater number of ideas. Civilization, starting from the earliest river valley settlements 8,000 years ago, can be considered a process by which possibilities and opportunities for the next generation are accumulated over time. The average middle-class person today working as a retail sales clerk has inherited far more choices than a king of old, just as the ancient king inherited more options than a subsistence nomad had before him.

While we amass possibilities, we do so because the very cosmos itself is on a similar expansion. As far as we can tell, the universe began as an undifferentiated point and steadily unfolded into the detailed nuancs that we call matter and reality. Over billions of years, cosmic processes created the elements, the elements birthed molecules, the molecules assembled into galaxies—each widening the realm of the possible.

The journey from nothing to the plentitudes of a materializing universe can be reckoned as the expansion of freedoms, choices, and manifest possibilities. In the beginning there was no choice, no free will, no thing but nothing. From the big bang onward, the possible ways matter and energy could be arranged increased, and eventually, through life, the freedom of possible actions increased. With the coming of imaginative minds, even possible possibilities increased. It is almost as if the universe was a choice assembling itself.

In general, the long-term bias of technology is to increase the diversity of artifacts, methods, and techniques of creating choices. Evolution aims to keep the game of possibilities going.

I began this book with a quest for a method, an understanding at least, that would guide my choices in the technium. I needed a bigger view to enable me to choose technologies that would bless me with greater benefits and fewer demands. What I was really searching for was a way to reconcile the technium's selfish nature, which wants more of itself, with its generous nature, which wants to help us to find more of ourselves. Looking at the world through the eyes of the technium, I've grown to appreciate the unbelievable levels of selfish autonomy it possesses. Its internal momentum and directions are deeper than I originally suspected. At the same time, seeing the world from the technium's point of view has increased my admiration for its transformative positive powers. Yes, technology is acquiring its own autonomy and will increasingly maximize its own agenda, but this agenda includes—as its foremost consequence—maximizing possibilities for us.

I've come to the conclusion that this dilemma between these two faces of technology is unavoidable. As long as the technium exists (and it must exist if we are), then this tension between its gifts and its demands will continue to haunt us. In 3,000 years, when everyone finally gets their jet packs and flying cars, we will still struggle with this inherent conflict between the technium's own increase and ours. This enduring tension is yet another aspect of technology we have to accept.

As a practical matter I've learned to seek the minimum amount of technology for myself that will create the maximum amount of choices for myself and others. The cybernetician Heinz von Foerster called this approach the Ethical Imperative, and he put it this way: "Always act to increase the number of choices." The way we can use technologies to increase choices for others is by encouraging science, innovation, education, literacies, and pluralism. In my own experience this principle has never failed: In any game, increase your options.

There are two kinds of games in the universe: finite games and infinite games. A finite game is played to win. Card games, poker rounds, games of chance, bets, sports such as football, board games such as Monopoly, races, marathons, puzzles, Tetris, Rubik's Cube, Scrabble, sudoku, online games such as *World of Warcraft,* and *Halo*—all are finite games. The game ends when someone wins.

An infinite game, on the other hand, is played to keep the game going. It does not terminate because there is no winner.

Finite games require rules that remain constant. The game fails if the rules change during the game. Altering rules during play is unforgivable, the very definition of unfairness. Great effort, then, is taken in a finite game to spell out the rules beforehand and enforce them during the game.

An infinite game, however, can keep going only by changing its rules. To maintain open-endedness, the game must play with its rules.

A finite game such as baseball or chess or Super Mario must have boundaries—spatial, temporal, or behavioral. So big, this long, do or don't do that.

An infinite game has no boundaries. James Carse, the theologian who developed these ideas in his brilliant treatise *Finite and Infinite Games,* says, "Finite players play within boundaries; infinite players play with boundaries."

Evolution, life, mind, and the technium are infinite games. Their game is to keep the game going. To keep all participants playing as long as possible. They do that, as all infinite games do, by playing around with the rules of play. The evolution of evolution is just that kind of play.

Unreformed weapon technologies generate finite games. They produce winners (and losers) and cut off options. Finite games are dramatic; think sports and war. We can think of hundreds of more exciting stories about two guys fighting than we can about two guys at peace. But the problem with those exciting 100 stories about two guys fighting is that they all lead to the same end—the demise of one or both of them—unless at some point they turn and cooperate. However, the one boring story about peace has no end. It can lead to a thousand unexpected stories—maybe the two guys become partners and build a new town or discover a new element or write an amazing opera. They create something that will become a platform for future stories. They are playing an infinite game. Peace is summoned all over the world because it births increasing opportunities and, unlike a finite game, contains infinite potential.

The things in life we love most—including life itself—are infinite games. When we play the game of life, or the game of the technium, goals are not fixed, the rules are unknown and shifting. How do we proceed? A good choice is to increase choices. As individuals and as a society we can invent methods that will generate as many new *good* possibilities as possible. A good possibility is one that will generate more good possibilities . . . and so on in the paradoxical infinite game. The best "open-ended" choice is one that leads to the most subsequent "open-ended" choices. That recursive tree is the infinite game of technology.

The goal of the infinite game is to keep playing—to explore every way to play the game, to include all games, all possible players, to widen what is meant by playing, to spend all, to hoard nothing, to seed the universe with improbable plays, and if possible to surpass everything that has come before.

In his mythic book *The Singularity Is Near,* Ray Kurzweil, serial inventor, technology enthusiast, and unabashed atheist, announces: "Evolution moves toward greater complexity, greater elegance, greater knowledge, greater intelligence, greater beauty, greater creativity, and greater levels of subtle attributes such as love. In every monotheistic tradition God is likewise described as all of these qualities, only without limitation. . . . So evolution moves inexorably toward this conception of God, although never quite reaching this ideal."

If there is a God, the arc of the technium is aimed right at him. I'll retell the Great Story of this arc again, one last time in summary, because it points way beyond us.

As the undifferentiated energy at the big bang is cooled by the expanding space of the universe, it coalesces into measurable entities, and, over time, the particles condense into atoms. Further expansion and cooling allows complex molecules to form, which self-assemble into self-reproducing entities. With each tick of the clock, increasing complexity is added to these embryonic organisms, increasing the speed at which they change. As evolution evolves, it keeps piling on different ways to adapt and learn until eventually the minds of animals are caught in self-awareness. This self-awareness thinks up more minds, and together a universe of minds transcends all previous limits. The destiny of this collective mind is to expand imagination in all directions until it is no longer solitary but reflects the infinite.

There is even a modern theology that postulates that God, too, changes. Without splitting too many theological hairs, this theory, called Process Theology, describes God as a process, a perfect process, if you will. In this theology, God is less a remote, monumental, gray-bearded hacker genius and more of an ever-present flux, a movement, a process, a primary self-made becoming. The ongoing self-organized mutability of life, evolution, mind, and the technium is a reflection of God's becoming. God-as-Verb unleashes a set of rules that unfold into an infinite game, a game that continually loops back into itself.

I bring up God here at the end because it seems unfair to speak about autocreation without mentioning God—the paragon of autocreation. The only other alternative to an endless string of creations triggered by previous creation is a creation that emerges from its own self-causation. That prime self-causation, which is not preceded but instead first makes itself before it makes either time or nothingness, is the most logical definition of God. This view of a mutable God does not escape the paradoxes of self-creation that infect all levels of self-organization, but rather it embraces them as necessary paradoxes. God or not, self-creation is a mystery.

In one sense, this is a book about continuous autocreation (with or without the concept of a prime autocreation). The tale told here tells how the ratcheting bootstrapping of increasing complexity, expanding possibilities, and spreading sentience—which we now see in the technium and beyond—is driven by forces that were inherent within the first nanospeck of existence and how this seed of flux has unfolded itself in such a manner that it can, in theory, keep unfolding and making itself for a very long time.

What I hope I have shown in this book is that a single thread of self-generation ties the cosmos, the bios, and the technos together into one creation. Life is less a miracle than a necessity for matter and energy. The technium is less an adversary to life than its extension. Humans are not the culmination of this trajectory but an intermediary, smack in the middle between the born and the made.

For several thousand years, humans have looked to the organic world, the world of the living, for clues about the nature of creation and even of a creator. Life was a reflection of the divine. Humans in particular were deemed to be made in the image of God. But if you believe humans are made in the image of God, the autocreator, then we have done well, because we have just birthed our own creation: the technium. Many, including many believers in God, would call that hubris. Compared to what has come before us, our accomplishments are puny.

"As we turn from the galaxies to the swarming cells of our own being, which toil for something, some entity beyond their grasp, let us remember man, the self-fabricator who came across an ice age to look into the mirrors and magic of science. Surely he did not come to see himself or his wild visage only. He came because he is at heart a listener and a searcher for some transcendent realm beyond himself." That's Loren Eiseley, anthropologist and author, ruminating on what he calls our "immense journey" so far under the stars.

The bleak message of the stars in their overwhelming infinitude is that we are nothing. It is hard to argue with 500 billion galaxies, each with a billion stars. In the mists of the endless cosmos, our brief blink in an obscure corner is nothing at all.

Yet the fact that there is something in one corner that sustains itself against the starry vastness, the fact that there is anything bootstrapping at all, is an argument against the nihilism of the stars. The smallest thought could not exist unless the entire universe and the laws of physics were in some way encouraging it. The existence of a single rosebud, a single oil painting, a single parade of costumed hominins strolling down a street of bricks, a single glowing screen waiting for input, or a single book on the nature of our creations requires life-friendly attributes baked deeply into the primeval laws of being. "The universe knew we were coming," says Freeman Dyson. And if the cosmic laws are biased to produce one bit of life and mind and technology, then one bit will flow after another. Our immense journey is a trace of tiny, improbable events stacked into a series of inevitabilities.

The technium is the way the universe has engineered its own self-awareness. Carl Sagan put it memorably: "We are starstuff pondering the stars." But by far humanity's greatest, most immense journey is not the long trek from star dust to wakefulness but the immense journey we have in front of us. The arc of complexity and open-ended creation in the last four billion years is nothing compared to what lies ahead.

The universe is mostly empty because it is waiting to be filled with the products of life and the technium, with questions and problems and the thickening relations between bits that we call *con scientia*—shared knowledge—or consciousness.

And whether we like it or not, we stand at the fulcrum of the future. We are in part responsible for the evolution of this planet proceeding onward.

About 2,500 years ago most of humanity's major religions were set in motion in a relatively compact period. Confucius, Lao-tzu, Buddha, Zoroaster, the authors of the Upanishads, and the Jewish patriarchs all lived within a span of 20 generations. Only a few major religions have been born since then. Historians call that planetary fluttering the Axial Age. It was as if everyone alive awoke simultaneously and, in one breath, set out in search of their mysterious origins. Some anthropologists believe the Axial Age awakening was induced by the surplus abundance that agriculture created, enabled by massive irrigation and waterworks around the world.

It would not surprise me if we saw another axial awakening someday, powered by another flood of technology. I find it hard to believe that we could manufacture robots that actually worked and not have them disturb our ideas of religion and God. Someday we will make other minds, and they will surprise us. They will think of things we never could have imagined, and if we give these minds their full embodiment, they will call themselves children of God, and what will we say? When we alter the genetics in our veins, will this not reroute our sense of a soul? Can we cross over into the quantum realm, where one bit of matter can be in two places at once, and still not believe in angels?

Look what is coming: Technology is stitching together all the minds of the living, wrapping the planet in a vibrating cloak of electronic nerves, entire continents of machines conversing with one another, the whole aggregation watching itself through a million cameras posted daily. How can this not stir that organ in us that is sensitive to something larger than ourselves?

For as long as the wind has blown and the grass grown, people have sat beneath trees in the wilderness for enlightenment—to see God. They have looked to the natural world for a hint of their origins. In the filigree of fern and feather they find a shadow of an infinite source. Even those who have no use for God study the evolving world of the born for clues to why we are here. For most people, nature is either a very happy long-term accident or a very detailed reflection of its creator. For the latter, every species can be read as a four-billion-year-long encounter with God.

Yet we can see more of God in a cell phone than in a tree frog. The phone extends the frog's four billion years of learning and adds the open-ended investigations of six billion human minds. Someday we may believe the most convivial technology we can make is not a testament to human ingenuity but a testimony of the holy. As the technium's autonomy rises, we have less influence over the made. It follows its own momentum begun at the big bang. In a new axial age, it is possible the greatest technological works will be considered a portrait of God rather than of us. In addition to holding spiritual retreats in redwood groves, we may surrender ourselves in the labyrinths of a 200-year-old network. The intricate, unfathomable layers of logic built up over a century, borrowed from rainforest ecosystems, and woven together into beauty by millions of active synthetic minds will say what redwoods say, only louder, more convincingly: "Long before you were here, I am."

The technium is not God; it is too small. It is not utopia. It is not even an entity. It is a becoming that is only beginning. But it contains more goodness than anything else we know.

The technium expands life's fundamental traits, and in so doing it expands life's fundamental goodness. Life's increasing diversity, its reach for sentience, its long-term move from the general to the different, its essential (and paradoxical) ability to generate new versions of itself, and its constant play in an infinite game are the very traits and "wants" of the technium. Or should I say, the technium's wants are those of life. But the technium does not stop there. The technium also expands the mind's fundamental traits, and in so doing it expands the mind's fundamental goodness. Technology amplifies the mind's urge toward the unity of all thought, it accelerates the connections among all people, and it will populate the world with all conceivable ways of comprehending the infinite.

No one person can become all that is humanly possible; no one technology can capture all that technology promises. It will take all life and all minds and all technology to begin to see reality. It will take the whole technium, and that includes us, to discover the tools that are needed to surprise the world. Along the way we generate more options, more opportunities, more connection, more diversity, more unity, more thought, more beauty, and more problems. Those add up to more good, an infinite game worth playing.

That's what technology wants.

Acknowledgments

This work is dedicated to my children: Kaileen, Ting, and Tywen. And also to my wife, Gia-Miin, who supplied the necessary love for the long trek.

I am grateful for Paul Slovak at Penguin, who supported this book through its many years of gestation. He never gave up on it, and his enthusiasm for the ideas in this book made its birth possible.

The best editor I have ever worked with, Paul Tough, rescued this book from verbosity. He streamlined its narrative to a readable form, carving a book out of an almost book. Paul gave this work both its outline and its polish.

Camille Cloutier was my chief collaborator. The tasks she contributed are almost too many to list. Camille tracked down experts, arranged interviews, prepared quotes and passages, found key charts, fact-checked the entire book, footnoted it, proofed it, managed its many versions, compiled the bibliography, kept software going, and in every way made sure what I said was true and accurate.

Research librarian Michele McGinnis performed most of the original research reported in this book. She spent months in the library and five years online searching for sources. Almost every page of this book has been improved by her work.

Jonathan Corum, master designer and illustrator, rendered the charts on these pages in his distinctive, radically clear style. The jacket of the hardcover book was designed by Ben Wiseman.

This is the sixth book that John Brockman, adviser and agent extraordinaire, has engineered into existence with me. I would not think of doing a book without him.

Behind the scenes Victoria Wright made exact transcriptions of my interviews, and with a few Zen-like koans, William Schwalbe, book coach, offered extremely helpful suggestions when I was stuck. The book's layout was done by Nancy Resnick and the index compiled by Cohen Carruth, Inc.

The following readers endured the first draft of this book and provided me with valuable and constructive feedback: Russ Mitchell, Michael Dowd, Peter Schwartz, Charles Platt, Andreas Lloyd, Gary Wolf, and Howard Rheingold.

During the course of researching this book I interviewed, spoke with, or corresponded with the smartest people I know. Listed in alphabetical order, each of these experts lent their valuable time and insight for my project. Of course, any errors in transmitting their thoughts are mine.

Chris Anderson
Gordon Bell
Katy Borner
Stewart Brand
Eric Brende
David Brin
Rob Carlson
James Carse
Jamais Cascio
Richard Dawkins
Eric Drexler
Freeman Dyson
George Dyson
Niles Eldredge
Brian Eno
Joel Garreau
Paul Hawken
Danny Hillis
Piet Hut
Derrick Jensen
Bill Joy
Stuart Kauffman
Donald Kraybill
Mark Kryder
Ray Kurzweil
Jaron Lanier
Pierre Lemonnier
Seth Lloyd
Lori Marino
Max More
Simon Conway Morris
Nathan Myhrvold
Howard Rheingold
Paul Saffo
Kirkpatrick Sale
Tim Sauder
Peter Schwartz
John Smart
Lee Smolin
Alex Steffen
Steve Talbot
Edward Tenner
Sherry Turkle
Hal Varian
Vernor Vinge
Jay Walker
Peter Warshall
Robert Wright

Annotated Reading List

Of the hundreds of books I consulted for this project, I found the following selected ones to be the most useful for my purposes. Listed in order of importance. (The rest of my sources are listed in the Source Notes.)

Autonomous Technology: Technics-Out-of-Control as a Theme in Political Thought. Langdon Winner. Cambridge: MIT Press, 1977.

Langdon Winner comes closest to my own notions about technology's autonomy, but his ideas predate mine by decades. Although he reaches very different conclusions, he's done tons of research, and I owe his book a lot. He is an elegant writer as well.

Technology Matters: Questions to Live With. David Nye. Cambridge: MIT Press, 2006.

Probably the best all-around overview of the scope, scale, and philosophy of the technium. Nye offers depth of scholarship and careful, evenhanded introductions to various theories, with lots of examples, yet in a short, readable book.

The Nature of Technology: What It Is and How It Evolves. W. Brian Arthur. New York: Free Press, 2009.

This is the clearest, most utilitarian description of technology that I've come across. Arthur reduces the complexity of technology to an almost mathematical purity. At the same time his is a very humane, artful view. I agree with Arthur's perspective 100 percent.

Visions of Technology: A Century of Vital Debate About Machines, Systems, and the Human World. Richard Rhodes, ed. New York: Simon & Schuster, 1999.

In this one-volume anthology, Rhodes collects writings about technology written over the past century or so. Critics, poets, inventors, authors, artists, and ordinary citizens present some of the most quotable passages and perspectives on technology. I found all kinds of insights I had not seen elsewhere.

Does Technology Drive History? The Dilemma of Technological Determinism. Merritt Roe Smith and Leo Marx, eds. Cambridge: MIT Press, 1994.

A fairly scholarly anthology of historians trying to answer this vexing question.

The Singularity Is Near. Ray Kurzweil. New York: Viking, 2005.

I call this book mythical because I think the Singularity is a brand-new myth for our age. It is unlikely to be true, but probably very influential. The Singularity is a myth much like Superman or Utopia; it is an idea that once born will never go away, but will be reinterpreted forever. This is the book that launched this indelible idea. You cannot ignore it.

Thinking Through Technology: The Path Between Engineering and Philosophy. Carl Mitcham. Chicago: University of Chicago Press, 1994.

An accessible entry into the history of technology, sometimes used as a textbook.

Life's Solution: Inevitable Humans in a Lonely Universe. Simon Conway Morris. Cambridge: Cambridge University Press, 2004.

Two big ideas swim in this rambling book: evolution is convergent, and life forms are inevitable. Written by the biologist who deciphered the Burgess Shale fossils, the same evidence on which Gould based his *Wonderful Life* book, but with a 180-degree different conclusion from Gould.

The Deep Structure of Biology: Is Convergence Sufficiently Ubiquitous to Give a Directional Signal? Simon Conway Morris, ed. West Conshohocken, PA: Templeton Foundation Press, 2008.

An anthology from many disciplines on convergent evolution.

Cosmic Evolution. Eric J. Chaisson. Cambridge: Harvard University Press, 2002.

A little-known exploration of the idea that evolution procedes on a continuum that began before life, written by a physicist.

Biocosm : The New Scientific Theory of Evolution: Intelligent Life Is the Architect of the Universe. James Gardner. Makawao Maui, HI: Inner Ocean, 2003.

At the center of this book is a very radical idea (that the cosmos is a living organism) that may be too far out for most people, but surrounding this core is a book filled with tons of evidence for the continuum between the inert universe, life, and the technosphere. *Biocosm* is the only other book I know about that covers the same cosmic trends I try to capture.

Cosmic Jackpot: Why Our Universe Is Just Right for Life. Paul Davies. Boston: Houghton Mifflin, 2007.

Davies uses his professional knowledge of physics to tie the processes of life, mind, and entropy together. He is the most original reporter today who is digging into the tough big philosophical questions, yet who is still grounded in experimental scientific results. He is my guide into the large-scale structure of existence. *Cosmic Jackpot* is his latest and best summary.

Finite and Infinite Games. James Carse. New York: Free Press, 1986.
This tiny book holds a universe of wisdom. Written by a theologian, you probably need to read only the first and last chapters, but that is enough. It altered my thinking about life, the universe, and everything.

The Riddle of Amish Culture. Donald B. Kraybill. Baltimore: The Johns Hopkins University Press, 2001.
Kraybill conveys the paradoxes of the Amish with both objective insight and warm sympathy. He's the expert on their use of technology. He was also my guide on one visit to the Amish.

Better Off: Flipping the Switch on Technology. Eric Brende. New York: HarperCollins, 2004.
This is a refreshing, fast read about the two years Brende lived off the grid near an Amish community. His book is the best way to get the feel—the warmth, the smell, the atmosphere—of the minimal lifestyle. Because Brende comes from a technological background, he anticipates your questions.

Laws of Fear: Beyond the Precautionary Principle. Cass Sunstein. Cambridge: Cambridge University Press, 2005.
Case studies on the faults of the Precautionary Principle and a suggested framework for an alternative approach.

Whole Earth Discipline. Stewart Brand. New York: Viking, 2009.
Many of my themes about progress and urbanization and constant vigilance were first developed by Brand. This book also celebrates the transformative nature of tools and technology.

Limited Wants, Unlimited Means: A Reader on Hunter-Gatherer Economics and the Environment. John M. Gowdy, ed. Washington, D.C.: Island Press, 1998.
Plenty of scholarly papers on the surprising research of anthropologists who found that hunter-gathering lifestyles were not as undesirable as moderns think. It is impossible to read this anthology without changing your mind several times.

The Foraging Spectrum: Diversity in Hunter-Gatherer Lifeways. Robert L. Kelly, ed. Washington, D.C.: Simthsonian Institution Press, 1995.
Solid, cross-cultural data on how hunter-gatherers actually spend their time, calories, and attention. The best scientific studies on the economics and sociality of preagricultural life.

Neanderthals, Bandits, and Farmers: How Agriculture Really Began. Colin Tudge. New Haven: Yale University Press, 1999.
A courageously tiny book that manages to sum up the reasons for the birth of agriculture into 52 small pages. It packs in about five volumes of insight a whole library of research and then distills the essence into one beautiful essay. I wish I could write a tiny brilliant book like this.

After Eden: The Evolution of Human Domination. Kirkpatrick Sale. Durham: Duke University Press, 2006.

An expose of early Sapiens' quick path to environmental domination and disruption, long before agriculture or industry.

The Ascent of Man. Jacob Bronowski. Boston: Little, Brown, 1974.

Based on the 1972 BBC-TV series of the same name, this book inspired the sweep and scope of my own and provided some key concepts. Part geek, part poet, part mystic, part scientist, Bronowski was way ahead of his time.

Source Notes

Sources for illustrations and charts are included below.

1. My Question

6 **Until 1939:** Franklin D. Roosevelt. (1939, January 4) "Annual Message to Congress." http://www.presidency.ucsb.edu/ws/index.php?pid=15684.

6 **until 1952:** Harry S. Truman. (1952, January 9) "Annual Message to the Congress on the State of the Union." http://www.presidency.ucsb.edu/ws/index.php?pid=14418.

6 **King Odysseus was a master of *techne*:** Steve Talbott. (2001) "The Deceiving Virtues of Technology." NetFuture, (125). http://netfuture.org/2001/Nov1501_125.html.

7 **the term *technology* essentially disappeared:** Carl Mitcham. (1994) *Thinking Through Technology: The Path Between Engineering and Philosophy.* Chicago: University of Chicago Press, pp. 128–129.

7 **glass, cement, sewers, and water mills:** Henry Hodges. (1992) *Technology in the Ancient World.* New York: Barnes & Noble Publishing.

7 **"and not just for technical reasons":** Carl Mitcham. (1994) *Thinking Through Technology: The Path Between Engineering and Philosophy.* Chicago; University of Chicago Press, p. 123.

8 **"primarily on non-human power":** Lynn White. (1940) "Technology and Invention in the Middle Ages." *Speculum*, 15 (2), p. 156. http://www.jstor.org/stable/2849046.

8 **that forgotten Greek word:** Johann Beckmann. (1802) *Anleitung zur Technologie* [*Guide to Technology*]. Gottingen: Vandenhoeck und Ruprecht.

9 **mathematical problems, just like a computer:** L. M. Adleman. (1994) "Molecular Computation of Solutions to Combinatorial Problems." *Science*, 266 (5187). http://www.sciencemag.org/cgi/content/abstract/266/5187/1021.

12 **the society and culture of tools:** David Nye. (2006) *Technology Matters: Questions to Live With.* Cambridge, MA: MIT Press, pp. 12, 28.

14 **one mega-scale computing platform:** Kevin Kelly. (2008) "Infoporn: Tap into the 12-Million-Teraflop Handheld Megacomputer." *Wired*, 16 (7). http://www.wired.com/special_multimedia/2008/st_infoporn_1607.

14 **eyes (phone and webcams) plugged in:** Ibid.

14 **searches at the humming rate of 14 kilohertz:** comScore. (2007) "61 Billion Searches Conducted Worldwide in August." http://www.comscore.com/Press_Events/Press_Releases/2007/10/Worldwide_Searches_Reach_61_Billion. Calculation based on comScore's figure for the number of searches performed in a month.

14 **5 percent of the world's electricity:** Kevin Kelly. (2007) "How Much Power Does the Internet Consume?" The Technium. http://www.kk.org/thetechnium/archives/2007/10/how_much_power.php. This figure was calculated by David Sarokin; see http://uclue.com//index.php?xq=724.

14 **from the system at large:** Reginald D. Smith. (2008, revised April 20, 2009) "The Dynamics of Internet Traffic: Self-Similarity, Self-Organization, and Complex Phenomena," *arXiv:0807.3374.* http://arxiv.org/abs/0807.3374.

14 **fractal pattern of self-organization:** Ibid.

2. Inventing Ourselves

21 **to give themselves claws:** Jay Quade, Naomi Levin, et al. (2004) "Paleoenvironments of the Earliest Stone Toolmakers, Gona, Ethiopia." *Geological Society of America Bulletin,* 116 (11/12). http://gsabulletin.gsapubs.org/content/116/11-12/1529.abstract.

21 **predigesting, with fire:** Richard Wrangham and NancyLou Conklin-Brittain. (2003) "Cooking as a Biological Trait." *Comparative Biochemistry and Physiology—Part A: Molecular & Integrative Physiology,* 136 (1). http://dx.doi.org/10.1016/S1095-6433(03)00020-5.

22 **red deer skeleton:** Kirkpatrick Sale. (2006) *After Eden: The Evolution of Human Domination.* Durham, NC: Duke University Press.

22 **the human form we see in ourselves:** Paul Mellars. (2006) "Why Did Modern Human Populations Disperse from Africa Ca. 60,000 Years Ago? A New Model." *Proceedings of the National Academy of Sciences,* 103 (25). http://www.pnas.org/content/103/25/9381.full.pdf+html.

22 **Some say 200,000:** Ian McDougall, Francis H. Brown, et al. (2005) "Stratigraphic Placement and Age of Modern Humans from Kibish, Ethiopia." *Nature,* 433 (7027). http://dx.doi.org/10.1038/nature03258.

22 **they were outwardly indistinguishable from us:** Paul Mellars. (2006) "Why Did Modern Human Populations Disperse from Africa Ca. 60,000 Years Ago? A New Model." *Proceedings of the National Academy of Sciences,* 103 (25). http://www.pnas.org/content/103/25/9381.full.pdf+html.

23 **"something was missing":** Jared M. Diamond. (2006) *The Third Chimpanzee: The Evolution and Future of the Human Animal.* New York: HarperPerennial, p. 44.

23 **Prehistory Explosion of Human Population:** Data from Quentin D. Atkinson, Russell D. Gray et al. (2008) "Mtdna Variation Predicts Population Size in Humans and Reveals a Major Southern Asian Chapter in Human Prehistory." *Molecular Biology and Evolution,* 25 (2), p. 472. http://mbe.oxfordjournals.org/cgi/content/full/25/2/468.

24 **before the dawn of agriculture 10,000 years ago:** United States Census Bureau. (2008) "Historical Estimates of World Population." United States Census Bureau. http://www.census.gov/ipc/www/worldhis.html.

24 they reached the edges of Asia: Kirkpatrick Sale. (2006) *After Eden: The Evolution of Human Domination.* Durham, NC: Duke University Press, p. 34.

24 to fill the whole of the New World: Jared M. Diamond. (1997) *Guns, Germs, and Steel: The Fates of Human Societies.* New York: W. W. Norton, pp. 50–51.

24 "the world's most rugged terrain": Ibid., p. 51.

24 tailored hides in graves: Kirkpatrick Sale. (2006) *After Eden: The Evolution of Human Domination.* Durham, NC: Duke University Press, p. 68.

24 woven net and loose fabrics on them: Ibid., p. 77.

25 individuals at one time: Juan Luis de Arsuaga, Andy Klatt, et al. (2002) *The Neanderthal's Necklace: In Search of the First Thinkers.* New York: Four Walls Eight Windows, p. 227.

26 "rapidly produced, articulate speech": Richard G. Klein. (2002) "Behavioral and Biological Origins of Modern Humans." California Academy of Sciences/BioForum, Access Excellence. http://www.accessexcellence.org/BF/bf02/klein/bf02e3.php. Transcript of a lecture, "The Origin of Modern Humans," delivered December 5, 2002.

26 "far beyond all other Earthly species": Daniel C. Dennett. (1996) *Kinds of Minds.* New York: Basic Books, p. 147.

27 rapid sequences of notions: William Calvin. (1996) *The Cerebral Code: Thinking a Thought in the Mosaics of the Mind.* Cambridge, MA: MIT Press.

27 the Neanderthal diet was mostly meat: Kirkpatrick Sale. (2006) *After Eden: The Evolution of Human Domination.* Durham, NC: Duke University Press, p. 51.

28 Hemple Bay tribe, 2,160: Marshall David Sahlins. (1972) *Stone Age Economics.* Hawthorne, NY: Aldine de Gruyter, p. 18.

28 days spent sleeping were not uncommon: Ibid., p. 23.

28 "without showing great fatigue": Ibid., p. 28.

29 was only half that of gathering: Mark Nathan Cohen. (1989) *Health and the Rise of Civilization.* New Haven, CT: Yale University Press.

29 "the cost of his own esteem": Marshall David Sahlins. (1972) *Stone Age Economics.* Hawthorne, NY: Aldine de Gruyter, p. 30.

30 the goodness of the forest: Nurit Bird-David. (1992) "Beyond 'The Original Affluent Society': A Culturalist Reformulation." *Current Anthropology,* 33(1), p. 31.

31 six to eight children in agricultural communities: Robert L. Kelly. (1995) *The Foraging Spectrum: Diversity in Hunter-Gatherer Lifeways.* Washington, DC: Smithsonian Institution Press, p. 244.

31 at 16 or 17 years old: Ibid., p. 245.

31 suckling for as long as 6 years: Ibid., p. 247.

31 "rate of population growth increases": Ibid., p. 254.

32 close encounters with large, angry animals: Juan Luis de Arsuaga, Andy Klatt, et al. (2002) *The Neanderthal's Necklace: In Search of the First Thinkers.* New York: Four Walls Eight Windows, p. 221.

33 "longevity in the modern humans" began about 50,000 years ago: Rachel Caspari and Sang-Hee Lee. (2004) "Older Age Becomes Common Late in Human Evolution." *Proceedings of the National Academy of Sciences of the United States of America,* 101 (30). http://www.pnas.org/content/101/30/10895.abstract.

34 **a shirt, a jacket, trousers, and moccasins:** Robert L. Kelly. (1995) *The Foraging Spectrum: Diversity in Hunter-Gatherer Lifeways.* Washington, DC: Smithsonian Institution Press.

35 **(in war between tribes):** Lawrence H. Keeley. (1997) *War Before Civilization.* New York: Oxford University Press, p. 89.

35 **Comparison of War Fatality Rates:** Data from Lawrence H. Keeley. (1997) *War Before Civilization.* New York: Oxford University Press, p. 89.

36 **in his survey of early warfare:** Ibid., pp. 174–75.

36 **how far you could go with a handful of tools:** Carl Haub. (1995) "How Many People Have Ever Lived on Earth?—Population Reference Bureau." Population Reference Bureau. http://www.prb.org/Articles/2002/HowManyPeopleHaveEverLivedonEarth.aspx.

37 **for the previous six million years:** Gregory Cochran and Henry Harpending. (2009) *The 10,000 Year Explosion: How Civilization Accelerated Human Evolution.* New York: Basic Books, p. 1.

38 **northernmost portions of the planet by now:** William F. Ruddiman. (2005) *Plows, Plagues, and Petroleum: How Humans Took Control of Climate.* Princeton NJ: Princeton University Press, p. 12.

39 **aristocratic feudalism in Europe:** John Sloan. (1994) "The Stirrup Thesis." http://www.fordham.edu/halsall/med/sloan.html.

40 **"and the process of Science began":** John Brockman. (2000) *The Greatest Inventions of the Past 2,000 Years.* New York: Simon & Schuster, p. 142.

41 **"structure, organization, information, and control":** Richard Rhodes. (1999) *Visions of Technology: Machines, Systems and the Human World.* New York: Simon & Schuster, p. 188.

3. History of the Seventh Kingdom

43 **time: four billion years:** Lynn Margulis. (1986) *Microcosmos: Four Billion Years of Evolution from Our Microbial Ancestors.* New York: Summit Books.

45 **"combinatorial evolution is foremost, and routine":** W. Brian Arthur. (2009) *The Nature of Technology: What It Is and How It Evolves.* New York: Free Press, p. 188.

46 **the major transitions in biological organization were:** John Maynard Smith and Eors Szathmary. (1997) *The Major Transitions in Evolution.* New York: Oxford University Press.

50 **punctuated, stepwise evolution:** Stephen Jay Gould and Niles Eldredge. (1977) "Punctuated Equilibria: The Tempo and Mode of Evolution Reconsidered." *Paleobiology,* 3 (2).

50 **some dating back to 1825:** Belinda Barnet and Niles Eldredge. (2004) "Material Cultural Evolution: An Interview with Niles Eldredge." *Fibreculture Journal,* (3). http://journal.fibreculture.org/issue3/issue3_barnet.html.

51 **Evolutionary Tree of Cornets:** Data from Ilya Temkin and Niles Eldredge. (2007) "Phylogenetics and Material Cultural Evolution." *Current Anthropology,* 48 (1). http://dx.doi.org/10.1086/510463.

51 **"bilaterally symmetrical conformation of the ancestral fish":** Belinda Barnet and Niles Eldredge. (2004) "Material Cultural Evolution: An Interview with Niles Eldredge." *Fibreculture Journal* (3). http://journal.fibreculture.org/issue3/issue3_barnet.html.

52 **A Thousand Years of Helmet Evolution:** Bashford Dean. (1916) *Notes on Arms and Armor.* New York: Metropolitan Museum of Art, p. 115.

53 **power delivered to them by overhead driveshafts:** David Nye. (2006) *Technology Matters: Questions to Live With.* Cambridge, MA: MIT Press, p. 57.

54 **Catalogs of Durable Goods:** Aaron Montgomery Ward and Joseph J. Schroeder, Jr. (1977) *Montgomery Ward & Co 1894–95 Catalogue & Buyers Guide, No. 56.* Northfield, IL: DBI Books, p. 562. Right-hand portion of this side-by-side comparison assembled by the author.

55 **brand-new spear and arrow points per year:** John Charles Whittaker. (2004) *American Flintknappers.* Austin: University of Texas Press, p. 266.

55 **to highlight the ephemeral nature of popular gadgetry:** Bruce Sterling. (1995, September 15) "The Life and Death of Media." Sixth International Symposium on Electronic Art ISEA, Montreal. http://www.alamut.com/subj/artiface/deadMedia/dM_Address.html. The Dead Media project is now defunct.

4. The Rise of Exotropy

58 **the beginning of time, billions of years ago:** National Aeronautics and Space Administration. (2009) "How Old Is the Universe?" http://map.gsfc.nasa.gov/universe/uni_age.html.

60 **Power Density Gradient:** Data from Eric J. Chaisson. (2002) *Cosmic Evolution.* Cambridge, MA: Harvard University Press, p. 139.

60 **shines the computer chip:** Eric J. Chaisson. (2005) *Epic of Evolution: Seven Ages of the Cosmos.* New York: Columbia University Press.

65 **Dominant Eras of the Universe:** Designed by the author.

66 **per genetic lineage (such as a parrot or a wallaby):** Motoo Kimura and Naoyuki Takahata. (1994) *Population Genetics, Molecular Evolution, and the Neutral Theory.* Chicago: University of Chicago Press.

67 **"precisely because it evades chemical imperatives":** Paul Davies. (1999) *The Fifth Miracle: The Search for the Origin and Meaning of Life.* New York: Simon & Schuster, p. 256.

67 **"financial and legal advice, and the like":** Richard Fisher. (2008) "Selling Our Services to the World (with an Ode to Chicago)." Chicago Council on Global Affairs, Chicago: Federal Reserve Bank of Dallas. http://www.dallasfed.org/news/speeches/fisher/2008/fs080417.cfm.

68 **The Dematerialization of U.S. Exports:** Data from "U.S. International Trade in Goods and Services Balance of Payments Basis, 1960–2004." U.S. Department of Commerce, International Trade Administration. http://www.ita.doc.gov/td/industry/OTEA/usfth/aggregate/H04t01.html.

69 **rather than manufactured goods (atoms):** Robert E. Lipsey. (2009) "Measuring International Trade in Services." *International Trade in Services and Intangibles in the Era of Globalization,* eds. Mathew J. Slaughter and Marshall Reinsdorf. Chicago: University of Chicago Press, p. 60.

5. Deep Progress

74 **"more good than evil in the world—but not by much":** Matthew Fox and Rupert Sheldrake. (1996) *The Physics of Angels: Exploring the Realm Where Science and Spirit Meet.* San Francisco: HarperSanFrancisco, p. 129.

75 **hoping to survive on those crowded shelves:** Barry Schwartz. (2004) *The Paradox of Choice: Why More Is Less.* New York: Ecco, p. 12.

75 **at least 30 million of them in use worldwide:** GS1 US. (2010, January 7) In discussion with the author's researcher. Jon Mellor, of GS1 US, explains that 1.2 million company prefixes have been issued worldwide. This is the first string of numbers used in both UPC and EAN bar codes. Based on this, he estimates the number of active UPC/EANs worldwide to be 30–48 million.

76 **contained 18,000 objects:** David Starkey. (1998) *The Inventories of King Henry VIII.* London: Harvey Miller Publishers.

77 **surrounded by all their possessions:** Peter Menzel. (1995) *Material World: A Global Family Portrait.* San Francisco: Sierra Club Books.

78 **objects in the entire estate:** Edward Waterhouse and Henry Briggs. (1970) "A declaration of the state of the colony in Virginia." The English experience, its record in early printed books published in facsimile, no. 276. Amsterdam: Theatrum Orbis Terrarum.

78 **now-classic study by Richard Easterlin in 1974:** Richard A. Easterlin. (1996) *Growth Triumphant.* Ann Arbor: University of Michigan Press.

78 **affluence brings increased satisfaction:** David Leonhardt. (2008, April 16) "Maybe Money Does Buy Happiness After All." *New York Times.* http://www.nytimes.com/2008/04/16/business/16leonhardt.html.

80 **only a small percentage of humans lived in cities:** United States Census Bureau. (2008) "Historical Estimates of World Population." http://www.census.gov/ipc/www/worldhis.html; George Modelski. (2003) *World Cities.* Washington, D.C.: Faros.

80 **now 50 percent do:** United Nations. (2007) "World Urbanization Prospects: The 2007 Revision." http://www.un.org/esa/population/publications/wup2007/2007wup.htm.

81 **Global Urban Population:** The author's calculations based on data from United States Census Bureau. (2008) "Historical Estimates of World Population." http://www.census.gov/ipc/www/worldhis.html; United Nations. (2007) "World Urbanization Prospects: The 2007 Revision." http://www.un.org/esa/population/publications/wup2007/2007wup.htm; Tertius Chandler. (1987) *Four Thousand Years of Urban Growth: An Historical Census.* Lewiston, N.Y.: Edwin Mellen Press; George Modelski. (2003) *World Cities.* Washington, D.C.: Faros.

82 **Paris in the Middle Ages:** Bronislaw Geremek, Jean-Claude Schmitt, et al. (2006) *The Margins of Society in Late Medieval Paris.* Cambridge, UK: Cambridge University Press, p. 81.

83 **"where they huddle around a fireplace":** Joseph Gies and Frances Gies. (1981) *Life in a Medieval City.* New York: HarperCollins, p. 34.

83 **slum at its peak in the 1880s:** Robert Neuwirth. (2006) *Shadow Cities.* New York: Routledge.

83 **"this serves all the purposes of the family":** Ibid., p. 177.

83 **"bona fide legal title to their land":** Ibid., p. 198.

83 **"half a dozen tents or shanties":** Ibid., p. 197.

84 **"Cities are wealth creators":** Stewart Brand. (2009) *Whole Earth Discipline.* New York: Viking, p. 25.

84 **"nearly 9 in 10 new patented innovations":** Ibid., p. 32.

84 "GNP growth occurs in cities": Ibid., p. 31.
84 "in the city at least six years": Mike Davis. (2006) *Planet of Slums*. London: Verso, p. 36.
85 but 94 percent of their kids were literate: Stewart Brand. (2009) *Whole Earth Discipline*. New York: Viking, pp. 42–43.
85 "Discomfort is an investment": Ibid., p. 36.
85 "get education for her children": Ibid., p. 26.
86 "more options for their future": Donovan Webster. (2005) "Empty Quarter." *National Geographic*, 207 (2).
88 "drawing that was pointed up": Gregg Easterbrook. (2003) *The Progress Paradox: How Life Gets Better While People Feel Worse*. New York: Random House, p. 163.
92 World Population in Civilization: Data from United States Census Bureau. (2008) "Historical Estimates of World Population." http://www.census.gov/ipc/www/worldhis.html.
93 only in expanding population: Niall Ferguson. (2009) In discussion with the author.
93 prime source of deep progress: Julian Lincoln Simon. (1996) *The Ultimate Resource 2*. Princeton, NJ: Princeton University Press.
94 World Population Forecasts: Data from United Nations Population Division. (2002) "World Population Prospects: The 2002 Revision." http://www.un.org/esa/population/publications/wpp2002/WPP2002-HIGHLIGHTSrev1.pdf.
94 that is, for the next 300 years: United Nations Department of Economic and Social Affairs Population Division. (2004) "World Population to 2300." http://www.un.org/esa/population/publications/longrange2/WorldPop2300final.pdf.
95 Estimated Long-Range World Population: Data from United Nations Department of Economic and Social Affairs Population Division. (2004) "World Population to 2300." http://www.un.org/esa/population/publications/longrange2/WorldPop2300final.pdf.
95 every country in Europe is below 2.0: Rand Corporation. (2005) "Population Implosion? Low Fertility and Policy Responses in the European Union." http://www.rand.org/pubs/research_briefs/RB9126/index1.html.
95 Japan is at 1.34: (2008, June 24) "Negligible Rise in Fertility Rate." *Japan Times Online*. http://search.japantimes.co.jp/cgi-bin/ed20080624a1.html.
96 Recent Fertility Rates in Europe: Data from Rand Corporation. (2005) "Population Implosion? Low Fertility and Policy Responses in the European Union." http://www.rand.org/pubs/research_briefs/RB9126/index1.html.
100 "would have turned out to be right": Julian Lincoln Simon. (1995) *The State of Humanity*. Oxford, UK: Wiley-Blackwell, pp. 644–45.
100 to 75.7 years in 1994: Kevin M. White and Samuel H. Preston. (1996) "How Many Americans Are Alive Because of Twentieth-Century Improvements in Mortality?" *Population and Development Review*, 22 (3), p. 415. http://www.jstor.org/stable/2137714.
101 "that we farm at the moment": Ronald Bailey. (2009, February) "Chiefs, Thieves, and Priests: Science Writer Matt Ridley on the Causes of Poverty and Prosperity." *Reason Magazine*. http://reason.com/archives/2009/01/07/chiefs-thieves-and-priests/3.

101 **"but simply part of our reality":** Simon Conway Morris. (2004) *Life's Solution: Inevitable Humans in a Lonely Universe.* New York: Cambridge University Press, p. xiii.

6. Ordained Becoming

104 **"may well have found them all":** Richard Dawkins. (2004) *The Ancestor's Tale: A Pilgrimage to the Dawn of Evolution.* Boston: Houghton Mifflin, p. 588.

105 **species coinhabiting Earth:** W. Hardy Eshbaugh. (1995) "Systematics Agenda 2000: An Historical Perspective." *Biodiversity and Conservation,* 4 (5). http://dx.doi.org/10.1007/BF00056336.

105 **"Evolution is remarkably reproducible":** Sean Carroll. (2008) "The Making of the Fittest DNA and the Ultimate Forensic Record of Evolution." *Paw Prints,* p. 154.

106 **evolution is hundreds long and counting:** (2009) "List of Examples of Convergent Evolution." Wikipedia, Wikimedia Foundation. http://en.wikipedia.org/w/index.php?title=List_of_examples_of_convergent_evolution&oldid=344747726.

107 **many of which evolved independently:** John Maynard Smith and Eors Szathmary. (1997) *The Major Transitions in Evolution.* New York: Oxford University Press.

107 **which uses a bubble to breathe:** Richard Dawkins. (2004) *The Ancestor's Tale: A Pilgrimage to the Dawn of Evolution.* Boston: Houghton Mifflin, p. 592.

109 **"independently reevolved fins":** George McGhee. (2008) "Convergent Evolution: A Periodic Table of Life?" *The Deep Structure of Biology,* ed. Simon Conway Morris. West Conshohocken, PA: Templeton Foundation, p. 19.

111 **Size Ratio in Life:** Data from K. J. Niklas. (1994) "The Scaling of Plant and Animal Body Mass, Length, and Diameter." *Evolution,* 48 (1), pp. 48–49. http://www.jstor.org/stable/2410002.

112 **"rates and times are remarkably similar":** Erica Klarreich. (2005) "Life on the Scales—Simple Mathematical Relationships Underpin Much of Biology and Ecology." *Science News,* 167 (7).

112 **"being invariant platonic forms":** Michael Denton and Craig Marshall. (2001) "Laws of Form Revisited." *Nature,* 410 (6827). http://dx.doi.org/10.1038/35068645.

113 **"pass examinations on Arcturus":** David Darling. (2001) *Life Everywhere: The Maverick Science of Astrobiology.* New York: Basic Books, p. 14.

115 **"largely incapable of self-replication":** Kenneth D. James and Andrew D. Ellington. (1995) "The Search for Missing Links Between Self-Replicating Nucleic Acids and the RNA World." *Origins of Life and Evolution of Biospheres,* 25 (6). http://dx.doi.org/10.1007/BF01582021.

116 **"strangest molecule in the universe":** Simon Conway Morris. (2004) *Life's Solution: Inevitable Humans in a Lonely Universe.* New York: Cambridge University Press.

117 **"functional groups used in life":** Norman R. Pace. (2001) "The Universal Nature of Biochemistry." *Proceedings of the National Academy of Sciences of the United States of America,* 98 (3). http://www.pnas.org/content/98/3/805.short.

117 **at least "one in a million":** Stephen J. Freeland, Robin D. Knight, et al. (2000)

"Early Fixation of an Optimal Genetic Code." *Moleculor Biology and Evolution*, 17 (4). http://mbe.oxfordjournals.org/cgi/content/abstract/17/4/511.

117 **several billion years of evolution have produced it?:** David Darling. (2001) *Life Everywhere: The Maverick Science of Astrobiology*. New York: Basic Books, p. 130.

118 **"where there is carbon-based life":** Michael Denton and Craig Marshall. (2001) "Laws of Form Revisited." *Nature*, 410 (6827). http://dx.doi.org/10.1038/35068645.

120 **"encoded implicitly in the genome":** Lynn Helena Caporale. (2003) "Natural Selection and the Emergence of a Mutation Phenotype: An Update of the Evolutionary Synthesis Considering Mechanisms That Affect Genomic Variation." *Annual Review of Microbiology*, 57 (1).

121 **from the same starting point:** (2009) "Skeuomorph." Wikipedia, Wikimedia Foundation. http://en.wikipedia.org/w/index.php?title=Skeuomorph&oldid=340233294.

122 **"the embodiment of contingency":** Stephen Jay Gould. (1989) *Wonderful Life: The Burgess Shale and Nature of History*. New York: W. W. Norton, p. 320.

123 **The Triad of Evolution:** Inspired by Stephen Jay Gould. (2002) *The Structure of Evolutionary Theory*. Cambridge, MA: Belknap Press of Harvard University Press, p. 1052; designed by the author.

124 **"walk through genetic drift":** Simon Conway Morris. (2004) *Life's Solution: Inevitable Humans in a Lonely Universe*. New York: Cambridge University Press, p. 132.

124 **"back hundreds of thousands of years":** Stephen Jay Gould. (2002) *The Structure of Evolutionary Theory*. Cambridge: Belknap Press of Harvard University Press, p. 1085.

124 **if the tape of life was rewound:** Michael Denton. (1998) *Nature's Destiny: How the Laws of Biology Reveal Purpose in the Universe*. New York: Free Press, p. 283.

125 **"are rigged in favor of life":** Paul Davies. (1998) *The Fifth Miracle: The Search for the Origin of Life*. New York: Simon & Schuster, p. 264.

125 **"predetermined by the interatomic forces":** Ibid., p. 252.

125 **"seem to direct the synthesis":** Ibid., p. 253.

126 **"but we the expected":** Stuart A. Kauffman. (1995) *At Home in the Universe*. New York: Oxford University Press, p. 8.

126 **"an inevitable process":** Manfred Eigen. (1971) "Self-organization of Matter and the Evolution of Biological Macromolecules." *Naturwissenschaften*, 58 (10), p. 519. http://dx.doi.org/10.1007/BF00623322.

126 **"into the fabric of the universe":** Christian de Duve. (1995) *Vital Dust: Life as a Cosmic Imperative*. New York: Basic Books, pp. xv, xviii.

126 **"becomes increasingly inevitable":** Simon Conway Morris. (2004) *Life's Solution: Inevitable Humans in a Lonely Universe*. New York: Cambridge University Press, p. xiii.

127 **with details left to chance:** Richard E. Lenski. (2008) "Chance and Necessity in Evolution." *The Deep Structure of Biology*, ed. Simon Conway Morris. West Conshohocken, PA: Templeton Foundation.

127 **"lines on similar phenotypes":** Sean C. Sleight, Christian Orlic, et al. (2008) "Genetic Basis of Evolutionary Adaptation by Escherichia Coli to Stressful Cycles of Freezing, Thawing and Growth." *Genetics*, 180 (1). http://www.genetics.org/cgi/content/abstract/180/1/431.

127 **"all outcomes would be different":** Sean Carroll. (2008) *The Making of the Fittest: DNA and the Ultimate Forensic Record of Evolution*. New York: W. W. Norton.

128 **precisely, but elegantly, backward:** Stephen Jay Gould. (1989) *Wonderful Life: The Burgess Shale and Nature of History*. New York: W. W. Norton, p. 320.

128 **Buckminster Fuller once said:** Richard Buckminster Fuller, Jerome Agel, et al. (1970) *I Seem to Be a Verb*. New York: Bantam Books.

7. Convergence

132 **strung across our countryside:** Christopher A. Voss. (1984) "Multiple Independent Invention and the Process of Technological Innovation." *Technovation*, 2 p. 172.

132 **"claimed by more than one person":** William F. Ogburn and Dorothy Thomas. (1975) "Are Inventions Inevitable? A Note on Social Evolution." *A Reader in Culture Change*, eds. Ivan A. Brady and Barry L. Isaac. New York: Schenkman Publishing, p. 65.

132 **the efficacy of vaccinations:** Bernhard J. Stern. (1959) "The Frustration of Technology." *Historical Sociology: The Selected Papers of Bernhard J. Stern*. New York: The Citadel Press, p. 121.

132 **came upon the same process:** Ibid.

133 **occurred within a month or so:** Dean Keith Simonton. (1979) "Multiple Discovery and Invention: Zeitgeist, Genius, or Chance?" *Journal of Personality and Social Psychology*, 37 (9), p. 1604.

133 **"is not the electric railroad inevitable?":** William F. Ogburn and Dorothy Thomas. (1975) "Are Inventions Inevitable? A Note on Social Evolution." *A Reader in Culture Change*, eds. Ivan A. Brady and Barry L. Isaac. New York: Schenkman Publishing, p. 66.

134 **known in statistics as a Poisson distribution:** Dean Keith Simonton. (1978) "Independent Discovery in Science and Technology: A Closer Look at the Poisson Distribution." *Social Studies of Science*, 8 (4).

134 **greatest discoverers buy lots of tickets:** Dean Keith Simonton. (1979) "Multiple Discovery and Invention: Zeitgeist, Genius, or Chance?" *Journal of Personality and Social Psychology*, 37 (9).

134 **Westinghouse laboratory in Paris:** John Markoff. (2003, February 24) "A Parallel Inventor of the Transistor Has His Moment." *New York Times*. http://www.nytimes.com/2003/02/24/business/a-parallel-inventor-of-the-transistor-has-his-moment.html.

135 **within months of each other in 1977:** Adam B. Jaffe, Manuel Trajtenberg, et al. (2000, April) "The Meaning of Patent Citations: Report on the NBER/Case-Western Reserve Survey of Patentees." Nber Working Paper No. W7631.

135 **"far above the accidents of personality":** Alfred L. Kroeber. (1917) "The Superorganic." *American Anthropologist*, 19 (2) p. 199.

135 **but never quite reached it:** Spencer Weart. (1977) "Secrecy, Simultaneous Discovery, and the Theory of Nuclear Reactors." *American Journal of Physics*, 45 (11), p. 1057.

136 the same body by different means: Dean Keith Simonton. (1979) "Multiple Discovery and Invention: Zeitgeist, Genius, or Chance?" *Journal of Personality and Social Psychology*, 37 (9), p. 1608.

136 "singleton discoveries are imminent multiples": Robert K. Merton. (1961) "Singletons and Multiples in Scientific Discovery: A Chapter in the Sociology of Science." *Proceedings of the American Philosophical Society,* 105 (5), p. 480.

136 "to investigate something else": Augustine Brannigan. (1983) "Historical Distributions of Multiple Discoveries and Theories of Scientific Change." *Social Studies of Science*, 13 (3), p. 428.

137 another 26 percent more than once: Eugene Garfield. (1980) "Multiple Independent Discovery & Creativity in Science." *Current Contents*, 44. Reprinted in *Essays of an Information Scientist: 1979–1980,* 4(44). http://www.garfield.library.upenn.edu/essays/v4p660y1979-80.pdf.

138 or even the patent office examiner: Adam B. Jaffe, Manuel Trajtenberg, et al. (2000) "The Meaning of Patent Citations: Report on the Nber/Case-Western Reserve Survey of Patentees." National Bureau of Economic Research, April 2000, p. 10.

138 "involve near-simultaneous invention": Mark Lemley and Colleen V. Chien. (2003) "Are the U.S. Patent Priority Rules Really Necessary?" *Hastings Law Journal*, 54 (5), p. 1300.

139 "a regular feature of innovation": Adam B. Jaffe, Manuel Trajtenberg, et al. (2000) "The Meaning of Patent Citations: Report on the Nber/Case-Western Reserve Survey of Patentees." National Bureau of Economic Research, April 2000, p. 1325.

139 inventors of incandescent bulbs prior to Edison: Robert Douglas Friedel, Paul Israel, et al. (1986) *Edison's Electric Light.* New Brunswick, NJ: Rutgers University Press.

139 Varieties of the Lightbulb: Collage by the author from archival materials.

141 "merely an efficient source of insight": Malcolm Gladwell. (2008, May 12) "In the Air: Who Says Big Ideas Are Rare?" *New Yorker*, 84 (13).

141 "one third of our ideas": Nathan Myhrvold. (2009) In discussion with the author.

141 "of when, not if": Jay Walker. (2009) In discussion with the author.

142 all had the same idea: W. Daniel Hillis. (2009) In discussion with the author.

142 The Inverted Pyramid of Invention: Inspired by W. Daniel Hillis; designed by the author.

144 "merits of both investigators as being comparable": Abraham Pais. (2005) *"Subtle Is the Lord . . .": The Science and the Life of Albert Einstein*. Oxford: Oxford University Press, p. 153.

144 "even *after* they read his paper": Walter Isaacson. (2007) *Einstein: His Life and Universe.* New York: Simon & Schuster, p. 134.

144 "ten years or more": Walter Isaacson. (2009) In discussion with the author.

144 "appear the most determined of all": Dean Keith Simonton. (1978) "Independent Discovery in Science and Technology: A Closer Look at the Poisson Distribution." *Social Studies of Science*, 8 (4), p. 526.

144 *K-9* and *Turner & Hooch*: Sean Dwyer. (2007) "When Movies Come in Pairs: Examples of Hollywood Deja Vu." Film Junk. http://www.filmjunk.com/2007/03/07/when-movies-come-in-pairs-examples-of-hollywood-deja-vu/.

145 **a device called Toto:** Tad Friend. (1998, September 14) "Copy Cats." *New Yorker.* http://www.newyorker.com/archive/1998/09/14/1998_09_14_051_TNY_LIBRY_000016335.

146 **simultaneous spontaneous creation:** (2009) "Harry Potter Influences and Analogues." Wikipedia, Wikimedia Foundation. http://en.wikipedia.org/w/index.php?title=Harry_Potter_influences_and_analogues&oldid=330124521.

148 **Parallels in Blow Gun Culture:** Collage by the author from archival materials.

149 **the exquisite timing of when to blow:** Robert L. Rands and Caroll L. Riley. (1958) "Diffusion and Discontinuous Distribution." *American Anthropologist,* 60 (2), p. 282.

149 **what we call abacus:** John Howland Rowe. (1966) "Diffusionism and Archaeology." *American Antiquity,* 31 (3), p. 335.

149 **"similar trajectories in various parts of the world":** Laurie R. Godfrey and John R. Cole. (1979) "Biological Analogy, Diffusionism, and Archaeology." *American Anthropologist,* New Series, 81 (1), p. 40.

151 **grains before root crops:** Neil Roberts. (1998) *The Holocene: An Environmental History.* Oxford: Blackwell Publishers, p. 136.

151 **an independent parallel discovery:** John Troeng. (1993) *Worldwide Chronology of Fifty-three Innovations.* Stockholm: Almqvist & Wiksell International.

151 **help of a statistician:** Andrew Beyer. (2009) In discussion with the author.

152 **"the telephone in the United States in 1876":** Alfred L. Kroeber. (1948) *Anthropology.* New York: Harcourt, Brace & Co., p. 364.

152 **simultaneous inventions in history:** Robert K. Merton. (1973) *The Sociology of Science: Theoretical and Empirical Investigations.* Chicago: University of Chicago Press, p. 371.

153 **are fueled by new technologies:** Dean Keith Simonton. (1979) "Multiple Discovery and Invention: Zeitgeist, Genius, or Chance?" *Journal of Personality and Social Psychology,* 37 (9), p. 1614.

153 **the then-obvious next step:** A. L. Kroeber. (1948) *Anthropology.* New York: Harcourt, Brace & Co.

154 **"a disadvantage when it comes to new ones":** (2008, February 9) "Of Internet Cafés and Power Cuts." *Economist,* 386 (8566).

8. Listen to the Technology

157 **Super Sabre doing 1,215 kilometers per hour:** (2009) "Flight Airspeed Record." Wikipedia, Wikimedia Foundation. http://en.wikipedia.org/w/index.php?title=Flight_airspeed_record&oldid=328492645.

158 **Speed Trend Curve:** Robert W. Prehoda. (1972) "Technological Forecasting and Space Exploration." *An Introduction to Technological Forecasting,* ed. Joseph Paul Martino. London: Gordon and Breach, p. 43.

158 **to the Moon quite soon after that:** Damien Broderick. (2002) *The Spike: How Our Lives Are Being Transformed by Rapidly Advancing Technologies.* New York: Forge, p. 35.

158 **"Arthur C. Clarke had expected it to occur":** Ibid.

159 **start-up making the integrated chips:** John Markoff. (2005) *What the Dormouse Said: How the 60s Counterculture Shaped the Personal Computer.* New York: Viking, p. 17.

160 **Plotting Moore's Law:** Data from Gordon Moore. (1965) "The Future of Integrated Electronics." *Understanding Moore's Law: Four Decades of Innovation,* ed. David C. Brock. Philadelphia: Chemical Heritage Foundation, p. 54. https://www.chemheritage.org/pubs/moores_law/; David C. Brock and Gordon E. Moore. (2006) "Understanding Moore's Law." Philadelphia: Chemical Heritage Foundation, p. 70.

160 **they would also become better:** David C. Brock and Gordon E. Moore. (2006) "Understanding Moore's Law." Philadelphia: Chemical Heritage Foundation, p. 99.

160 **"drops as a result of the technology":** Gordon E. Moore. (1995) "Lithography and the Future of Moore's Law." *Proceedings of SPIE,* 2437, p. 17.

161 **"Moore's Law is really about economics":** David C. Brock and Gordon E. Moore. (2006) "Understanding Moore's Law." Philadelphia: Chemical Heritage Foundation.

161 **"to make it come to pass":** Bob Schaller. (1996) "The Origin, Nature, and Implications of 'Moore's Law.'" http://research.microsoft.com/en-us/um/people/gray/moore_law.html.

161 **what you're allowed to believe:** University Video Corporation. (1992) *How Things Really Work: Two Inventors on Innovation, Gordon Bell and Carver Mead.* Stanford: University Video Corporation.

162 **"it sort of drives itself":** Bob Schaller. (1996) "The Origin, Nature, and Implications of 'Moore's Law.'" http://research.microsoft.com/en-us/um/people/gray/moore_law.html.

163 **"while scaling of disk drives continues":** Mark Kryder. (2009) In discussion with the author.

163 **"not identical as might be expected":** Lawrence G. Roberts. (2007) "Internet Trends." http://www.ziplink.net/users/lroberts/IEEEGrowthTrends/IEEE Computer12-99.htm.

163 **"sequence of the physical DNA":** Rob Carlson. (2009) In discussion with the author.

164 **Four Other Laws:** Data from National Renewable Energy Laboratory Energy Analysis Office. (2005) "Renewable Energy Cost Trends." cost_curves_2005.ppt. www.nrel.gov/analysis/docs/cost_curves_2005.ppt; Ed Grochowski. (2000) "IBM Areal Density Perspective: 43 Years of Technology Progress." http://www.pcguide.com/ref/hdd/histTrends-c.html; Rob Carlson. (2009, September 9) "The Bio-Economist." *Synthesis.* http://www.synthesis.cc/2009/09/the-bio-economist.html. Deloitte Center for the Edge. (2009) "The 2009 Shift Index: Measuring the Forces of Long-Term Change," p. 29. http://www.edgeperspectives.com/shiftindex.pdf.

165 **"operative even when people disbelieved it":** Rob Carlson. (2009) In discussion with the author.

165 **more than an industry road map:** Ray Kurzweil. (2005) *The Singularity Is Near.* New York: Viking.

165 **Kurzweil's Law:** Data from Ray Kurzweil. (2005) "Moore's Law: The Fifth Paradigm." The Singularity Is Near (January 28, 2010). http://singularity.com/charts/page67.html.

167 **Doubling Times:** Data from Ray Kurzweil. (2005) *The Singularity Is Near.* New York: Viking; Eric S. Lander, Lauren M. Linton, et al. (2001) "Initial Sequencing and Analysis of the Human Genome." *Nature*, 409 (6822). http://www.ncbi.nlm.nih.gov/pubmed/11237011; Rik Blok. (2009) "Trends in Computing." http://www.zoology.ubc.ca/~rikblok/ComputingTrends/; Lawrence G. Roberts. (2007) "Internet Trends." http://www.ziplink.net/users/lroberts/IEEEGrowthTrends/IEEEComputer12-99.htm; Mark Kryder. (2009) In discussion with the author; Robert V. Steele. (2006) "Laser Marketplace 2006: Diode Doldrums." *Laser Focus World*, 42 (2). http://www.laserfocusworld.com/articles/248128.

169 **which it passed in 1997:** David C. Brock and Gordon E. Moore. (2006) "Understanding Moore's Law." Philadelphia: Chemical Heritage Foundation.

169 **The Continuum of Kryder's Law:** Data from Clayton Christensen. (1997) *The Innovator's Dilemma: When New Technologies Cause Great Firms to Fail*. Boston: Harvard Business School Press, p. 10.

170 **need to "listen to the technology":** (2001) "An Interview with Carver Mead." *American Spectator*, 34 (7). http://laputan.blogspot.com/2003_09_21_laputan_archive.html.

170 **Compound S Curves:** Data from Clayton Christensen. (1997) *The Innovator's Dilemma: When New Technologies Cause Great Firms to Fail*. Boston: Harvard Business School Press, p. 40.

9. Choosing the Inevitable

176 **First Glimpse of the Picture Phone:** AT&T archival photograph via "Showcasing Technology at the 1964–1965 New York World's Fair." http://www.westland.net/ny64fair/map-docs/technology.htm.

177 **everyone recognized the vision:** (2010) "Videophone." Wikipedia, Wikimedia Foundation. http://en.wikipedia.org/w/index.php?title=Videophone&oldid=340721504.

177 **"self-sustaining, ineluctable flow":** Langdon Winner. (1977) *Autonomous Technology: Technics-Out-of-Control as a Theme in Political Thought*. Cambridge, MA: MIT Press, p. 46.

181 **"much that can still be chosen":** Ibid., p. 55.

182 **"moments in the progressions":** Ibid., p. 71.

182 **The Triad of Biological Evolution:** Inspired by Stephen Jay Gould. (2002) *The Structure of Evolutionary Theory*, Cambridge, MA: Belknap Press of Harvard University Press, p. 1052; designed by the author.

183 **The Triad of Technological Evolution:** Designed by the author.

184 **North Korea at Night:** Paul Romer. (2009) "Rules Change: North vs. South Korea." Charter Cities (January 28, 2010). http://chartercities.org/blog/37/rules-change-north-vs-south-korea.

185 **that is sparse and minimal:** Paul Romer. (2009) "Paul Romer's Radical Idea: Charter Cities." TEDGlobal, Oxford.

185 **a matter of what it is designed for:** Robert Wright. (2000) *Nonzero: The Logic of Human Destiny*. New York: Pantheon.

187 **technology is our "second self":** Sherry Turkle. (1985) *The Second Self*. New York: Simon & Schuster.

188 **"but we hope in technology":** W. Brian Arthur. (2009) *The Nature of Technology: What It Is and How It Evolves*. New York: Free Press, p. 246.

10. The Unabomber Was Right

191 **"tendency to make war impossible":** Richard Rhodes. (1999) *Visions of Technology: A Century of Vital Debate About Machines, Systems, and the Human World.* New York: Simon & Schuster, p. 66.

191 **"will be beyond computation":** Christopher Cerf and Victor S. Navasky. (1998) *The Experts Speak: The Definitive Compendium of Authoritative Misinformation.* New York: Villard, p. 274.

191 **"war will become impossible":** Ibid.

191 **"a thousand world conventions":** Ibid., p. 273.

191 **"it will make war impossible":** Havelock Ellis. (1926) *Impressions and Comments: Second Series 1914–1920.* Boston: Houghton Mifflin.

191 **"it will make war ridiculous":** Ivan Narodny. (1912) "Marconi's Plans for the World." *Technical World Magazine* (October).

191 **"Good Will Toward Men a reality":** Christopher Cerf and Victor S. Navasky. (1998) *The Experts Speak: The Definitive Compendium of Authoritative Misinformation.* New York: Villard, p. 105.

192 **"good will towards men":** Janna Quitney Anderson. (2006) "Imagining the Internet: A History and Forecast." Elon University/Pew Internet Project. http://www.elon.edu/e-web/predictions/150/1870.xhtml.

192 **"will bring peace and harmony on Earth":** Nikola Tesla. (1905) "The Transmission of Electrical Energy Without Wires as a Means for Furthering Peace." *Electrical World and Engineer.* http://www.tfcbooks.com/tesla/1905-01-07.htm.

192 **"the unfettered movement of ideas":** David Nye. (2006) *Technology Matters: Questions to Live With.* Cambridge, MA: MIT Press, p. 151.

192 **"dream of participatory democracy":** Stephen Doheny-Farina. (1995) "The Glorious Revolution of 1971." *CMC Magazine,* 2 (10). http://www.december.com/cmc/mag/1995/oct/last.html.

192 **"seen as a sacrament":** Joel Garreau. (2009) In discussion with the author.

192 **"answers to solutions":** W. Brian Arthur. (2009) *The Nature of Technology: What It Is and How It Evolves.* New York: Free Press, p. 153.

192 **die in automobile accidents:** M. Peden, R. Scurfield, et al. (2004) "World Report on Road Traffic Injury Prevention." World Health Organization. http://www.who.int/violence_injury_prevention/publications/road_traffic/world_report/en/index.html.

192 **kills more people than cancer:** Melonie Heron, Donna L. Hoyert, et al. (2006) "Deaths, Final Data for 2006." National Vital Statistics Reports, Centers for Disease Control and Prevention, 57 (14).

193 **"round of technological innovation?":** Theodore Roszak. (1972) "White Bread and Technological Appendages: I." *Visions of Technology: A Century of Vital Debate About Machines, Systems, and the Human World,* ed. Richard Rhodes. New York: Simon & Schuster, p. 308.

194 **"the mind-forg'd manacles":** William Blake. (1984) "London." *Songs of Experience,* New York: Courier Dover Publications, p. 37.

194 **"changing our habits of mind":** Neil Postman. (1994) *The Disappearance of Childhood.* New York: Vintage Books, p. 24.

195 **50,000 per year:** John H. Lawton and Robert M. May. (1995) *Extinction Rates.* Oxford: Oxford University Press.

196 **the future with a short distance:** Paul Saffo. (2008) "Embracing Uncertainty: The Secret to Effective Forecasting." Seminars About Long-term Thinking. San Francisco: The Long Now Foundation. http://www.longnow.org/seminars/02008/jan/11/embracing-uncertainty-the-secret-to-effective-forecasting/.

196 **some kind of consequence of that:** Kevin Kelly and Paula Parisi. (1997) "Beyond Star Wars: What's Next for George Lucas." *Wired*, 5 (2). http://www.wired.com/wired/archive/5.02/fflucas.html.

197 **"they may never acknowledge the void":** Langdon Winner. (1977) *Autonomous Technology: Technics-Out-of-Control as a Theme in Political Thought.* Cambridge: MIT Press, p. 34.

197 **"self-respecting members of the former":** Eric Brende. (2004) *Better Off: Flipping the Switch on Technology.* New York: HarperCollins, p. 229.

198 **not by ideology but by technical necessity:** Theodore Kaczynski. (1995) "Industrial Society and Its Future." http://en.wikisource.org/wiki/Industrial_Society_and_Its_Future.

201 **civilization would collapse by 2020:** Kevin Kelly. (1995) "Interview with the Luddite." *Wired*, 3 (6). http://www.wired.com/wired/archive/3.06/saleskelly.html.

201 **readings focused on the theme called *Against Civilization*:** John Zerzan. (2005) *Against Civilization: Readings and Reflections.* Los Angeles: Feral House.

201 **gas lines and the information infrastructure:** Derrick Jensen. (2006) *Endgame, Vol. 2: Resistance.* New York: Seven Stories Press.

201 **"reductions in freedom":** Theodore Kaczynski. (1995) "Industrial Society and Its Future." http://en.wikisource.org/wiki/Industrial_Society_and_Its_Future.

202 **"defeatism, guilt, self-hatred, etc.":** Ibid.

202 **"There was even a waterfall there":** Theresa Kintz. (1999) "Interview with Ted Kaczynski." *Green Anarchist* (57/58). http://www.insurgentdesire.org.uk/tedk.htm.

202 **"that sort of thing became a priority for me":** Ibid.

204 **find themselves FORCED to use it:** Theodore Kaczynski. (1995) "Industrial Society and Its Future." http://en.wikisource.org/wiki/Industrial_Society_and_Its_Future.

205 **no rational and effective public resistance:** Ibid.

206 **eventually wipe out all of our freedom:** Ibid.

207 **Inside the Unabomber's Shack:** Federal Bureau of Investigation photograph via (2008) "Unabom Case: The Unabomber's Cabin." http://cbs5.com/slideshows/unabom.unabomber.exclusive.20.433402.html.

210 **"strong health and robusticity":** Green Anarchy. (n.d.) "An Introduction to Anti-Civilization Anarchist Thought and Practice." Green Anarchy Back to Basics (4). http://www.greenanarchy.org/index.php?action=viewwritingdetail&writingId=283.

210 **"a few of its devastating derivatives":** Ibid.

210 **"thousands and tens of thousands of years":** Derrick Jensen. (2009) In discussion with the author.

212 **"any radicals facing up to":** Theresa Kintz. (1999) "Interview with Ted Kaczynski." *Green Anarchist* (57–58). http://www.insurgentdesire.org.uk/tedk.htm.

11. Lessons of Amish Hackers

225 **adopted by the rest of America:** Stephen Scott. (1990) *Living Without Electricity: People's Place Book No. 9.* Intercourse, PA: Good Books.

228 **a tale he recounts in his book:** Eric Brende. (2004) *Better Off: Flipping the Switch on Technology.* New York: HarperCollins.

229 **"I am satisfied":** Wendell Berry. (1982) *The Gift of Good Land: Further Essays Cultural & Agricultural.* San Francisco: North Point Press.

232 **'Start your own business'":** Stewart Brand. (1995, March 1) "We Owe It All to the Hippies." *Time*, 145 (12). http://www.time.com/time/magazine/article/0,9171,982602,000.html.

234 **more work, but not better:** Wendell Berry. (1982) *The Gift of Good Land: Further Essays Cultural & Agricultural.* San Francisco: North Point Press, p. 180.

236 **"to imagine being somebody else":** Brink Lindsey. (2007) *The Age of Abundance: How Prosperity Transformed America's Politics and Culture.* New York: HarperBusiness, p. 4.

237 **"doesn't quite work yet":** W. Daniel Hillis. (2009) In discussion with the author.

12. Seeking Conviviality

239 **"master what the human mind has made?":** Langdon Winner. (1977) *Autonomous Technology: Technics-Out-of-Control as a Theme in Political Thought.* Cambridge, MA: MIT Press, p. 13.

241 **Duration of Prohibitions:** Data compiled from research gathered by Michele McGinnis and Kevin Kelly in 2004; originally presented at http://www.kk.org/thetechnium/archives/2006/02/the_futility_of.php.

242 **"in the defense of fortifications and on ships":** David Bachrach. (2003) "The Royal Crossbow Makers of England, 1204–1272." *Nottingham Medieval Studies* (47).

243 **numerals in their accounts:** Bernhard J. Stern. (1937) "Resistances to the Adoption of Technological Innovations." Report of the Subcommittee on Technology to the National Resources Committee.

243 **increasing at 9 percent per year globally:** Applications International Service for the Acquisition of Agri-Biotech. (2008) "Global Status of Commercialized Biotech/Gm Crops: 2008; The First Thirteen Years, 1996 to 2008." ISAAA Brief 39-2008: Executive Summary. http://www.isaaa.org/resources/publications/briefs/39/executivesummary/default.html.

243 **increasing globally by 2 percent a year:** International Atomic Energy Agency. (2007) "Nuclear Power Worldwide: Status and Outlook." International Atomic Energy Agency. http://www.iaea.org/NewsCenter/PressReleases/2007/prn200719.html.

243 **peaked at 65,000 units in 1986 and is now at 20,000:** National Resources Defense Council. (2002) "Table of Global Nuclear Weapons Stockpiles, 1945–2002." http://www.nrdc.org/nuclear/nudb/datab19.asp.

247 **"to prevent environmental degradation":** United Nations Environment Program. (1992) "Rio Declaration on Environment and Development." Rio de Janeiro: United Nations Environment Program. http://www.unep.org/Documents.multilingual/Default.asp?DocumentID=78&ArticleID=1163.

247 **such as Portland, Oregon, and San Francisco:** Lawrence A. Kogan. (2008) "The Extra-WTO Precautionary Principle: One European 'Fashion' Export the United States Can Do Without." *Temple Political & Civil Rights Law Review*, 17 (2). p. 497. http://www.itssd.org/Kogan%2017%5B1%5D.2.pdf.

247 **"it leads in no direction at all":** Cass Sunstein. (2005) *Laws of Fear: Beyond the Precautionary Principle*. Cambridge: Cambridge University Press, p. 14.

248 **DDT around the insides of homes:** Lawrence Kogan. (2004) "'Enlightened' Environmentalism or Disguised Protectionism? Assessing the Impact of EU Precaution-Based Standards on Developing Countries," p. 17. http://www.wto.org/english/forums_e/ngo_e/posp47_nftc_enlightened_e.pdf.

248 **EU agreed to phase out DDT altogether:** Tina Rosenberg. (2004, April 11) "What the World Needs Now Is DDT." *New York Times*. http://www.nytimes.com/2004/04/11/magazine/what-the-world-needs-now-is-ddt.html.

249 **"carried away in a single night":** Richard Rhodes. (1999) *Visions of Technology: A Century of Vital Debate About Machines, Systems, and the Human World*. New York: Simon & Schuster, p. 145.

250 **"hidden paths in the systems":** Charles Perrow. (1999) *Normal Accidents: Living with High-Risk Technologies*. Princeton NJ: Princeton University Press, p. 11.

251 **"totally unlike the world we now inhabit":** Langdon Winner. (1977) *Autonomous Technology: Technics-Out-of-Control as a Theme in Political Thought*. Cambridge, MA: MIT Press, p. 98.

252 **imagined a horseless carriage:** Arthur C. Clarke. (1984) *Profiles of the Future*. New York: Holt, Rinehart and Winston.

252 **subsets of technologies that permeate society:** M. Rodemeyer, D. Sarewitz, et al. (2005) *The Future of Technology Assessment*. Washington, D.C.: The Woodrow Wilson International Center.

253 ***eternally provisional:*** Stewart Brand. (2009) *Whole Earth Discipline*. New York: Viking, p. 164.

253 **but with any luck it works:** Edward Tenner. (1996) *Why Things Bite Back: Technology and the Revenge of Unintended Consequences*. New York: Knopf, p. 277.

255 **radical transhumanist, in 2004:** Max More. (2005) "The Proactionary Principle." http://www.maxmore.com/proactionary.htm.

256 **"overweighting human-technological risks":** Ibid.

261 **moratorium on all nanotechnological research:** James Hughes. (2007) "Global Technology Regulation and Potentially Apocalyptic Technological Threats." *Nanoethics: The Ethical and Social Implications of Nanotechnology*, ed. Fritz Allhoff. Hoboken, NJ: Wiley-Interscience.

261 **used in antimicrobial coatings:** Dietram A. Scheufele. (2009) "Bund Wants Ban of Nanosilver in Everyday Applications." http://nanopublic.blogspot.com/2009/12/bund-wants-ban-of-nanosilver-in.html; Wiebe E. Bijker, Thomas P. Hughes, et al. (1989) *The Social Construction of Technological Systems*. Cambridge, MA: MIT.

264 **were destructive no matter who ran them:** Ivan Illich. (1973) *Tools for Conviviality*. New York: Harper & Row.

13. Technology's Trajectories

275 **equally inadequate in real life:** Seth Lloyd. (2006) *Programming the Universe: A Quantum Computer Scientist Takes on the Cosmos.* New York: Knopf.

275 **any direction to evolution whatsoever:** Stephen Jay Gould. (1989) *Wonderful Life: The Burgess Shale and Nature of History.* New York: W. W. Norton.

275 **display effective complexity:** Seth Lloyd. (2006) *Programming the Universe: A Quantum Computer Scientist Takes on the Cosmos.* New York: Knopf, p. 199.

275 **"the cosmological origins of biology":** James Gardner. (2003) *Biocosm: The New Scientific Theory of Evolution.* Makawao Maui, HI: Inner Ocean.

276 **transitions in organic evolution:** John Maynard Smith and Eors Szathmary. (1997) *The Major Transitions in Evolution.* New York: Oxford University Press.

277 **Each step was also irreversible:** John Maynard Smith and Eors Szathmary. (1997) *The Major Transitions in Evolution.* New York: Oxford University Press, p. 9.

278 **Complexity of Software:** Data from Vincent Maraia. (2005) *The Build Master: Microsoft's Software Configuration Management Best Practices.* Upper Saddle River, NJ; Addison-Wesley Professional.

279 **Vista contained 50 million lines of code:** Vincent Maraia. (2005) *The Build Master: Microsoft's Software Configuration Management Best Practices.* Upper Saddle River, NJ: Addison-Wesley Professional.

279 **Complexity of Manufactured Machines:** Data from Robert U. Ayres. (1991) *Computer Integrated Manufacturing: Revolution in Progress.* London: Chapman & Hall, p. 3.

281 **say a UNIX kernel:** W. Daniel Hillis. (2007) In discussion with the author.

282 **varieties of stable elements were created:** George Wallerstein, Icko Iben, et al. (1997) "Synthesis of the Elements in Stars: Forty Years of Progress. *Reviews of Modern Physics,* 69 (4), p. 1053. http://link.aps.org/abstract/RMP/v69/p995.10.1103/RevModPhys.69.995.

283 **mineral species we find today:** Robert M. Hazen, Dominic Papineau, et al. (2008) "Mineral Evolution." *American Mineralogist,* 93 (11/12). http://ammin.geoscienceworld.org/cgi/content/abstract/93/11-12/1693.

283 **only 200 million years ago:** Dale A. Russell. (1995) "Biodiversity and Time Scales for the Evolution of Extraterrestrial Intelligence." *Astronomical Society of the Pacific Conference Series* (74). http://adsabs.harvard.edu/full/1995ASPC...74..143R.

283 **Total Diversity of Life:** J. John Sepkoski. (1993) "Ten Years in the Library: New Data Confirm Paleontological Patterns." *Paleobiology,* 19 (1), p. 48.

284 **articles has exploded in the last 50 years:** Stephen Hawking. (2001) *The Universe in a Nutshell.* New York: Bantam Books, p. 158.

284 **seven million patents issued in the United States alone:** Brigid Quinn and Ruth Nyblod. (2006) "United States Patent and Trademark Office Issues 7 Millionth Patent." United States Patent and Trademark Office.

284 **Total Patent Applications and Scientific Articles:** United States Patent and Trademark Office. (2009) "U.S. Patent Activity, Calendar Years 1790 to Present: Total of Annual U.S. Patent Activity Since 1790." http://www.uspto.gov/web/offices/ac/ido/

oeip/taf/h_counts.htm; Stephen Hawking. (2001) *The Universe in a Nutshell.* New York: Bantam Books, p. 158.

285 artificial learning machine can recognize: Irving Biederman. (1987) "Recognition-by-Components: A Theory of Human Image Understanding." *Psychological Review,* 94 (2), p. 127.

286 over 480,000 products in its catalog: "McMaster-Carr." http://www.mcmaster.com/#.

286 about one million TV episodes: "IMDB Statistics." Internet Movie Database. http://www.imdb.com/database_statistics.

286 different songs have been recorded: "iTunes A to Z." Apple Inc. http://www.apple.com/itunes/features/.

286 cataloged 50 million different chemicals: Paul Livingstone. (2009, September 8) "50 Million Compounds and Counting." *R&D Mag.* http://www.rdmag.com/Community/Blogs/RDBlog/50-million-compounds-and-counting/.

286 "it literally was one of a kind": David Nye. (2006) *Technology Matters: Questions to Live With.* Cambridge, MA: MIT Press, pp. 72–73.

286 paralyzing consumers: Barry Schwartz. (2004) *The Paradox of Choice: Why More Is Less.* New York: Ecco, pp. 9–10.

286 "the less likely they are to make a choice": Barry Schwartz. (2005, January 5) "Choose and Lose." *New York Times.* http://www.nytimes.com/2005/01/05/opinion/05schwartz.html.

287 "even be said to tyrannize": Barry Schwartz. (2004) *The Paradox of Choice: Why More Is Less.* New York: Ecco, p. 2.

287 U.S. Patent Office by 2060!: Kevin Kelly. (2009) Calculation extrapolated by the author based on historic U.S. Patent data. http://www.uspto.gov/web/offices/ac/ido/oeip/taf/h_counts.htm.

287 different books in English: Library of Congress. (2009). "About the Library." http://www.loc.gov/about/generalinfo.html.

290 "mere day's walk away": Pierre Lemonnier. (1993) *Technological Choices: Transformation in Material Cultures Since the Neolithic.* New York: Routledge, p. 74.

290 "logic of material efficiency or progress": Ibid., p. 24.

290 released by a passing pig: Ibid.

291 the physics of waterwheels are constant: Ibid.

292 bone cells, skins cells, and brain cells: Stuart Kauffman. (1993) *The Origins of Order: Self-Organization and Selection in Evolution.* New York: Oxford University Press, p. 407.

293 Specialized Cell Types: Data from James W. Valentine, Allen G. Collins, et al. (1994) "Morphological Complexity Increase in Metazoans." *Paleobiology,* 20 (2), p. 134. http://paleobiol.geoscienceworld.org/cgi/content/abstract/20/2/131.

297 mind and hand of humans: Peter M. Vitousek, Harold A. Mooney, et al. (1997) "Human Domination of Earth's Ecosystems." *Science,* 277 (5325).

298 Accelerating Pace of Technology Adoption: Peter Brimelow. (1997, July 7) "The Silent Boom." *Forbes,* 160 (1). http://www.forbes.com/forbes/1997/0707/6001170a.html.

300 throughout the manufacturing industry: "Electric Motor." Wikipedia, Wikimedia Foundation. http://en.wikipedia.org/w/index.php?title=Electric_motor&oldid=344778362.

301 Ubiquitous Motors: David A. Hounshell. (1984) *From the American System to Mass Production 1800–1932: The Development of Manufacturing Technology in the United States.* Baltimore: Johns Hopkins University Press, p. 232.

301 Ad for the Home Motor: Donald Norman. (1998) *The Invisible Computer: Why Good Products Can Fail, the Personal Computer Is So Complex, and Information Appliances Are the Solution.* Cambridge: MIT Press, p. 50.

302 to whip cream and beat eggs: Donald Norman. (1998) *The Invisible Computer: Why Good Products Can Fail, the Personal Computer Is So Complex, and Information Appliances Are the Solution.* Cambridge, MA: MIT Press.

304 "two Americas" emerging: Don Tapscott. (1999) *Growing up Digital.* New York: McGraw-Hill, p. 258. Referring to Brad Fay's research for the 1996 Roper Starch report "The Two Americas: Tools for Succeeding in a Polarized Marketplace."

307 must be described as an act of free will: Freeman J. Dyson. (1988) *Infinite in All Directions.* New York: Basic Books, p. 297.

307 shift of spin direction in cosmic particles: J. Conway. (2009) "The Strong Free Will Theorem." *Notices of the American Mathematical Society,* 56 (2).

308 quantum effects happen in all matter: Stuart Kauffman. (2009) "Five Problems in the Philosophy of Mind." Edge: The Third Culture, (297). http://www.edge.org/3rd_culture/kauffman09/kauffman09_index.html.

308 the ultimate explanation of our own: Conway. "The Strong Free Will Theorem."

309 "with more freedom we can be more human": Richard Rhodes. (1999) *Visions of Technology: Machines, Systems and the Human World.* New York: Simon & Schuster Inc., p. 266.

310 brain/arm combo: (1998) "'Quick-Thinking' Robot Arm Helps MIT Researchers Catch on to Brain Function." MITnews. http://web.mit.edu/newsoffice/1998/wam.html.

311 on this planet are parasitic: Peter W. Price. (1977) "General Concepts on the Evolutionary Biology of Parasites." *Evolution,* 31 (2). http://www.jstor.org.libaccess.sjlibrary.org/stable/2407761.

314 collaboratively write and edit material: Ward Cunningham. "Publicly Available Wiki Software Sorted by Name." http://c2.com/cgi/wiki?WikiEngines.

314 by YouTube each month in the United States alone: comScore. (2009) "YouTube Surpasses 100 Million U.S. Viewers for the First Time." ComScore. http://www.comscore.com/Press_Events/Press_Releases/2009/3/YouTube_Surpasses_100_Million_US_Viewers.

314 stories deposited on fan-fiction sites: M. E. Curtin. (2007) In discussion with the author's researcher. See M. E. Curtin's Alternate Universes for her earlier stats: http://www.alternateuniverses.com/ffnstats.html.

315 with categories, labels, and keywords: Heather Champ. (2008) "3 Billion!" Flickr Blog. http://blog.flickr.net/en/2008/11/03/3-billion/.

316 the release of the Fedora Linux 9 software: Amanda McPherson, Brian Proffitt, et al. (2008) "Estimating the Total Development Cost of a Linux Distribution." The Linux Foundation. http://www.linuxfoundation.org/publications/estimatinglinux.php.

316 different open-source projects: Ohloh. (2010) "Open Source Projects." http://www.ohloh.net/p.

316 **but without any bosses:** General Motors Corporation. (2008) "Form 10-K." http://www.sec.gov.

319 **innate attraction to living things:** Stephen R. Kellert and Edward O. Wilson. (1993) *The Biophilia Hypothesis.* Washington, D.C.: Island Press.

319 **Ergonomic Scissors:** Generic tailor's scissors, unknown origin.

320 **one began to pray to it:** Langdon Winner. (1977) *Autonomous Technology: Technics-Out-of-Control as a Theme in Political Thought.* Cambridge, MA: MIT Press, p. 44.

321 **to a world where no one is:** Joan Didion. (1990). *The White Album.* New York: Macmillan, p. 198.

322 **"passion and utility" of clocks snag some:** Mark Dow. (June 8, 2009) "A Beautiful Description of Technophilia [Weblog comment]." Technophilia. The Technium. http://www.kk.org/thetechnium/archives/2009/06/technophilia.php#comments.

322 **"we love the objects we think with":** Sherry Turkle. (2007) *Evocative Objects: Things We Think With.* Cambridge, MA: MIT Press, p. 3.

325 **so small as to be invisible:** Nigel R. Franks and Simon Conway Morris. (2008) "Convergent Evolution, Serendipity, and Intelligence for the Simple Minded." *The Deep Structure of Biology,* ed. Simon Conway Morris. West Conshohocken, PA: Templeton Foundation Press.

325 **only discovered by humans in 1733:** J. F. Ramley. (1969) "Buffon's Needle Problem." *American Mathematical Monthly,* 76 (8).

326 **flies away from their hind parts:** Donald R. Griffin. (2001) *Animal Minds: Beyond Cognition to Consciousness.* Chicago: University of Chicago Press, p. 12.

326 **so as not to discourage them:** Ibid.

326 **our definitions of animal intelligence:** Anthony Trewavas. (2008) "Aspects of Plant Intelligence: Convergence and Evolution." *The Deep Structure of Biology,* ed. Simon Conway Morris. West Conshohocken, PA: Templeton Foundation Press.

326 **"one of the lower animals":** Ibid., p. 80.

327 **food in a maze, much like a rat:** Ibid.

327 **other dolphins new to the pool:** Donald R. Griffin. (2001) *Animal Minds: Beyond Cognition to Consciousness.* Chicago: University of Chicago Press, p. 229.

328 **"not by an ornithologist but by a psychiatrist":** Anthony Trewavas. (2008) "Aspects of Plant Intelligence: Convergence and Evolution." *The Deep Structure of Biology,* West Conshohocken, PA: Templeton Foundation Press, p. 131.

329 **minds contain a billion transistors each:** Jim Held, Jerry Bautista, et al. (2006) "From a Few Cores to Many: A Tera-Scale Computing Research Overview." http://download.intel.com/research/platform/terascale/terascale_ovierview_paper.pdf

329 **brain of a sperm whale:** Lori Marino. (2004) "Cetacean Brain Evolution: Multiplication Generates Complexity." *International Journal of Comparative Psychology,* 17 (1).

330 **about 100 quadrillion (10^{17}) transistors:** Kevin Kelly. (2008) "Infoporn: Tap into the 12-Million-Teraflop Handheld Megacomputer." *Wired,* 16 (7). http://www.wired.com/special_multimedia/2008/st_infoporn_1607.

330 **generates nearly 20 exabytes of data:** Ibid.

331 approximately 2.7 billion cell phones: Portio Research. (2007) "Mobile Messaging Futures 2007–2012." http://www.portioresearch.com/MMF07-12.html.

331 1.3 billion land phones: Central Intelligence Agency. (2009) "World Communications." *World Factbook.* https://www.cia.gov/library/publications/the-world-factbook/geos/xx.html.

331 27 million data servers: Jonathan Koomey. (2007) "Estimating Total Power Consumption by Servers in the U.S. and the World." Oakland: Analytics Press. www.amd.com/us-en/assets/content_type/DownloadableAssets/Koomey_Study-v7.pdf.

331 80 million wireless PDAs: eMarketer. (2002) "PDA Market Report: Global Sales, Usage and Trends," p. 1. Citing Gartner Dataquest. http://www.info-edge.com/samples/EM-2058sam.pdf. Based on the cumulative total of 2003–2005.

331 holds about a trillion pages: Marcus P. Zillman. (2006) "Deep Web Research 2007." LLRX. http://www.llrx.com/features/deepweb2007.htm.

331 synaptic links to thousands of other neurons: David A. Drachman. (2005) "Do We Have Brain to Spare?" *Neurology,* 64 (12). http://www.neurology.org.

331 links to 60 other pages: Andrei Z. Broder, Marc Najork, et al. (2003) "Efficient URL Caching for World Wide Web Crawling." *Proceedings of the 12th International Conference on the World Wide Web,* Budapest, Hungary, May 20–24, p. 5. http://portal.acm.org/citation.cfm?id=775152.775247.

332 several quintillion online transistors: Semiconductor Industry Association. (2007) "SIA Hails 60th Birthday of Microelectronics Industry." Semiconductor Industry Association. http://www.sia-online.org/cs/papers_publications/press_release_detail?pressrelease.id=96.

332 hundreds of exabytes of real-life data: John Gantz, David Reinsel, et al. (2007) "The Expanding Digital Universe: A Forecast of Worldwide Information Growth Through 2010." http://www.emc.com/collateral/analyst-reports/expanding-digital-idc-white-paper.pdf.

334 by a few million bits: Stephen Hawking. (1996) "Life in the Universe." http://hawking.org.uk/index.php?option=com_content&view=article&id=65.

334 new information to the technium each year: Bret Swanson and George Gilder. (2008) "Estimating the Exaflood." Discovery Institute. http://www.discovery.org/a/4428.

334 an exponential curve for over 100 years: Andrew Odlyzko. (2000) "The History of Communications and Its Implications for the Internet." SSRN eLibrary. http://papers.ssrn.com/sol3/papers.cfm?abstract_id=235284.

335 when science began: Derek Price. (1965) *Little Science, Big Science.* New York: Columbia University Press.

337 the overthrow of scientific paradigms: Freeman J. Dyson. (2000) *The Sun, the Genome, and the Internet: Tools of Scientific Revolutions.* New York: Oxford University Press, p. 15.

14. Playing the Infinite Game

347 "could the steam-engine make men better?": Wendell Berry. (2000) *Life Is a Miracle: An Essay Against Modern Superstition.* Washington, D.C.: Counterpoint Press, p. 74.

350 **Missing Technologies:** Collage by the author.

352 **"act to increase the number of choices":** Heinz von Foerster. (1984) *Observing Systems.* Seaside, CA: Intersystems Publications, p. 308.

353 **"infinite players play with boundaries":** James Carse. (1986) *Finite and Infinite Games.* New York: Free Press, p. 10.

354 **"never quite reaching this ideal":** Ray Kurzweil. (2005) *The Singularity Is Near.* New York: Viking, p. 389.

355 **called Process Theology:** John B. Cobb Jr. and David Ray Griffin. (1977) *Process Theology: An Introductory Exposition.* Philadelphia: Westminster Press.

356 **so far under the stars:** Loren Eiseley. (1985) *The Unexpected Universe.* San Diego: Harcourt Brace Jovanovich, p. 55.

357 **"universe knew we were coming":** Freeman J. Dyson. (2001). *Disturbing the Universe.* New York: Basic Books, p. 250.

357 **"starstuff pondering the stars":** Carl Sagan. (1980) *Cosmos.* New York: Random House.

Index

Page numbers in *italics* refer to illustrations.